Productivity and Quality Through Science and Technology

Productivity and Quality Through Science and Technology

**Edited by
Y. K. Shetty and Vernon M. Buehler**

*Foreword by Donald E. Petersen,
Chairman of the Board and CEO,
Ford Motor Company*

Q

QUORUM BOOKS
New York • Westport, Connecticut • London

Library of Congress Cataloging-in-Publication Data

Productivity and quality through science and technology / edited by
   Y. K. Shetty and Vernon M. Buehler.
     p.   cm.
   Bibliography: p.
   Includes index.
   ISBN 0-89930-344-7 (lib. bdg. : alk. paper)
   1. Industrial management—United States. 2. Industrial
productivity—United States. 3. Labor productivity—United States.
4. Technological innovations—Economic aspects—United States.
5. Competition—United States. I. Shetty, Y. Krishna.
II. Buehler, Vernon M.
HD70.U5P69   1988
658—dc19   87-32595

British Library Cataloguing in Publication Data is available.

Library of Congress Catalog Card Number: 87-32595
ISBN: 0-89930-344-7

First published in 1988 by Quorum Books

Greenwood Press, Inc.
88 Post Road West, Westport, Connecticut 06881

Printed in the United States of America

The paper used in this book complies with the
Permanent Paper Standard issued by the National
Information Standards Organization (Z39.48-1984).

10 9 8 7 6 5 4 3 2 1

Gratefully Dedicated
to W. EDWARDS DEMING

As the twentieth century's most eminent consultant
on quality and productivity, Dr. Deming lectured at USU
in 1983 and 1984 to thousands of students and managers
on his management principles that he taught, first to
the Japanese in the early 1950s, and later to managers
around the globe.   His teachings have revolutionized
management thinking and practices that will endure for
future generations in the classroom, the boardroom, and
the workplace.

Contents

Foreword
by Donald E. Petersen

The U.S. today is struggling to protect and sustain the position of industrial supremacy that it enjoyed so broadly and securely from the years following World War II into the late 1960's. Its once undisputed leadership in quality, efficiency, technology, and innovation is under constant and increasing challenge from competitors around the world. Slowing productivity, erratic product quality, faltering of some of our basic industries, and persisting sizeable trade deficits have become major sources of concern.

The editors of this book believe that the U.S. can and will retain its strong competitive position. Further, they believe that our competitiveness will be improved by the growing use and benefits of teamwork throughout the various sectors of our economy. Such collaboration is typified by the many successful meetings and exchanges over the years between representatives of industry and Utah State University. The manuscripts produced by these exchanges--not yet available in textbooks--cover innovative approaches for managing human resources, information systems, quality, productivity, technology, and trends in world trade.

I believe that the title should not be interpreted to mean that technology and science are the sole ingredients to achieving productivity and quality improvements. There is also the ever-important human element. Science and technology must be introduced at a pace that is right for your company and people. It should not be blindly implemented without first confirming the compatibility of the technology with the people who will be using it. There is a pace to that, and too rapid a pace can lead to confusion, waste, and failure.

The editors point to successes that have bolstered our competitive prowess including: the labor and management problem-solving teams on the factory floor; and business-government cooperation in formulating public policy on trade, anti-trust, consumerism, and other issues; and university researchers and industry technologists joining at government-sponsored technology centers for expediting the commercialization of new ideas. These promising team efforts are gradually and meaningfully easing adversarial relationships and disharmonies that impeded our competitiveness in the past.

Our ability to generate and apply new knowledge for improving productivity, quality, and competitiveness is related directly to the vigor of our research and educational enterprise. By compiling and sharing the results of these exchanges, the editors remind us that our survival in this competitive world relies increasingly on efforts to unlock and apply new knowledge by teams of players from academe, government, and industry. The future competitive ability of the U.S. depends importantly on how well we step up to this challenge.

The Utah State University Partnership Program represents an outstanding endeavor to bring together people representing the many facets of our economy and society. It serves as a useful and practical model for other institutions interested in programs directly toward maintaining America's position as the principal industrial power of the world.

Donald E. Petersen
Chairman and CEO
Ford Motor Company

Preface

This is our fifth volume that reports on company experiences and public policy measures for improving U.S. competitiveness. Like previous volumes, this book describes successful practices in managing people, information, innovation, quality, productivity, and manufacturing strategy.

This time we have emphasized also the role of science and technology in enhancing competitiveness. Studies show that since World War II technological innovation, spawned from new knowledge, has been the source of over 40 percent of the improvements in U.S. productivity. Experiences related here show how new ideas are being commercialized and how the government is encouraging this process. Firms are leapfrogging each other with new ideas as they launch new generations of products having shorter and shorter life cycles.

Lessons from these experiences give us an optimistic outlook for U.S. competitiveness. They counter the claims of those who fear deindustrialization and prolonged exportation of jobs. Instead the experiences tell of enlightened and adaptive management practices that are promoting competitiveness through actions such as downsizing, teamwork, pay for performance, fast-track development processes, computer integrated manufacturing, and fostering innovation.

Utah State University's Partners Program of the College of Business is committed to improving U.S. competitiveness through academic-business exchanges. Dating from 1970, this student-driven program provides a forum for interaction between practitioners and our entrepreneurial-minded students. Each year some 100 business experts from top U.S. firms come at company expense to our scenic campus. They share their experiences on evolving global management practices with thousands of young aspiring managers and business friends. Realism is injected into the classroom by discussing creative approaches that are headlined in

the Wall Street Journal but are not yet in textbooks.
Several experts are world-renowned such as Harvard
professor Rosabeth Moss Kanter, corporate raider
T. Boone Pickens, and Harvard economist John Kenneth
Galbraith. So, in seven annual student-managed seminars, USU's Partners Program enables students and
friends from the community to discuss cutting-edge
issues in banking, accounting, real estate, world
trade, human resources, information systems, quality,
and productivity.

This volume contains the proceedings of three
Partners Program Seminars held in 1987 on managing
productivity, human resources, and information systems.
The manuscripts reported here were selected from over
100 presentations.

Has the Partners Program been successful in
helping students and executives understand how to
rebuild U.S. competitiveness? This can be partially
answered by examining the extent of annual participation by students and industry:

As many as 3,000 students with business and
other majors attend selected sessions of the seminars. Some students earn academic credit by sub-
mitting research papers and passing examinations
on the seminar topics.

Some 150 to 500 executives attend each of the
seven seminars and hear world-class keynoters.
Attendees may earn seventy units of Continuing
Professional Education and/or eight Continuing
Education Units.

One hundred speakers nationwide participate
by traveling to USU at their own expense. They
are hosted by 300 to 400 students and several
student associations while discussing career and
job opportunities and speaking at class sessions
as well as at seminars.

Close coordination is maintained with 500 to
600 dues-paying Partners by plant visits, newsletters, and luncheon meetings. Partners often participate as seminar panelists, lecturers, and
moderators.

Over thirty firms and individuals nationwide
qualify as Life Members by making significant
financial contributions and/or speaking at Partners events annually.

Some thirty professional societies representing approximately 5,000 members in the area serve
as co-sponsors of seminars and assist in planning,
promoting, and conducting these Partners events.

Seminar proceedings similar to this volume
are published bi-annually. Collections of distinguished lectures are available annually.

Fifteen to twenty students plan and conduct Partners events with minimal supervision, involving an annual $250,000 budget. Over 100,000 student-produced program brochures, semi-annual newsletters, and other Partners literature are distributed annually.

Partners seminars generate net revenues for awarding over twenty modest scholarships annually. Twenty-five $500 assistantships are granted annually from earnings on Partners endowments that exceed $150,000.

Over $150,000 has been endowed from donations and seminar revenues to support distinguished executive lectures annually. These lectures are endowed in the names of such leaders as W. Edwards Deming, Henry Kaufman, Robert Noyce, Joseph M. Juran, Philip Crosby, and William Hewlett.

Acknowledgments

The materials for this volume come from USU's student-driven Partners Program. The success of the Partners Program stems from many sources, but the predominant factor is its small, industrious student staff, supported by several student associations. The student staff works with minimal direction in developing and conducting a comprehensive program for infusing cutting-edge issues into the classroom. These new ideas are not yet in texts but they are causing major changes in the workplace and the boardroom. Students have a bone-deep desire to better understand how managers cope with such complex forces as fast-changing information and other technologies, shifting work force demographics, global competitive pressures for improved quality and productivity, innovative manufacturing strategies, and the evolving government-business relations. Meeting and exchanging views with executives fresh from the trenches of the competitive battlefield provide a lasting learning experience. These managers are eager to share their insights in this life-span learning endeavor and to close the gap between the worlds of business and academics.

Our Partners, who provide the life blood for these programs, are our industrious faculty and students, managers throughout the nation, and our loyal alumni. They generously contribute their time, advice, and resources. A few deserve special recognition.

The late George S. Eccles, and his charming wife, Dolores Dore Eccles, founded our prestigious George S. Eccles Distinguished Lecture Series in 1974. It has financed over fifty lectures by national leaders and is continuing under the generous sponsorship of Spencer F. Eccles, who is chairman, president, and CEO of First Security Corporation, Salt Lake City.

To recognize the exceptional generosity of those who have helped promote the interaction of academia and business, Partner Life memberships and memorial awards were recently presented to the following:

LIFE MEMBERS

Joseph A. Anderson, ZCMI, SLC, UT.

Russell V. Anderson, RVA Service Corp., Logan, UT.

Samuel Barker, Attorney at Law, Ogden, UT.

Val Browning, Industrialist, Ogden, UT.

Charles Bullen, Bullen's Inc., Logan, UT.

Reed Bullen, Utah Legislator/Businessman,
   Logan, UT.

G. H. "Herb" Champ, Mortgage Banker, Logan, UT .

John E. Clay, CPA, Logan, UT.

Philip B. Crosby, Quality Consultant, Winter
   Park, FL.

Thomas D. Dee II, Dee Co., Ogden, UT.

W. Edwards Deming, Consultant in Statistical
   Studies, Washington, DC.

Spencer F. Eccles, First Security Corp., SLC, UT.

Ronald S. Hanson, Zion's First National Bank,
   SLC, UT.

Jay Dee Harris, Harris Truck and Equipment,
   Tremonton, UT.

William R. Hewlett, Hewlett-Packard Corp., Palo
   Alto, CA.

Joseph M. Juran, Juran Institute, Wilton, CT.

Henry Kaufman, Salomon Brothers, NYC, NY.

John W. Kendrick, George Washington University,
   Washington, DC.

Jack Lampros, First Security Bank, Ogden, UT.

J. W. Marriott, Jr., Marriott Corp., Bethesda, MD.

D. B. "Bud" Ozmun, Selway Foundation,
   Wheeling, IL.

Jack B. Parson, Jack B. Parson Construction,
  Logan, UT.

T. Boone Pickens, Mesa Petroleum Co.,
  Amarillo, TX.

Rex G. Plowman, Lewiston State Bank, Lewiston, UT.

Ralph H. Redford, Tetrotech Int'l, Arlington, VA.

Lynn A. Richardson, Banker, Ogden, UT.

Kenneth O. Sorensen, Cache Mortgage Corp.,
  Logan, UT.

Fred H. Thompson, Banker, Logan, UT.

Morris H. Wright, Wall Street Banker, NYC, NY.

POSTHUMOUS MEMORIAL AWARDS

George S. Eccles, President, First Security
  Corporation.

D. Wade Mack, President, Mack Foundation.

Samuel C. Powell, Attorney at Law.

Bert L. Thomas, President, Winn-Dixie Stores.

Utah Manufacturers Association has assisted in the
promotion and conduct of the annual productivity seminars for twelve years. It is one of some thirty professional and industrial societies that co-sponsor
annual seminars. Most noteworthy of these is the Utah
Bankers Association which has co-sponsored highly successful annual banking seminars starting in 1970 and
has endowed $16,000 for student assistantships.
  Generous assistance was received from over 100
seminar speakers representing the following
organizations:

3M Company	St Paul, MN	* Beneficial Life	SLC, UT
AFL-CIO	Washington, DC	* BF Goodrich	Cleveland, OH
Allen-Bradley	Milwaukee, WI	* Boeing Aerospace	Seattle, WA
American Express	SLC, UT	* Borg-Warner	Chicago, IL
Amgen	Thousand Oaks, CA	* Boston University	Boston, MA
* Analog Devices	Norwood, MA	Bourns Networks	Logan, UT
ARCO Oil & Gas	Dallas, TX	Burroughs	Camarillo, CA
* Arthur Andersen	SLC, UT	Celanese	NYC, NY
Ashton-Tate	Chicago, IL	Cetus	Emeryville, CA
ASK Comuter Syst	Altos, CO	* Cincom Systems	Cincinnati, OH
A. T. Kearny	Chicago, IL	* CitiBank	NYC, NY
* AT & T	NYC, NY	Coasts	
* Bank of America	San Francisco, CA	Productions	Hollywood, CA

Collagen	Palo Alto, CA	McDonald's	Oak Brook, IL
Comshare	Ann Arbor, MI	* McDonnell	
Contexture	SLC, UT	Douglas	St Louis, MO
* Control Data	Minneapolis, MN	McKesson	San Francisco, CA
CRSS	Sterling, VA	* McKinsey & CO	Los Angeles, CA
* Cullinet	Atlanta, GA	Mead	Dayton, OH
* Dana Corp	Toledo, OH	Metropolitan Life	NYC, NY
Dayna Comm	SLC, UT	Micro DB	
* DEC	Maynard, MI	Systems	W Lafayette, IN
Deere &		MIT	Cambridge, MA
Company	Moline, IL	Mitchell Fein	Hillsdale, NJ
Delta Air Lines	SLC, UT	Molecular	
Dialogic		Computer	San Jose, CA
Systems	Sunnyville, CA	Monsanto	St Louis, MO
DMS	SLC, UT	* Moore Bus Forms	Glenview, IL
Do-It Systems	Albuquerque, NM	Moore Financial	Boise, ID
Dow Chemical	Midland, MI	* Morton-Thiokol	Brigham City, UT
DuPont	Wilmington, DE	Motorola	Scottsdale, AZ
Eastman Kodak	Rochester, NY	* Mutual Benefit Life	Newark, NJ
* Eaton Corp	Cleveland, OH	Mutual of Omaha	Omaha, NE
Epsilon Data Mgt	Burlington, MA	NASA	Washington, DC
* Federal Express	Memphis, TN	* Ntn'l Adv Syst	Mtn View, CA
FHP	SLC, UT	* Ntn'l Science	
* First Interstate	Torrance, CA	Foundation	Washington, DC
Florida,		Ntn'l Semicon-	
University of	Gainesville, FL	ductor	West Jordan, UT
FMC	Chicago, IL	* Nestar Systems	Palo Alto, CA
* Ford Motor	Dearborn, MI	Nordstroms	SLC, UT
* Gandalf Tech	Ontario, Canada	* N Telecomm Inc	Nashville, TN
Gartner Group	Stamford, CT	* NW Mutual Life	Milwaukee, WI
* General Dynamics	St Louis, MO	* Novel Data Syst	Orem, UT
General Electric	Selkirk, NY	Nucor Steel	Charlotte, NC
General Foods	White Plains, NY	Omnidata	Logan, UT
* General Motors	Detroit, MI	* Ore-Ida Foods	Boise, ID
Goetze	Muskegon, MI	Org Consultants	Ann Arbor, MI
Golder Thoma		Otis Elevators	Farmington, CT
& Cressey	Chicago, IL	Pacific Bell	San Francisco, CA
* Hercules	Magna, UT	* Peat Marwick	SLC, UT
* Hewlett-Packard	Palo Alto, CA	Penwalt	Philadelphia, PA
Honeywell	Minneapolis, MN	Pillsbury	Minneapolis, MN
Hughes	Los Angeles, CA	Pizza Hut	Wichita, KS
* IBM	San Jose, CA	* PPG Industries	Pittsburgh, PA
Info Builders	NYC, NY	Price Waterhouse	NYC, NY
Inland Steel	Chicago, IL	Procter & Gamble	Cincinnati, OH
Integrated Circuits	Scottsdale, AZ	Productivity Inc	Stamford, CT
* Intel	Santa Clara, CA	Psych Cnslts	Richmond, VA
Jacquelyn Wonder	Denver, CO	Rainier Bank	Seattle, WA
* JC Penney	NYC, NY	Rockwell Int'l	Pittsburgh, PA
Johnson Wax	Racine, WI	RTE Corp	Waukesha, WI
Kellogg	Battle Creek, MI	Rutgers University	New Brunswick, NJ
Kimberly-Clark	Neenah, WI	* Ryder Systems	Miami, FL
Leasway		Salomon Brothers	NYC, NY
Transportation	Cleveland, OH	Sanders	
Lee Scientific	SLC, UT	Associates	Nashua, NH
Lincoln Electric	Cleveland, OH	* Satellite Software	Orem, UT
* Litton Systems	SLC, UT	* Schreiber Foods	Green Bay, WI
LTV Steel	Cleveland, OH	Sec Pacific	
* Marriott	Washington, DC	Automation	Los Angeles, CA
Mary Kay	Dallas, TX	Sec Pacific Nt'l	Los Angeles, CA

Software	
AG of NA	Reston, VA
Southwestern	
Bell Laboratories	St Louis, MO
Sperry	Blue Bell, PA
SRI	Menlo Park, CA
Steelcase	Grand Rapids, MI
* Tandem Computer	Cupertino, CA
Tandy	Fort Worth, TX
TeleVideo Systems	San Jose, CA
Tennant CO	Minneapolis, MN
* Texas Instruments	Dallas, TX
TIAA	NYC, NY
Toshiba	Tustin, CA
Travelers	Hartford, CT

* TRW	Cleveland, OH
Twin City Fed Bank	Cleveland, OH
* Unisys	Detroit, MI
United Energy Res	Houston, TX
US Congress	Washington, DC
US West Info Sys	SLC, UT
* Utah Tech-	
Financial Corp	SLC, UT
* Wang	
Laboratories	Lowell, MA
Westin Hotel	Seattle, WA
* Weyerhaeuser	Tacoma, WA
* Xerox	Stamford, CT

* Indicates multiple visits

Assistance in arranging for these Partners seminars was provided by many individuals and organizations, including USU's President Stanford Cazier, Provost Peter E. Wagner, and Vice President William Lye. Our College of Business Dean, David Stephens, and his faculty provided guidance and support in planning and conducting the seminars.

Assistance was willingly and conscientiously provided by numerous students. Kevin Baugh typifies the work ethic orientation of our students; he prepared the manuscripts with flawless accuracy and exceptional timeliness. Invaluable assistance in the planning and execution of the seminars was provided by the student staff including Brett Bagley, Greg Bassett, Jim Birch, Ray Buttars, David Bland, Mark Fitzgerald, Kris Hammer, Mark Hazelgren, Kristen Henrie, Shaun Henrie, Kristy Hoffmann, Jennifer Jenkins, Victoria Jurinak-Long, Julie Liday, Glenn Morris, Daryl Nielson, Sheila Ossowski, Kent Parker, Roger Preece, Paula Rosson, Connie Vance, Jay Ward, and Larry Ward. Several have graduated and are expertly applying their new understanding.

Part I

GLOBAL COMPETITIVE CHALLENGE

1

Overview

Y. K. Shetty and Vernon M. Buehler

In the last two decades, American business has undergone a crisis of declining competitiveness. Growing international competition and widespread globalization of industry have made many branches of U.S. industry that were once healthy and profitable moribund and ineffective today. It is of deep concern to national and business leaders. At the top of many corporate agendas now rests the determination to restore U.S. competitiveness.

A variety of proposals have been suggested, at the national as well as business level, to confront the challenge. They include: increased capital investment, reduced government regulation, stepped-up research and development, reduced labor costs, new labor-management initiative and the like. Though many of these measures are macro in nature, they have the potential for regaining the competitive advantage by directly focusing attention on improving productivity and quality through science and technology at the national and business level. The ability to compete and be profitable in business depends increasingly on a successful and sustained effort at improving productivity and product quality.

PRODUCTIVITY AND QUALITY

Productivity growth, a measure of how fast output increases relative to inputs (labor, capital, and natural resources), in the United States has been slowing since the mid-1960s and declined in the 1970s. In spite of its recent improvement, the lag in productivity growth is putting the United States behind other industrialized countries. Likewise, the ability of U.S. industry to provide quality products--defined as satisfying the levels of relevance, uniformity, and dependability required by customers--has also deteriorated seriously in recent years.

Slower productivity growth coupled with deteriorating product quality have placed the United States at serious competitive disadvantage in global markets. American business is slowly but surely beginning to recognize that restoring productivity growth and quality is critical to the very survival of the nation's business system. Signs of their genesis and the willingness to accept this challenge are appearing at the national as well as business level. The results include definite moves toward careful enhancement of productivity and product quality.

President Reagan's Commission on Industrial Competitiveness made several recommendations including: increasing incentive for research and development; commercializing new technologies through improved manufacturing; balancing regulation with the needs of industrial competitiveness; increasing effective dialogue among government, business, and labor; establishing new cooperative relationships between labor and management; strengthening employee incentives; improving worker skills; and enhancing business and engineering education. All the recommendations are aimed at improving U.S. competitiveness. Quality and productivity improvements are gaining momentum in the corporate sector also. John Young, president and CEO of Hewlett-Packard and Chairman of the Commission says, "In today's competitive environment, ignoring the quality issue is tantamount to corporate suicide. The private sector must improve the quality and cost competitiveness of its products through productive investments and innovations in technology and human resources management." F. James McDonald, president of General Motors, says, "If quality is not the number one operating priority at GM, there may be a time when there is no GM." Productivity and quality improvement are gaining momentum in companies such as General Motors, Westinghouse, Polaroid, Motorola, Ford, Hewlett-Packard, Northrup, Honeywell, and Xerox.

SOURCES OF COMPETITIVE ADVANTAGE

Productivity and quality are two major sources of competitive advantage. Improved productivity clearly reduces costs and enhances competitiveness and profitability. Such an advantage allows a firm to use its cost advantage to increase profit margins, lower prices, or both. Traditionally, in the United States, quality and productivity are often considered as distinctly different issues, even though the two concepts are closely linked. Quality is concerned with both the outputs and the inputs of the production process. Since productivity is the ratio of outputs to inputs, an increase in inputs and a decline in outputs lowers productivity. A deterioration in the quality of the products or services can disrupt schedules, delay deliveries, increase rework, increase scrap, waste manpower and materials and machine time, and increase warranty cost. Reworking products, inspecting parts, and the product lost due to scrap all lower productivity.

Quality of products and services not only increases costs but also affects sales. For example, poor quality may result in rejection and substantially diminish the chance of selling the customer a replacement product, thus precipitating a decline in the level of output. Furthermore, the damage to the company reputation can eventually cause incalculable losses due to reduced sales and market share lost to competitors. On the other hand, businesses that improve quality acquire a competitive advantage through quality-induced product differentiation, the creation of something that is perceived as being unique throughout the industry. Product differentiation on the basis of quality creates a defensible competitive position and insulates a firm against inroads of rival firms. Customers who prefer quality products are willing to pay more for the product. Customer loyalty and the uniqueness associated with quality are difficult barriers for new competing firms to surmount. In short, productivity and quality improvement are important sources of competitive advantage. In order to have improved productivity and quality we must have technology and we need to generate new knowledge through basic and applied research. This has to be pursued at the national as well as the corporate level.

NATIONAL SEMINARS ON COMPETITIVE CHALLENGES

This book grows out of the 1987 seminars on improving productivity and quality--through the management of human resources, information, and science and technology--within the context of U.S. competitiveness in the global market. Sponsored by the College of Business at Utah State University as part of the Partners Program, the nationally recognized annual series of seminars is conducted in collaboration with business constituencies of the college including Utah Manufacturers Association, American Society of Industrial Engineers, Utah Chamber of Commerce, U.S. Department of Commerce, American Production and Inventory Control Society, Utah Economic and Industrial Development Division, World Trade Association of Utah, Utah Association of Quality Circle Facilitators, American Society for Training and Development, Utah Bankers Association, Utah Association of Realtors, Utah

Association of Certified Public Accountants, and many
others. Objectives of the program are to:

1. Provide a forum in which members of business and
    academia can discuss the critical issues of produc-
    tivity, quality, science and technology, innova-
    tion, global competition, and human resource
    management.

2. Facilitate analysis of the experiences of well-
    managed companies that have successfully developed
    and implemented programs for improving produc-
    tivity, quality, innovation, and human resource
    management.

3. Give students an opportunity to meet and hear
    experienced executives discuss emerging management
    issues that generally have not yet appeared in
    textbooks.   Additionally, these seminars are
    designed to help our students observe business
    executives in action, interact and learn from them,
    and get advice on coursework and career planning
    for the workplace.

4. Help other firms learn from the experiences of
    well-managed firms.

These seminars demonstrate that encouraging
progress is being made in bridging the gap between
business and education.   Business and education must
collaborate closely to help prepare our managers, current and prospective, to meet the challenges of the
expanding global marketplace. As Roger Smith, Chairman
of GM puts it: "USU's Partners Program is a model
program for promoting such coordinated endeavors."

MAJOR THEMES

The major theme of this volume is improving
productivity and quality through science and
technology--being innovative in management of people,
information, and manufacturing technology. Part I
introduces and outlines a perspective for studying
productivity, quality, and science and technology
within the context of global competitive challenge.
This perspective examines selected public policy as
well as corporate challenges that will complement and
reinforce the changes outlined in Parts II, III, and
IV.   This material is drawn from lectures by several
seminar participants including John Kenneth Galbraith
of Harvard, John H. Moore of the National Science
Foundation, T. Boone Pickens of Mesa Petroleum, Mark

Shepherd, Jr., of Texas Instruments, Rosabeth Moss
Kanter of Harvard, and F. Kenneth Iverson of Nucor
Steel.

Part II analyzes human resource strategies for
improved productivity and quality, and includes the
experiences of several firms including Intel, Lincoln
Electric, Eaton Corporation, General Foods, National
Semiconductor, Monsanto, and others.

Part III discusses certain aspects of information
technology. Information technology is emerging as an
important tool for sustained improvement of productivity and quality. Experiences of several firms
including AT&T, Mead Corporation, Unisys, Arthur
Anderson & Company, McDonnell Douglas, Digital
Equipment, and others are analyzed here.

Part IV focuses on three major issues--
productivity, quality, and manufacturing technology.
Experiences in a wide variety of industries and compa-
nies are explored. Companies analyzed include Xerox,
GM, Hewlett-Packard, Ford, IBM, DuPont, Metropolitan
Life, and others.

Part V develops some general guidelines for
practice in developing effective strategies to gain
competitive advantage. The volume's central focus is
the problems of global competition and the ways U.S.
companies can position themselves in order to gain
sustainable competitive advantage.

PART I: NATIONAL AND BUSINESS INITIATIVES FOR GLOBAL
COMPETITION

The chapters in this section provide perspectives
on national measures as well as certain business initiatives aimed at improving U.S. competitiveness
through science and technology.

"Economic Priorities in Our Time: The Larger
Context," by John Kenneth Galbraith, traces the
experience of the 1930s and says that one lesson we
have learned is clear: governments of whatever political persuasion must be responsible for the overall
behavior of their economic system. The basic instruments of management of the economy are: (1) government
budget or fiscal policy, and (2) central banks or what
has come to be called monetary policy.

This system was successful for a number of
reasons. For one thing, it left the basic structure of
macroeconomics untouched--the market mechanisms, business enterprise, corporations, property rights, labor
relations, and, in basic measure, the distribution of
income. Second, in the quarter century following World
War II, this system worked wonderfully. However, since
the early 1970s the economic performance has been far

less satisfactory, resulting in inflation combined with rising unemployment and sometimes negative growth. Then in the early 1980s, the United States suffered a great slump. Recently, inflation ceased to be a problem but unemployment, particularly in parts of the country, has been high. Our economy has not responded well, resulting in the huge budget deficit in these last years. Also, we have had problems in mass production industries, farming, oil, and commodities. We have been running a huge deficit in our international trade account. It is this trade deficit that has largely canceled out the expansive effect of the budget deficit.

Part of the problem was inherent in the Keynesian system, says Galbraith. It assumes that while macroeconomic policy requires strong action, microeconomics can be left to the magic of the market. Keynesian measures to deal with deflation and employment are more politically feasible, while measures for dealing with inflation (higher taxes and reduced public expenditures) are politically unpleasant. This situation encouraged a resort to monetary policy which seemed to be a politically more agreeable line of action. However, emphasis on monetary policy has weakened our industrial base via higher interest rates.

Another factor contributing to the weakening of our mass production industrial base is what Galbraith calls the corporate senility. This manifests itself in the great increase in corporate staff and a multiplication of the layers of command leading to bureaucratic inflexibility.

What public policies are needed to correct the problem? First of all, Galbraith notes, we must recognize the limitations of monetary policy. That is, it is far better to restrain the economy, when that is required, by higher taxes or by a responsible fiscal policy than by high interest rates. The latter has a direct effect on capital investment that is most needed for the growth, modernization, and competitive competence of our industry. We must also develop a system that will encourage restraints between employer and union on the wage-price spiral. Other needed lines of remedy include the recognition of the aging tendencies of the great corporate enterprise. Galbraith also feels that tariffs for protecting the weaker firms or government bailouts are not the appropriate measures for improving our industrial strength. These may be short-term measures. But we must concentrate on long-term solutions including good education, public support for investment in advanced technology, and strong support for arts and design.

John Moore's essay, "U.S. Competitiveness and the National Perspective," discusses the emerging trends in science and technology and their relation to productivity. He believes that while public policy measures aimed at enhancing international competitiveness are essential, we still have to produce high quality products and services that are competitive. To do that, we must have the technology that is required. In order to develop technology required in today's--and tomorrow's--world, we need two things: (1) new knowledge and (2) people to produce and use it. This means that preconditions of competitiveness require new knowledge and education of science and engineering. The United States must greatly improve its performance in these two areas in order to compete in the future.

Over the past two decades, the world economy has changed and the market for nearly all significant manufacturing industries has become worldwide. Seventy percent of the goods we produce today must compete against the goods from abroad. These developments, Moore feels, have had a pervasive effect on the U.S. economy: the position of world leadership it took for granted during the decades immediately following World War II is no longer assured. Key areas of strength in industries such as steel and automobiles have eroded, and we are having serious problems even in high tech industries where we once held an impressive lead.

One key to improving our competitive position is enhancing our productivity--making better products at lower cost. The generation of the knowledge for productivity and, ultimately, for competitiveness is directly related to the health of our research and educational effort.

However, Moore says that our research and education base has weakened over the past two decades. In the 1960s, the United States spent more on research & development as a percent of GNP than any other nation. But since that time we have stood still while others have caught up. More specifically, in civilian research, Japan, West Germany, and France are now ahead of the United States. The amount of federal R & D spending devoted to basic research has remained relatively flat over the past two decades. While we have been diminishing our relative effort in basic research, other nations are emphasizing it, challenging one of our most important long-standing advantages. Furthermore, other nations have recognized that economic leadership depends on the quality and the availability of scientific and engineering personnel. They are making all-out efforts to gain an advantage in this area.

At this time when the United States is facing rising competition and its leadership in research and development and scientific and technical personnel is being threatened, the science and engineering fields are changing profoundly resulting in tremendous growth in the generation of knowledge in biotechnology, materials, information, science, and others. Also, research in all fields increasingly relies on complex and expensive research equipment, and requires the pooling of diverse interdisciplinary resources.

Understanding the importance of the above trends and changes, the federal government is responding to the problems by focusing on three major areas: (1) education and human resources, (2) science and technology centers, and (3) interdisciplinary programs. The National Science Foundation is playing a major role in these efforts. Moore believes that our survival in the competitive world is going to require great effort from all of us--in our schools, colleges, and universities, in our state and federal governments, and in our industrial enterprises. Productivity improvements depend on people and how they develop and use new knowledge.

In "Takeovers and Mergers: A Function of the Free Market," T. Boone Pickens, Jr., defends those who agree that the market is exercising a healthy discipline on managers. The proponents of this view say that the merger lawyers, investment bankers, and raiders such as Pickens are simply acting as brokers in a massive redeployment or restructuring of assets that will help U.S. business to enhance its competitive strength. If the Pickenses don't force the restructuring, the Japanese will do it for us by putting our companies out of business and taking what's left of our markets.

The opponents in this great debate over deal mania say that the raiders and investment bankers are chopping and reassembling U.S. business and forcing corporate managers to focus on the short term as they try to preserve themselves. In other words, they are too busy fighting Wall Street to fight international competition and are neglecting vital, long-term spending on plant, R & D, and job training.

In supporting the proponents of the market forces in this debate, Pickens says that everyone benefits from these transactions except possibly a few management egos. He contends that he works within the rules and the system, but despite this the "good ol' boys," as he calls 85 percent of the corporate managers, are really on his case. He calls their charges ridiculous when they claim he has a scheme to bust up U.S. industry. Pickens says that these professional managers are upset because they are vulnerable for having done a lousy job for the stockholders. The stockholders cannot afford the risk and expense required to make managers account for their actions. On the other hand, corporate executives are paying huge fees from stockholders funds for legal takeover defense. Furthermore, Pickens states that managers of targeted takeover firms

are pressuring bankers to deny credit requests from
Pickens.

Time will tell whether restructuring has been
beneficial. Pickens and other raiders have been able
to redeploy undissolved corporate assets and cut costs.
This can bring a quick flow of cash to stockholders and
enforce "lean and mean" organizations. This is what
Galbraith calls correcting corporate senility. Others
call it bloated corporate bureaucracy, or "corpocracy."
International competition will be the real judge of
this debate as to whether restructuring turns out to be
simply a reshuffling of the economic decks, or a source
of economic efficiency and competitive strength.

"Current Challenges for American Industry," by
Mark Shepherd, Jr., analyzes the massive changes taking
place in U.S. industry and makes a number of recommendations for improving its competitiveness. While the
United States is struggling to regain its competitiveness, the nations of the Asia/Pacific Region have
skillfully exploited their labor-cost and productivity
advantages to steadily increase their share of world
markets.

Productivity is the key to a country's
competitiveness. Lower productivity growth in the
United States has contributed to reduced profit
margins, flat production, downsized capacity in terms
of both capital and labor, and an increasing loss of
market share to imports. However, America's competitive strategy should not rely on forcing our trading
partners to give. up their legitimate advantages.
Instead, we should develop a strategy that builds on
America's strengths to tip the competitive scales in
our favor, says Shepherd. To this end, he suggests:

1. Balancing the financial scales between the United
   States and its international competitors. This
   would necessitate actions aimed at: increasing the
   availability of capital and reducing its cost and,
   reducing the distortions in exchange rates.

2. Improving the basic skills of the work force.
   Technical manpower shortage should be remedied
   through encouraging more students to pursue
   engineering and science education.

3. Accelerating the development of advanced
   manufacturing processes that facilitate the transition from R & D prototype to full-scale commercial
   production. The government should encourage R & D--
   and help provide firms with the cash flow necessary
   to develop advanced manufacturing technologies.

Shepherd believes that in these changing times our traditional ideals of freedom, patriotism, and spirit must be blended with new values: a zeal for winning; a firm belief in fiscal responsibility; a determined effort to tilt our nation's resources toward productivity and investment; a revival of the American work ethic; and a firm determination to manage our destiny. It is time for American ingenuity to come to bear on the problem of international competition.

Lawrence Schein's chapter, "Changing National Business Culture," deals with two major areas. First, it analyzes the impact of the changing business culture on managing people. Second, it provides some ideas about how individual companies change their own culture or use their cultures to adapt to the painful process of change. The forces that are causing change in the national business culture include shifts in the world market place, political conditions, political alliances, demography, and the educational level of the work force.

Schein focuses his attention on the massive corporate restructuring in America. This has been triggered by several forces coming together simultaneously: lagging productivity, very intense competition in many industries, and a very substantial increase in our trade deficit. Restructuring is being carried out by downsizing, reduction of labor forces, stripping out of layers of management, and unprecedented activity in mergers and acquisitions. These actions have many implications for management, particularly human resource management. First, we will see increased emphasis for promotion and advancement being placed on the performance criteria. Second, more attention will be given to job rotation, lateral transfers, cross training, learning multiple skills, and making work more challenging. Third, there will be a shift in managerial culture from controlling the worker to getting the worker involved and committed.

Schein notes that corporate culture, which is influenced by the changing national business culture, can be used to manage crisis, to create an identity, and to provide the "glue" to hold a company together. However, cultural changes are not easy and require a great deal of time, patience, and managerial effort.

In her essay, "Managing Change in Innovative Organizations," Rosabeth Kanter, author of one of the most talked about books, The Change Masters, discusses the environment that makes it possible for people to master the change adequately and develop corporate support for winning the new competitive game in the business world. This is the new game: everything is in flux, technology is changing rapidly, customers, employees, and users are more demanding and willing to shop around, and the whole structure of the industry is changing. Consequently, it is very difficult to win a game like this using conventional methods and innovation is critically important.

Kanter says that in order to win the competitive game, business has to do a number of things. First, be focused on just a few things that you know how to do very well, to build on competence and avoid being distracted to trying to keep up with too many things. Second, develop the ability to move quickly to gain the benefit of "first mover" advantage. Third, be flexible to keep bending and redirecting to take advantage of the new opportunities that are arising in an environment of constant change, where technology provides an opportunity a minute. Behind all of the above three actions, the need for partnership is essential. The best strategy here is to befriend all the groups in the environment on whom the company depends--customers, employees, and suppliers. Finally, in order to make these happen, we need a particular corporate environment. Stimulating innovation and entrepreneurship is a method of having an integrative organization, where people pull together rather than apart, joint projects and joint developments are encouraged, communication and information knit everything together, and shared goals and focus for a business create unanimity of purpose. The real issue in managing an innovative organization, according to Kanter, is to create more opportunity for people themselves to take charge of change.

In his chapter, "Effective Leadership: The Key Is Simplicity," Kenneth Iverson provides some background on Nucor Steel, a nationally recognized company that is known for its successful competitive strategy, and outlines his views about what makes a good manager in business.

Nucor used a low-cost competitive advantage by employing more efficient mini-mill technology. The prices of their products have been equal or less than the price of the same products produced by foreign steel companies. The company has been highly profitable and in a number of years its return on stockholders' equity exceeded 20 percent.

When many of the U.S. steel companies are continuously losing money, why has Nucor been successful? Nucor's success is directly related to its ability to achieve a low-cost advantage largely resulting from its organization structure and employee relations program. Following Iverson's "lean management" philosophy, only four levels of management separate Iverson from the hourly employees. Nucor is a decentralized corporation and managers make day-to-day decisions that determine the success of the company. It also has very few staff personnel because of top management's great confidence in operating personnel.

Nucor is also known for its employee relations program. A major part of the employee motivation system at Nucor is its employee incentive system. Many workers earn very attractive wages and this has made the company attractive to job seekers. The company has never laid off or fired an employee for lack of work in the past fifteen years. In addition, the company maintains an open communications system so that employees are familiar with the company's plans and problems. It provides attractive benefits to employees and makes an all-out effort to eliminate the distinction between management and employees in its management practices. What is the result of this type of program? The result has been a very productive organization. The average integrated steel company in the United States last year produced about 350 tons per employee. Nucor produced about 980! The company prides itself on being the most productive steel company in the world.

In short, Iverson's message is simple: even in a highly competitive smokestack industry, it is possible to achieve high productivity and compete successfully. The ingredients of success are: modern technology, highly motivated employees, and an enlightened environment.

Major themes of these seven chapters are clear. Our public agenda should include encouraging capital investment, building healthy labor-management relationships, improving the skills of the work force, supporting investment in research and development, facilitating healthy redeployment of corporate assets, and improving education, particularly technical education--all aimed at improving productivity and quality for regaining our competitive advantage. The private sector must improve quality and cost competitiveness, create corporate environment for innovation and entrepreneurship, and improve its products through the use of the most modern technology, enlightened human resource policies, and creative management.

2

Economic Priorities in Our Time: The Larger Context

John Kenneth Galbraith

LESSONS OF THE KEYNESIAN ERA

History, as it is often practiced, can be an escape from the inconveniences of modern reality. But history is also an indispensable guide to the present, and to the questions of the present policy with which you are today concerned. Anyone seeking to understand modern economic life and behavior, or frequent misbehavior, must begin by going back in time. Specifically, one must now go back half a century to the early months of 1936, and to the publication in that year of The General Theory of Employment, Interest, and Money by John Maynard Keynes. That year marked the beginning of what has ever since been called the Keynesian Era.

The great step that was taken fifty years ago has been variously and often erratically described. The essence was clear. It was that governments of whatever political coloration--Republican, Democratic, right or left--would thereafter be responsible for what economists call the macroeconomic behavior of their economic systems. Until the Age of Keynes, it was assumed that the performance of the economic system was autonomous, something that could be left to itself. It reflected its own employment and its own price-stabilizing tendencies. The level of employment and the stability of prices were thought to have a general tendency to be benign. There will be fluctuations--business cycles-- but they were self-correcting. Government served best by keeping its hands off. Depressions, in the metaphor of Joseph Schumpeter, one of the great figures in the early part of this century and a onetime colleague of mine at Harvard, served to extrude the accumulated poison in the economic system.

In the devastating context of the Great Depression of the 1930s, Keynes brought this relaxed view to an end. Thereafter, in all countries, governments of all political complexion would take responsibility for the level of economic activity, for having

an increase or growth in that activity, and for maintaining what would be thought an acceptable stability
of prices.   Taking responsibility   for all these
things, I have no hesitation in adding, does not mean
that government always successfully fulfilled that
responsibility.

The basic instruments of this overall or macroeconomic management of the economy are not well
recognized.   They were and continue to be: 1) the
government budget, what has come to be called fiscal
policy, and 2) central bank action or what has come to
be called monetary policy or monetarism.   And there is
no mystery as to how these work or, in any case, how
they were meant to work.   Given insufficient demand for
goods and services in the economy with the consequence
of underemployment of manpower and other production
resources, demand would be fiscally enhanced by an
excess of public expenditure over taxation or income.
There would be a deliberate resort to deficit financing.   And with overemployment and inflation, the
reverse policy--higher taxes and less public spending--
would be employed to dry up the excess purchasing power
that was pressing on markets and causing prices to
rise.

Complementing this would be monetary action.   Low
interest rates and consequent encouragement of easy
lending and easy investment in industrial enterprises
and in housing would enlarge demand.   There would be
more consumer borrowing and spending or investing.
These would expand the economy.   A restrictive monetary
policy, the opposite policy, would be pursued when
demand was pressing on capacity and threatening inflation.   There were an infinity of requirements and
qualifications associated with the policy that I have
just mentioned.   There was varying emphasis on the
different steps, and the timing of these steps.   There
is, however, no serious disagreement in what has come
to be called the economic mainstream as to the central
thrust of Keynesian policy and of the Keynesian
Revolution.

This revolution, as I've said, lodged itself
deeply in the thinking of the industrial countries.
Three factors contributed to this.   The first of these
factors that gave credence to this system was the
public suffering, and indeed the very threat to a
capitalist system itself, from economic depression.
Marx saw depression, the capitalist crisis, as sounding
the death knell of capitalism.   During the depression
years of the 1930s, that bell was at least distantly
heard.   There was deep alienation in the industrial
countries and an undercurrent of revolutionary senti-
ment.   Socialism and communism seemed a viable
alternative.

The second factor lodging the Keynesian system in the public policy of the industrial countries was, oddly enough, its highly agreeable conservatism. It took responsibility for the overall behavior of the economy--macroeconomics. But it left the basic structure of microeconomics untouched--the market mechanisms, business enterprises, corporations, property rights, labor relations and, in basic measure, the distribution of income. The Keynesian Revolution, in other words, left the essentials of capitalism unchanged. There were some, indeed many, who thought at the time that Keynesianism was a very radical thing. We should always be aware that the conservative fears change as endemic and is by no means discriminating. No revolution, in fact, could have been more protective of what those with the greatest stake in the system had the greatest concern to preserve.

The third reason for the success of the Keynesian Revolution was that in the quarter century following World War II, a little more than a quarter century, or let us say the years from 1945 to around 1972, this system worked. Indeed, it worked wonderfully. These years were a most remarkable time in the history of capitalism. The scars of war were soon healed and forgotten, and the industrial economy settled down to an extended period of political tranquility, stable prices, low unemployment and steady economic growth.

In only two years between 1950 and 1970 did the U.S. economy fail to register expansion. In most years the expansion was very substantial. Over the whole period real output of goods and services about doubled, and, as I have noted, prices were generally stable, inflation was not a problem, and unemployment was very low. Economists of my generation were not reluctant to take credit for this achievement. We did. As compared to later generations, we were very shrewd in our selection of the particular time in which we chose to practice our profession. The next generation would not be so wise.

PROBLEMS OF THE 1970s AND 1980s

I come now to the present or the near present. In the decade and a half since the early 1970s, the economic record has been far less satisfactory. At the very beginning of the 1970s, inflation showed signs of becoming a problem. And after 1972, inflation was combined with rising unemployment and sometimes negative growth. This led to the invention of the least lovely word in the economic lexicon. These, it was said, were years of "stagflation."

Then in the early 1980s we suffered here in the United States the greatest slump since the Great Depression itself. And our depression pulled much of the rest of the industrial world down with us.   In 1981, 1982, and 1983, inflation ceased to be a problem, but unemployment in these years was at distressing levels. So also were small business failures and agricultural distress. Economic growth was negative. Since then there has been some recovery, but. it has been generally slow and uncertain, and unemployment, in certain parts of the country, has been, by all earlier standards, very high.

Our economy has not responded as we might have expected to the huge budget deficits in these last years nor, as would have been presumed by the Keynesian analysis and expectation, have the deficits had a very large expansive effect. (These deficits must be wonderfully impressive to John Maynard Keynes from wherever he is now watching. He must be quite extraordinarily surprised at the politics of his modern disciples.) In these last years we have had yet further problems. The older mass-production industries, the farmers and the commodity producers generally (as I am reminded to say here in Utah) and other industries, not excluding oil, have been singled out for continued suffering. And through what has been called by some--I hope there is no adverse ethnic overtone here--the Mexicanization of our economic policy, we have been running a huge deficit in our international trade account--buying far more than we sell and borrowing heavily from abroad to cover the difference.   It is this trade deficit that has canceled out the expansive effect of the budget deficit. If money is spent abroad for goods, it does not expand our domestic economy, and what we have spent abroad has, by and large, offset the excess of Keynesian spending by the federal government through the budget deficit. In further consequence, we have converted ourselves from being the largest creditor nation in the world to being the largest debtor nation. The debt service on international accounts and the stability of the dollar itself are problems that are either with us or are strongly in prospect.

POLITICAL ASYMMETRY OF KEYNESIAN MODEL

We must ask ourselves what went wrong, what lies back of this reversal of what was previously a period of great good fortune. The control of nuclear arms and getting out from under the threat of nuclear war apart, this is the most compelling question in our time.

The answer is not simple, but neither, I think, is the answer beyond the reasonably accessible view. Part of the problem was inherent in the Keynesian system. More arises from some new, obtrusive and neglected problems in the market system, in microeconomics, that part of the economic system to which Keynes gave only passing attention. We have been suffering very severely from the Keynesian assumption that while macroeconomic policy requires strong and intelligent action, microeconomics can be left to the magic of the market, to which we can apply the ancient doctrine of laissez-faire.

The microeconomic failure arises from what we may call the political asymmetry of the Keynesian model. That is a rather formidable expression, and I use it to prove, as economists do, that I have some language, some insight, that is not available to the public at large. By the political asymmetry of the Keynesian system I mean that the policy is politically very different when applied to inflation as compared to when it is applied to unemployment. It was politically easy and politically rewarding in the Keynesian system to act against unemployment and deflation. That required lower taxes, higher public expenditures, and lower interest rates. These were politically very agreeable actions. At least for the years until roughly 1970, they were generally the required measures. There was, in those years in all the industrial countries, a broad tendency of the economy to deflation and unemployment, the conditions that had characterized and caused the Great Depression. But after 1970, inflation became a central and compelling concern. For inflation, the relevant Keynesian action was higher taxes and reduced public expenditures. This was politically unpleasant; thus the political asymmetry. Deflation and unemployment were politically pleasant to deal with; inflation remedies were politically very disagreeable.

This practical manifestation I saw at first hand. When a President saw that he had economists on his appointment schedule in Washington and he knew that they were proposing lower taxes and more ample public spending, he did not feel depressed. He went ahead with the appointment because he knew they were talking about pleasant action. But with inflation, when he saw that he had economists on his calender, he knew they were coming in to talk about higher taxes, lower public expenditures. In consequence, he postponed the appointment certainly for a week, possibly for a month and in some cases forever.

Along with this political asymmetry we also in these years had a new intransigent cause of inflation that came out of the microeconomic sector. This was the increasing seriousness of the interaction of wages and prices. We had wages shoving up prices and prices pulling up wages, something that did not yield easily

to the modest restraints that were implicit in fiscal
policy and a phenomenon that the Keynesian system did
not wholly foresee.

MONETARY POLICY CAUSED SUFFERING

     The political asymmetry of Keynesian fiscal or
budget policy encouraged another step. It encouraged a
resort to monetary policy, which seemed a politically
more agreeable line of action. Instead of facing up to
the need to increase taxes, the need to reduce public
expenditure, as measures against inflation, the government in the late 1970s and early 1980s sought an escape
from that unpleasantness. It escaped from the world of
John Maynard Keynes into the world of the next most
influential economic figure of this century, Professor
Milton Friedman. With monetary policy--firm and intelligent control of the money supply--Professor Friedman
promised a relatively painless end to inflation.
     Monetary policy worked through high interest rates
and restrictive lending. And we learned that this
worked only as it imposed severe restrictions on business investment as well as on housing construction and
also on consumer spending. In other words, we learned
in 1981, 1982 and 1983 that monetary policy worked only
as it produced a great deal of unemployment, a great
deal of idle plant capacity and a great many failures
of small businessmen, the highest rate of failures
since the Great Depression.
     Professor Friedman, however, had held that the
suffering from such constraint would be brief. He went
so far as to say that monetarism had a kind of special
magic of its own that could not be fully explained. In
fact, we escaped the political asymmetry of the
Keynesian fiscal policy through a remedy that was worse
than the disease it sought to control.
     There was another source of support for monetary
action that is not discussed among my economic colleagues. High interest rates with their operative
restraints on investment and on consumer borrowing
reward people who have money to lend. As a broad
proposition, those who lend money are likely to have
more money than those who do not have money to lend.
There is a certain internal logic about that proposition that I hope will appeal to you. (There are some
inescapable conclusions in economics; one of them, my
favorite, is attributed, perhaps falsely, to Calvin
Coolidge. He is said to have observed that when many
people are out of work, unemployment results.) Monetary policy, accordingly, is not, as often supposed,
neutral. It has a strong tilt in favor of the

financially favored. Monetary policy, in consequence, has a strong affluent support, especially as compared with fiscal policy, which has always in it the threat of higher taxes. Professor Friedman, an old friend of mine, is avowedly and proudly a conservative, and in his service to the affluent he wholly keeps his faith.

So, in summary, an astringent monetary policy beginning under President Carter in the late 1970s and continuing with great severity in the early 1980s with the advent of the Reagan administration did break the force of inflation. But it also produced, as I have noted, the most severe recession, let us say depression, since the 1930s.

Here the role of microeconomic factors also enters. Modern inflation, as I have just noted, is caused not only by an excess of aggregate demand; it is caused also by the direct pressure of wages, trade union wages in particular, on prices, and in turn the upward pull of prices on wage contracts. This microeconomic process is not arrested by a slight or modest reduction in demand. We have learned that it is arrested only by severe cutbacks in employment, particularly in the mass-production industries. And it is also only arrested as employer power is curtailed, as the ability to pay higher wages is sharply weakened. Only when that happens do wage increases come to an end. Only as employers are weakened do we have wage stability or wage reductions--givebacks--as the union takes responsibility for keeping the employer in business. An astringent monetary policy arrests wage and price inflation only by drastically weakening the strength and vitality of the employing industries.

One of the great and somber lessons of these last few years is that a strong trade union movement requires strong employing industries. The companion lesson is that monetary policy escapes from the political asymmetry of fiscal-policy controls on inflation by drastically and even permanently damaging the industrial base of the countries that pursue it. The rust-belt cities of the American Middle West with their abandoned shops and mills are a monument to the magic of monetary policy.

There was a further and final effect of monetary policy: its high interest rates attracted investment funds to the United States. This attraction bid up the dollar and had the effect of subsidizing imports. These imports further damaged the already weakened mass-production industries and also lodged foreign goods firmly in the American markets, where they still remain. We have learned that once foreign products are solid in the market, a changed level of the dollar does not cause them to be removed easily.

CORPORATE SENILITY

   Our tendency in the post-Keynesian years to associate all public policy with macroeconomics has thus
caused important developments to escape our notice.
There are still some further factors that have not come
fully to our attention.
   We have not in these last years seen the aging
tendencies that are at work in great corporate structures. We have long recognized that, with the passage
of time, feeblement and eventual senility are the tendency of the human frame. We agree on that. We have
not recognized that intellectual senility is also the
tendency of the great corporation. In the older industrial countries, and especially here in the United
States, we are suffering from the bureaucratic and
intellectual sclerosis that affects the older firms in
the mass-production industries.
   This manifests itself in a great increase in corporate staff and a multiplication of the layers of
command. We attribute importance to individuals in a
corporation in accordance with the number of people
they command. "How many does he have under him?" And
there is a powerful tendency in older industrial estab-
lishments for intelligence to be measured by what is
most companionable to those already there, what somebody who is already there already believes. And we
approve as wisdom whatever most closely accords with
what is already being done. This sclerotic tendency we
have not recognized.
   We have also not recognized until very recently
the singular flexibility, the singular competence, of
the new countries in the industrial world--of Japan and
the even newer industrial countries of the Pacific
Basin. We see here being played out the longest-
running and best-established scenario of international
economic development. That is the advantage in all
industry that accrues to youth. This was the advantage that the United States and Germany once had in
relation to Britain. It is the advantage that Japan
has over us, and it is the advantage that Korea,
Taiwan, and Brazil are coming to have vis-a-vis Japan.
A year ago I went with my wife to Nagasaki. While
there, we were taken on a tour of the great harbor as
well as the rather grim terrain where one of the first
atomic bombs was exploded. As one goes along that
great harbor for a mile, one passes the huge Mitsubishi
ship-building and ship-repair plants. The work has
gone on to Korea and Taiwan.
   Not all of this advantage is from younger and
better management. Some, perhaps most of it, comes

from having a labor force that is new to industrial
employment. In all countries at all times, traditional
industry has performed best with labor recently
recruited from the even more demanding, even more ill-
paid toil of rural life. The first generation of
workers in industry is the most efficient industrial
generation. It is so because to that generation industrial work seems easy in comparison with agricultural
toil. Earnings also seem abundant, and the disciplines
of industrial toil are a slight thing as compared with
the much more severe oppression of a hostile and
demanding nature. As I have noted, this vigor of the
younger countries is a threat that Japan now faces.
Far more, needless to say, it is a threat to us here in
the United States.

NEEDED POLICY CHANGES

Now having reviewed the problems that we have
accumulated, I must turn to the answers. What should
be the policy in the years ahead? I do not have any
great hope for the next twenty months. Perhaps I could
be tempted to ask what should be the policy in the
post-Reagan era? This should not be taken as any sort
of a political comment. I, of course, adhere rigidly
to one of our severe conventions, which is that no one
should ever allow his politics to be revealed in the
course of a public lecture.

As to macroeconomic policy the answer is clear.
We must, first of all, as will be evident, see the
magic of monetary policy as one of the more grievously
destructive policies of modern times. We must see that
it is far better to restrain the economy, when that is
required, by higher taxes or by an otherwise responsible fiscal policy than by high interest rates. The
latter have a direct and inescapable negative effect on
capital investment, the very expenditure which is most
needed for the growth, modernization and the competitive competence of our industry.

I do not urge fiscal action against social expenditure. Such expenditure is important for its contribution to social tranquility, another markedly
conservative goal. We do have a large opportunity in
the United States for expenditure reduction--and for
making capital and manpower available to civilian
industry--by getting our defense expenditures under
control. Our defense expenditures are not now a function of military need; to a very substantial extent, as
President Dwight D. Eisenhower warned as he retired
from office, they are the result of the exercise of
military power on its own behalf.

But also let us be willing now to accept the need for higher taxes to close the budget deficit. This we should then combine with lower interest rates. This will ease the transfers on the public debt account. It will also help our Third World debtors. And, a matter of emerging importance, it will also reduce the amount of money we will have to pay to our foreign creditors on the foreign debt account.

But we must not in the future rely exclusively on fiscal policy to keep inflation under control. We have seen that fiscal policy also works against inflation by restricting demand, creating unemployment. With wage/price interaction as a cause of inflation, fiscal policy, if it is to be effective, must be severe. That severity we have seen also has a deeply damaging effect, not only on employees but on employee and employer alike. It works only as it weakens the employing firm. So the further answer, the only answer, is to have a system of agreed restraints on the wage/price spiral between employer and union. In one form or another, this kind of restraint is now accepted in most of the industrial countries of the world. It has reached a very high level of sophistication in the Social Market Policy of the Austrians. And it is accepted policy in Germany, Switzerland, Scandinavia, and of course Japan. The English-speaking countries have lagged in putting such a social contract into effect. In the English-speaking countries, we still relish the destructive enjoyments of old-fashioned class struggle, old-fashioned class conflict. We are almost the last to do so, the last to adhere to the Marxian commitment to class struggle.

The prospect as regards the other needed lines of remedy is perhaps a bit more clouded. Perhaps we are coming to recognize the aging tendencies of the great corporate enterprise. Something is gained when we do. We are certainly escaping to some extent from the self-satisfied euphoria that characterized much of our corporate culture in the aftermath of World War II. We are now having a healthy reexamination of the nature and efficiency of that culture. I do not, however, believe that what is now being done in Wall Street is at all satisfactory. We are now having a great episode of asset shuffling--the take-over and acquisition mania. This will not make our industrial establishment more efficient. It is diverting attention from the necessity to improve productivity. That is our greatest need. No one should suppose that the financially eager young men who are now making fortunes on Wall Street are contributing anything to the improved efficiency of industry. They are extracting personal income from the system to no purpose; they will continue to do so until, perhaps, a few more become public charges in our minimum security jails.

I do not rise every morning, as economists generally are required to do, to make obeisance to free trade. But I am not enthusiastic about tariffs for protecting the less confident firms of the older industrial countries from the younger and more effective producers of the New World. I think we should go slow on this. If some of the measures now before the Congress are enacted into law, we will be putting protective fences around the corporate senility and incompetence to which I have already adverted.

I do not respond well to another solution that we have followed in the past and that has become very common in other of the older industrial countries. That is to have the government bail out or take over the faltering industrial enterprise. It has been our policy vis-a-vis a large number of banks and thrift organizations. We should be aware that socialism in our time consists largely of the failed children of capitalism. First we have speeches on the eternal verities and values of free enterprise, then the corporate jet to Washington. This is a very costly solution, and it leaves the government saddled with the congenital losers, including a certain number of firms that deserve to lose.

However, to ease the pain of short-term transitions, retraining, relocation assistance, pay until retirement for older workers, and even temporary tariff protection are all meritorious steps. We should, as a longer-run solution, have good education, public support to investment in advanced technology and, a much neglected point, strong public support for arts and design. Arts and design are very much a part of the future. After things work well, people want them to look well. After people have enough things, they want entertainment and enjoyments. Here is an area where we still have a substantial advantage. (Nobody can compete with us in the production of socially depraved television programs.) These areas of advantage we somehow or other decline to recognize. We do talk about the possibility of high-tech industry; we very rarely talk about our advantages in highly accomplished arts, design, entertainment.

I am not wholly optimistic about the economic prospect. I would like to send you away with an amiable feeling of confidence. One would be more sanguine if we did not have the present formula by which we evade the problem and evade the need to act on it. During these last years we have made it a minor act of religious observance to avoid thought and action on specific measures of the sort that I have urged, both macroeconomic and microeconomic. During these last years we have   returned our faith to the eighteenth-century belief in laissez-faire, to the belief that markets are comprehensively benign and planning for the future, somehow subversive.   We have turned our faith to the belief that God is for free enterprise and will always provide.   I, myself, do not believe that God is that much involved with American economics and politics.   What we have is a wonderful design for evading what should be painful thought and, on occasion, quite painful action.   This is not the way by which we should ensure the future of what we, in our courageous moments, still call the capitalist system.   If we are concerned with that system, we must address ourselves to it with far more attention and imagination than we have yet been inclined to use.

3

U.S. Competitiveness and the National Perspective

John H. Moore

In Washington, the word of the day is economic competitiveness--you might say the buzzword of the day. As just one indication of its prevalence on the Washington scene, at least sixty bills professing to deal with competitiveness have been introduced in Congress in 1987.

What do we mean by competitiveness? A couple of years ago, a presidential commission chaired by John Young produced a definition of competitiveness that says it well. It goes like this:

> Competitiveness is the degree to which a nation can, under free and fair market conditions, produce goods and services that meet the test of international competition while simultaneously maintaining or expanding the real incomes of its citizens.

Clearly, seeking fair conditions for marketing our goods abroad is consistent with that definition. But even under the best of international trade conditions, we still have to produce high quality goods and services that compete. To do that, we must have the technology that is required in today's--and tomorrow's--world. And to succeed at that, we need two things: (1) new knowledge and (2) people to produce and use it. Many of the bills I mentioned have this as an objective.

So at the NSF we see competitiveness as consisting of two principal elements: new knowledge and education of science and engineering personnel.

We must greatly improve our performance in these areas if we hope to compete in the future. Why is this? What has changed in the world that leads to our concerns about knowledge and education?

THE AGE OF KNOWLEDGE AND GLOBAL MARKETS

Over the last two decades, the world economy has changed dramatically. Along with a rapid growth in the volume of world trade--a sevenfold increase since 1970--there has been a profound change in its structure. The market for nearly all significant manufacturing industries has become worldwide. In a total reversal of the pattern of thirty years ago, 70 percent of the goods we produce today must compete against merchandise from abroad.

These developments have had a pervasive effect on the U.S. economy. The position of world leadership we took for granted during the decades immediately following World War II is no longer assured. The trade figures give a rough idea. An examination of U.S. trade balances in high technology and other manufactured product groups shows that:

1. Key areas of strength, such as steel and automobile manufacture, have eroded.

2. We are having serious problems in areas where we once held an impressive lead--high tech industries like semiconductors and software.

The situation is not improving--it's getting worse. Obviously, one key to improving our competitive stance is enhancing our productivity--making better products at lower cost. And the generation of new knowledge is key to that, as our history shows. A comparison of the contributors to U.S. productivity reveals the following:

Technological Innovation	44%
Tangible Capital	16%
Scale Economies	16%
Education	12%
Better Resource Allocation	12%

This clearly shows technological innovation, achieved by applying new knowledge created through basic research, has been responsible for nearly half of all U.S. productivity gains since World War II.

STATUS OF U.S. SPENDING FOR RESEARCH

Our ability to generate and apply new knowledge for productivity and, ultimately, for competitiveness is directly related to the health of our research and education enterprise.   That's what lies behind productivity.

But our research and education base has weakened in relative terms over the last two decades.   True, there has been a great increase in spending on R & D in the last few years.   Federal spending for R & D has almost doubled since 1980, from $33 billion to $63 billion.   But these increases must be put in context by making a country comparison of national expenditures for performance of R & D as a percent of Gross National Product.   In the 1960s the U.S. spent more on R & D as a percent of GNP than any other nation.   But since that time, we have stood still while others have caught up. And while other nations have been placing more and more emphasis on civilian research, the bulk of recent spending increases in U.S. R & D has gone to military research.   As a result, in civilian research, Japan, West Germany, and France are now ahead of the United States. When comparing civilian R & D as percent total federal R & D we see that, in 1988, defense functions will account for about 70 percent of the federal R & D budget.   At one time, investments in military R & D fueled advances in the civilian sector.   That is no longer the case in many critical areas.   In fields like biotechnology, computers and semiconductors, civilian research is ahead of the defense sector.

These observations pertain to both research and development. But the picture in basic research, which underlies our technological developments, isn't any better. A comparison of total federal and basic research spending, 1960-1988, shows the amount of federal R & D spending devoted to basic research has remained relatively flat over the last two decades.   So obviously it has fallen as a percent of the total.

A comparison of R & D at universities as percent total federal R & D shows that our universities are at the heart of our basic research effort.   Federal research at universities has risen significantly since 1980, increasing from about $4.2 billion to $6.4 billion.   But, as basic research has received a shrinking share of the federal R & D funding pool, the proportion of federal R & D spending supporting work at colleges and universities has experienced a decline.

While we have been diminishing our relative effort in basic research, other nations are emphasizing it, challenging one of our most important long-standing advantages. A look at the Japanese effort on research shows:

1. There is a new drive in basic research in Japan, which supplements their long-standing efforts to apply science and develop technology. The image of low-wage workers and derivative technology--once quite accurate--has yielded to one of technological sophistication and enlightened management.

2. And there is another transformation coming: Japan has begun a reform of its educational system that will place much more emphasis on individual creativity, a characteristic that its leaders expect will help it advance to the front rank in basic research. Lots of people are skeptical about this--and especially about creativity in Japan.

3. Lest anyone doubt Japan's ability to do creative fundamental work, look at what they have achieved in electronic materials. A 1986 National Research Council study concludes that they are ahead of us in these critical areas:

    - Microwave plasma processing

    - Lithographic sources

    - Electron and ion microbeams

    - Laser-assisted processing

    - Compound semiconductor processing

    - Optoelectronic integrated circuits

    - Three-dimensional device structures

    Publications in scientific journals also show a clear strengthening in the Japanese research effort.
    Japan is not the only example. There is a renewed effort in the European Economic Community countries. To judge by publication data, France is making important progress in basic science. And there are numerous community projects, such as Esprit, Race, Brite, Alvey, and Eureka.
    We should also not forget that we can expect serious challenges to our research leadership from the newly industrializing countries of the Pacific Rim. There is a shift in capabilities occurring in the world, a shift to the Pacific. Japan is just the first to represent serious inroads on our competitiveness in this high tech way. Others are following.

STATUS OF U.S. SCIENTISTS/ENGINEERS IN R & D

These are discussions of spending. But that's not
the only indicator of the increased capabilities of our
competitors. Recent trends in human resources also
reflect the sweeping change that is taking place around
the world. Other nations are recognizing what we have
long known--that economic leadership depends on the
quality and availability of scientific and engineering
personnel.

We see important trends by comparing country data
on the scientists and engineers in R & D per 10,000
labor force population. What's the importance of this?
It shows that twenty years ago, the United States led
the world in S & E personnel as a percent of the labor
force. Since then, others have made substantial gains,
while we have essentially stood still. Japan, for
example, has doubled its technical workforce in the
last two decades. By studying the U.S. average annual
growth in scientist and engineering employment versus
other manpower and economic variables, we see the
increased importance of the scientific and engineering
talent pool. The data show that in the last decade the
employment of scientists and engineers increased three
times faster than total U.S. employment, and twice as
fast as total professional employment. There is no
sign of change in this trend. But despite these
trends, the proportion of our students pursuing science
and engineering bachelor's degrees has dropped. Science and engineering B.S. degrees account for a lower
percent of all degrees than they have at any time since
1960.

When looking at science and engineering
baccalaureates attaining Ph.D.s after seven years, we
see problems at the graduate level. The proportion of
B.S. graduates who stay on through the Ph.D. has
experienced a long, steady decline over the last two
decades. While fewer U.S. students are in the Ph.D.
track, more and more foreign nationals are. Nearly 60
percent of all engineering doctoral students and 40
percent of all mathematics Ph.D. candidates are now
foreign nationals. Overall, foreign students have
accounted for almost 85 percent of the growth in graduate education in the United States in recent years. We
are creating a situation where we are highly dependent
on a resource over which we have little control.

The lack of U.S. student interest in attaining
Ph.D.s is especially disconcerting in light of recent
employment trends. Since 1975, industry has been
hiring Ph.D. scientists at a rate nearly double that of
other sectors, expanding its scientific Ph.D. pool by

almost 110 percent. And because of security requirements, many of these jobs cannot be filled by foreign
nationals.

These trends are even more alarming given
demographic changes. Our data show that the number of
twenty-two-year-olds in the U.S. population has already
peaked. We have begun a long decline in the number of
young people from whom future scientists and engineers
can be drawn.

The composition of the college-age population is
changing as well. Minorities and women, who have historically had low participation rates in science and
engineering, are becoming larger fractions. These are
the largest pools of available personnel that are not
now involved in science and engineering. We need to
expand their participation, not just because that is
good policy, but because we cannot afford to waste
their talent.

PROFOUND CHANGES IN SCIENCE/ENGINEERING IMPACT INSTITUTIONS

At the same time that competition is rising,
science and engineering are changing profoundly. These
changes influence our ability to respond effectively to
the challenges posed by economic competition. Let me
describe these trends. Nearly all fields are experiencing tremendous growth in the generation of
knowledge.

1. Advances in established disciplines are paralleled
    by the emergence of new disciplines, like bio-
    technology, new materials, and information science,
    which are providing some of the most   exciting
    opportunities for the growth of knowledge.

2. New instrumentation is making possible
    unprecedented advances in nearly every field.
    Remote sensing, high-speed lasers, scanning tun-
    neling electron microscopes, and the automatic gene
    sequencer are prominent examples, but they are not
    the only examples.

3. Large-scale computing is fundamentally altering the
    way we do science. The modeling, computation, and
    simulation it makes possible provide a new research
    methodology that takes its place alongside theory
    and empirical research.

As a result, institutional changes in the science
and engineering research enterprise are occurring.

1. Increasingly, research in all fields relies on expensive and complex research equipment, which makes sharing and cooperation inevitable. Even mathematics, whose instrumentation needs were once satisfied by a reasonable supply of paper and pencils, has come to rely on computers.

2. Some of the most promising research, like biotechnology, information science, and materials research, is now located in the intersections of traditional disciplines. These fields require the pooling of diverse disciplinary resources, influencing the culture of science.

The traditional disciplinary basis of university organization is not optimal for meeting the needs of this multidisciplinary research. As a result, centers and research institutions are complementing established academic departments. At NSF, we are actively promoting this approach to research. But even within the disciplines themselves, the sharing of resources--such as equipment--is requiring an increase in group activities.

The organization of our research effort needs modification in another way. Most basic research in the U.S. is conducted in universities. Development is left to industry. We need to improve the linkage between the two sectors, so new knowledge generated in the universities finds its way to application faster.

SUPPORT FOR SCIENCE AND ENGINEERING

What do these international changes and new demands mean to how we support science and engineering research?

1. The explosion of knowledge and the rising cost of providing adequate facilities and instrumentation are putting a strain on limited university resources.

2. We need to provide a more hospitable research environment for promising multidisciplinary and cooperative work.

3. While cooperation among universities and industries is improving, we still have a distance to go before we achieve a full partnership.

In the last six years we have begun to move in the right direction. We have a better grip on the problem and made substantial changes in funding and attitudes.

In this respect, we may now be at a watershed in our willingness to deal with the problems seriously. The president's 1988 budget message to Congress proposes "continued increases in federally supported basic research that lead to longer term improvements in the nation's productivity and global competitiveness."

A key aspect of this initiative centers on NSF. To put it in perspective, a few facts about the foundation are needed.

NSF, with an FY 1987 budget of $162 billion, is small in total, compared to all federal R & D support. But it is relatively large as a supporter of basic research. It is relatively larger still as a supporter of research at universities and colleges--receiving 25.5 percent--after National Institute Health, the largest.

The president's legislative budget message provides for an increase of nearly 17 percent in the NSF budget for FY 88--far above the government average and probably the largest for any agency. We are now trying to see that budget through Congress.

In addition, and in some respects even more significant, the president proposed that the NSF budget be doubled over the next five years. This proposal is a centerpiece of the administration's Trade, Employment, and Productivity Act of 1987, its main proposal in the effort to improve competitiveness. Clearly the administration understands the importance of science and technology and is taking it very seriously. The FY 88 budget proposal responds to the problems I have mentioned by focusing on three major themes: (1) Education and human resources, (2) science and technology centers and groups, and (3) disciplinary programs and facilities.

EDUCATION AND HUMAN RESOURCES

The first major theme is education and human resources. The nation's science and engineering base-- the source of our ability to do basic research--is made up of people, equipment, and facilities. Of the three, people are the most important. The number of scientists and engineers cannot be increased quickly, because their education takes many years. And it is particularly important that we identify the most promising students early, and encourage them in every way possible to stay with science and engineering.

I emphasized the need for building better links between universities and industry. That is one of the key features of NSF's Engineering Research Centers, an initiative begun three years ago. The Center for Advanced Combustion Research--a cooperative endeavor of

Brigham Young University and the University of Utah--is an example.

The centers share a number of characteristics:

- Universities

- Multidisciplinary

- Industry

- Education

We have established thirteen of these centers to date. So far, they are performing well.

In FY 1988, we plan to extend this concept outside engineering in the Science and Technology Centers, and accomplish other tasks shown below:

1. Establish new science centers:

    Biology and biotechnology

    Social and behavioral sciences

    Computer and information sciences

    Materials sciences

2. Increase and enhance engineering research centers

3. Strengthen advanced scientific computing centers

4. Establish minority resource centers for research and education

Like the Engineering Research Centers, the Science and Technology Centers will be university-based, multidisciplinary research centers, each with strong industry involvement, working on problems that are scientifically important and relevant to industrial technology. The close coupling of industry with academic research will substantially improve transfer of new knowledge from research labs to factories and to the marketplace.

I want to point out especially the Minority Research Centers of Excellence Project, which will serve both a research and a human resources development goal by building research capabilities at minority institutions.

The steady improvement of the foundation's traditional broad support for disciplinary programs and specialized research facilities is the third major theme of the FY 1988 budget as shown below:

1. Increase grant size:

    Support of students and postdocs

    Instrumentation

2. Start new research programs in:

    Materials processes and interfaces

    Cognitive science and architecture

    Large systems engineering

    Supercomputer applications

    Low temperature chemistry

    Storm-scale meteorology

3. Assess science and engineering education and career
    opportunities:

    Career pattern research and analysis

    Education policy studies

    Personnel data bases

    While multidisciplinary research is critical, it
must rest on a solid disciplinary base. And research
facilities must be adequate to support the research in
both disciplinary and multidisciplinary work. So the
FY 88 budget proposal includes an increase of about
$100 million in these areas.

CONCLUSION

    NSF is just one element in the huge national R & D
complex. It is being singled out now because of its
critical strategic role in the generation of new knowledge and the education of scientific and engineering
personnel. The NSF budget initiative is being taken
very seriously in the administration and the Congress.
And I should emphasize that this is a bipartisan issue.
Both sides of the aisle realize how important this is
to the nation's future.
    But NSF is only one part of the picture, a picture
that is much larger. This is not the place for a
comprehensive survey, but clearly, our survival in
this competitive world is going to require great

efforts from all of us--in our schools, in our colleges and universities, in our state and federal governments, and in our industrial enterprises. Productivity improvements depend on people and how they develop and use new knowledge.

The world has changed. It's not that rather comfortable postwar world in which American economic and technological dominance could be taken for granted, and it isn't going to be that way in the future. We must recognize that fact and we must respond to the challenges it presents. Our future depends on it.

4

Takeovers and Mergers: A Function of the Free Market

T. Boone Pickens, Jr.

INTRODUCTION

I have a clear-cut view of takeovers, mergers, and acquisitions. They are a basic function of America's free market system. I believe that once companies elect to become publicly owned, their stock is for sale to anyone who wants to buy it. If the founders or executives of a company did not want it to be that way, they should not have gone public in the first place.

MERGERS ARE VIABLE MEANS OF GROWTH

I feel the same about Mesa Petroleum, the company I founded in 1956. If today anybody wants to make an offer for Mesa, I promise you we will drop all the golden parachutes and shark repellent. We will not spend a dime of the stockholders' money to save our jobs. If you want to take over the company, your offer will go straight to the stockholders. If the company is sold to you, I will be a professional until I am released. And on the day you release me, you will not have to pay me past 5 o'clock. I will not even take the desk set when I leave.

Sure, it will hurt when I leave the company that I founded. But I am realistic enough to know that I do not own that company. I am a stockholder. I am one of the owners, but there are 25,000 other owners out there. They are entitled to see any offer that is made for the company, and we will do nothing to stop it.

I also think that mergers and acquisitions are a viable means of achieving corporate growth. For example, in one of the ten largest transactions in the oil industry in 1984, we acquired Mesa Royalty Trust, a group of producing wells we had spun off to shareholders. We spun it off in 1979, and we bought it back in 1984. All we had to do was make an attractive offer to shareholders.

I had a good friend who owned over a million shares of Mesa Royalty Trust, and he asked me, "What do you want me to do?"

I replied, "You do whatever you want to. It is your call."

He said, "If you believe it is worth $35 a share, I think I will keep it."

I responded, "That is fine."

He still has a million shares. We bought back 89 percent of the Trust, but there was very little communication involved because it was a straightforward offer to the shareholders. They could decide for themselves what they wanted to do about it. Eighty-nine percent of them took the offer. The other 11 percent did not, and they still have the shares.

Our stockholders have certainly prospered from some of the things we have done. Stockholders for other companies have prospered, too. In the five large transactions we have been involved in--Gulf, General American, Cities Service, Phillips, and Unocal--some 750,000 stockholders made about $6.5 billion in profits. That $6.5 billion would have never been realized by those 750,000 stockholders if there had not been an activist stockholder who made something happen.

Moreover, the government collected some $2 billion in taxes on that $6.5 billion. Now, my job is not to round up taxes for the federal government, but after the Chevron/Gulf deal, I saw Senator Paul Laxalt, and he said, "Congratulations on your deal."

And I responded, "Congratulations on your deal."

He inquired, "I do not understand."

And I replied, "Well, of the $6.5 billion profit made in that particular transaction, about 10 percent was made by Mesa, as the largest stockholder. But the federal government got about 30 percent, more than $2 billion dollars in taxes. So you guys did better than we did."

MERGERS ARE HARD ON MANAGEMENT EGOS

Personally, I can not see that anybody gets hurt in these transactions, with the exception of a few management egos--and I really do not worry too much about that. We work within the system. We always follow the rules. We do not cheat to win. Even my harshest critics say that we play by the rules. Stockholders own companies, and they should be allowed to decide for themselves if they want to accept an offer to sell their shares.

When Mesa has tried to take over a company, we have really wanted to acquire it and run it. We felt

that we could operate these oil companies more
efficiently than the previous management. Ninety
percent of our experience is in running exploration and
production operations. Our record is excellent. We
started in 1956 with $2,500 in capital. In 1964, when
we went public, our balance sheets reflected less than
$2 million in assets. Today, our company is the lar-
gest independent oil company in the United States, with
assets of about $4 billion. A $10,000 investment in
Mesa stock in 1964 would be worth $300,000 today. We
are proud of that record.

Nevertheless, the good ol' boys are really on my
case. The good ol' boys, to me, are about 85 percent
of the managements of the large companies in corporate
America. They are professional managers. In fact, the
managements and the boards of directors of the largest
200 companies in America, otherwise known as the
Business Roundtable, own less than three-tenths of one
percent of those companies.

But the good ol' boys say that "Pickens has a
scheme to bust up American industry--to liquidate,
dismember, and destroy companies." That is ridiculous.

If you look at the three largest deals in 1984--
Chevron/Gulf, Mobil/Superior, and Texaco/Getty--you
will find a common denominator: a large stockholder
who did not like the way management was running the
company. In the Gulf deal, of course, we were
involved. We invested $1 billion in Gulf and became
their largest stockholder. We were unhappy. We asked
Gulf management to spin off a portion of the company's
cash flow. We did not ask them to liquidate the
company, or sell the company. We did not try to take
over the company. We only asked them to take a fourth
of the cash flow and put it in a royalty trust for the
stockholders--that was all.

Why did we ask them to share part of their cash
flow with stockholders? Because the record had been
horrible at Gulf. From 1978 to 1983, they actually
lost off their reserve base--meaning they had produced
and not replaced-- the equivalent of 634 million
barrels of oil.

We were only asking them to distribute about a
fourth of their cash flow, about $800 million a year,
which would have left them with plenty of money for
their operation budget. Their immediate response, as
reported in the newspapers, was, "We're going to strap
on our sixguns and go after this guy."

Now, that is unusual. We were their largest
stockholder, and they were going to come after us.
And all we were trying to tell them was, "Look, don't
continue to make poor investments with the cash flow--
distribute some of it to the owners." And this was

just absolutely abhorrent to them.   They could not
believe anybody would be stupid enough to propose that
they should actually distribute anything more than what
they were required to through dividends.

My father, although blind and eighty-seven years
old, follows things very closely.   And he told me,
"Son, I notice that Mr. Getty at Getty Oil and Mr. Keck
at Superior Oil haven't been called anything deroga-
tory, but you've been called everything but a woodchuck
in this deal.   [I had been called a pirate, a raider, a
predator, you name it.]   And the only difference was
how long you owned your stock and how long Getty and
Keck owned their stock.   That was the only difference."

Does length of ownership make any difference when
you buy a house?   Do you have fewer rights of ownership
because you bought your house today from a guy who had
it for fifty years?   Don't you have the same rights
that he does?   Of course you do.   It is the same for
stockholders, as far as I am concerned.

So, the good ol' boys are really on my case.   But
the main reason they are upset is because they are
vulnerable, and they are vulnerable precisely because
they have done a lousy job for the stockholders.

Think of it.   They have no risk.   They have no
ownership to speak of.   And yet they have full control
of the assets.   They often try to push the stockholders
aside and take away the assets so they can have the
empire all to themselves, to do with as they please.

When stockholders look upon themselves as
"investors," executives love it.   I had one executive
tell me one time,   "I would like to have a million
stockholders who all own a hundred shares each."   And
then he said, "I would not have to fool with any of
them."   Well, of course not.   Nobody with a hundred
shares can afford the risk and expense it would take to
get something done.

I cannot believe the amount of money that American
executives spend to protect themselves and to keep
their jobs.   It is not their money, of course; it is
the stockholders' money.   Yet they spend it like it was
their own.

For instance, the law firm that gets most of the
takeover defense action is Wachtell Lipton in New York.
Wachtell Lipton is the highest paid law firm in
America.   The average for its partners is $795,000 a
year.   (The second best firm receives about $350,000.)
You can imagine what is happening here.   Executives
absolutely panic when someone makes an offer for their
company, and they will pay anything.   The Wachtell
Lipton billing for nineteen days of work in December
1984 for Phillips Petroleum was $5.5 million.

PRESSURE ON BANKS TO OPPOSE TAKEOVERS

And the good ol' boys put pressure on the banks. One of our lead banks has four members of the Business Roundtable on its board. During a ten-day period, in one-on-one conversations with the chairman, they said, "You either drop Boone Pickens, or we're going to resign from the board and remove our accounts from the bank." Now, that is hardball. They had been customers of the bank for several years.

The chairman talked to me about it. He said, "Boone, you are a gold-plated customer. Your repayment record is excellent. You don't quibble about fees. You have excellent collateral....But I'm sorry," he said, "I have reached a point..."

And I interrupted, "Let me stop you."

He said, "Okay."

I said, "...where you can't do business with me."

And he agreed, "That's right."

I was sorry about it. I thought that the bank had a policy of loaning money to creditworthy customers. The four companies that were putting the heat on the bank were not oil companies or oil industry-related businesses. They had nothing to do with me, but they applied pressure from the outside. They said, "You have to help us out," and the bank was helping them out.

I want to share the following quote because it is an incredible admission by Harold Hammer, the chief financial officer of Gulf Oil, to a Pittsburgh-area newspaper (June 1984):

> Hammer admits surprise at the high credibility which Boone Pickens achieved with the news media during his own personal canvassing in last December's proxy contest against the Texan. Hammer ran into unusually hostile proxy committees at several major banks. Invariably, they seemed to be manned by 28-year-old MBAs, a lot of them women. These folks seem to share with Pickens a propensity to believe the worst of so-called big business. On some occasions, Gulf executives had to phone the bank chairman to reverse the proxy committee.

Would you say that is pressure? The chairman of Gulf calls the chairman of the bank and says, "You've got to reverse your proxy committee." When somebody says there is a "Chinese wall" between the trust department and the loan department of a bank, they are kidding somebody.

But Hammer did not just stop there. He revealed how he assesses whether a banker is on management's

side in a May 10, 1985, issue of <u>USA Today</u>. He said,
"If the banker looks at his shoes or the ceiling, you
change banks--that forces a bank to listen." You can
see why Chevron did not keep him on the payroll after
they took over Gulf.

One more published example of pressure placed on
banks was in the <u>Los Angeles Times</u>, May 9, 1985:

> The Fred Hartley's are shouting, "Stop the
> T. Boone Pickenses of the world." And they have
> some banks listening. Said one banker who has done
> business with Pickens, "I don't think banks
> ought to be making public policy based on a few
> phone calls from big business." Said another
> banker, "We are all paranoid. Are banks supposed
> to be the policeman for corporate America? That
> is clearly the signal we are getting."

And the answer is no. That is not the proper role
of banks. They are not to make moralistic judgments.
Banks have a job to do; and if a customer is creditworthy, there should be no problem with loaning him the
money.

We were identified as a good customer by the bank
that had to drop us. They said, "You are gold-plated.
You are the best oil account we have in the bank. But
we are sorry--we cannot do business with you any more
because we have had tremendous pressure put on us."

The issue is simple. Shareholders own companies,
and they should be allowed to decide whether or not to
accept a tender offer. If they want to sell their
stock, fine. If they do not want to sell it, that is
fine, too.

5

Current Challenges for American Industry

Mark Shepherd, Jr.

The U.S. economy is in the midst of the third longest-running expansion in the postwar period. Inflation is at its lowest level in twenty-five years, and the stock market reaches new highs almost weekly.

Beneath the glitter, however, we still face the threat of large and growing deficits: budget deficit, whose solution appears out of reach, and a trade deficit, whose impact on our industrial landscape is approaching seismic proportions. In industry after industry, manufacturers are going out of business, curtailing their operations, or giving up on the United States as a suitable place for making their products.

EROSION OF U.S. PRODUCTIVITY

The erosion of American competitiveness in recent years is only a part of the changes now sweeping the global economy, including a massive shift in the focus of world economic activity toward the Pacific Basin. While the United States is struggling to regain its position of leadership, the nations of the Asia/Pacific Region have skillfully exploited their labor-cost and productivity advantages to steadily increase their share of world markets.

Productivity is the key to a country's competitiveness. Since 1960, Japan's productivity gains in manufacturing have increased sixfold relative to the United States, and Europe has outstripped the United States by a factor of two. The United States still enjoys an advantage over our competitors in the absolute level of total productivity. Much of this advantage, however, was accumulated over the long period from the Civil War to World War II--and has been eroded by the low relative rate of U.S. productivity growth in recent years.

This lower U.S. productivity growth has contributed to reduced profit margins, flat production, downsized capacity in terms of both capital and labor, and an increasing loss of market share to imports.

No single statistic exemplifies the troubles of American industry as much as our manufacturing trade balance. After seventy years of nearly uninterrupted surpluses, the U.S. trade balance for manufactured goods plunged to a $40 billion deficit in 1983, $80 billion in 1984, and to $142 billion in 1986.

The loss of market share has been extensive even in high technology industries--a sector in which we have always taken our leadership for granted. Foreign penetration of the U.S. computer market has increased from 3 percent to 29 percent in less than six years. Communications and instruments have also suffered large losses in market share. These trends are worrisome for a nation accustomed to carrying the banner of world economic leadership.

America's competitive strategy should not rely on forcing our trading partners to give up their legitimate advantages. Instead, we should be tough with those who take advantage of our good will; and we should develop a strategy that builds on America's strengths to tip the competitive scales in our favor. To this end, we should aim at: 1) balancing the financial scales between the U.S. and its international competitors; 2) improving the skill levels--not only the technical skills, but more importantly, the basic skills--of the work force; and 3) accelerating the development of advanced manufacturing processes.

COMPETITIVE STRATEGY: BALANCE FINANCIAL SCALES

The first step in balancing the financial scales is to increase the availability of capital for American companies. The cost of capital for a typical Japanese firm is less than half of that for its U.S. competitors (8.6 percent versus 17.4 percent). High savings rates in Japan, and close ties between banks and industry, have favored highly leveraged financial structures. Despite recent trends toward less reliance on debt, Japanese firms continue to exploit the low cost of debt with debt-to-equity ratios that are at least twice as high as comparable ratios in the United States.

Additionally, Japanese companies benefit from a tax system that effectively exempts from taxation much dividend income, interest income, and capital gains on stock. The result is that even if debt/equity ratios were the same in Japan and the United States, the Japanese would still enjoy a cost-of-capital advantage.

This lower cost of capital--combined with different earning standards of the financial community--have allowed Japanese firms to satisfy their investors with only 1 or 2 percent after-tax profit on sales, as opposed to the 5 or 6 percent required in the United States. This difference in acceptable profit margins for a given sales level means more cash available to our Japanese competitors for additional capital investment and research.

Instead of narrowing this cost-of-capital gap, the Tax Reform Act of 1986 increased it still further. According to a recent study by two economists at Stanford University--Douglas Bernhein and John Shoven--the effect of this legislation will be to increase the inflation-adjusted cost of capital in the United States by at least 30 percent. This will worsen the cost-of-capital disadvantage of U.S. firms with respect to their Japanese competitors from 2:1 to 2.6:1. If this disparity continues, U.S. companies will not be able to keep up with Japanese investments, and America's technological leadership will continue its erosion.

Future legislation should consider ways of enhancing the incentives for personal saving and for capital formation: 1) by repealing taxes on the interest from savings; 2) by eliminating the double taxation of dividends; 3) by lowering tax rates on capital gains; and above all, 4) by reinstating the investment tax credit. Consideration should be given to a consumption tax as an offset to tax revenues lost through these reforms.

While tax reform could help increase the availability of capital, the federal budget deficit continues to threaten our economy. A return to fiscal discipline on the part of the U.S. government is essential to our continued recovery. The United States must reduce the size of the budget deficit by reducing government spending. In the late 1960s, total government spending was about 20 percent of GNP despite defense requirements that absorbed more than 9 percent of GNP. In fiscal year 1986, government spending totalled nearly 24 percent of GNP, and only 7 percent of that was in defense.

This is not meant to suggest that defense spending should go back to 9 percent of GNP. But we must go back to the kind of restraint on non-defense spending that has served us so well throughout most of our history.

While fiscal reform is necessary to restore U.S. competitiveness, it is not sufficient. Chronic distortions in exchange rates have played a large role in the deterioration of America's trade balance. Ending these distortions must be an important part of a new competitive strategy.

The extreme undervaluation of the yen in the first half of the 1980s is well documented. Today, Taiwanese and Korean exports to the United States are boosted by currencies that are grossly undervalued--in Korea's case by more than 60 percent--with respect to the dollar. These relative currency values do not reflect the true positions of Taiwan and Korea in world markets. The use of more realistic trade weights in computing the value of their currencies, instead of the current policy of pegging them exclusively to the dollar, would go a long way toward eliminating this distortion.

But Korea and Taiwan are only part of much larger problems inherent in the present system of flexible exchange rates--problems that include increased short-term volatility and lack of predictability for planning purposes. To address these problems, the United States should continue working to establish a system of target zones for major international currencies. The success of the European Monetary System suggests that the idea of target zones might be appropriate on a broader scale.

This system would require the identification (and periodic revision) of a zone, perhaps 20 percent wide, outside of which rates would be considered in clear disequilibrium. Concerted intervention on the part of major economic powers, followed by changes in monetary and/or fiscal policies if necessary, would discourage exchange rates from straying outside their target zones. This system would: 1) require governments to consider explicitly the effects of national economic policies on international competitiveness; 2) encourage productive investment by providing a more stable financial environment; and 3) prevent the "beggar thy neighbor" policies employed by countries with chronically undervalued currencies.

Our major trading partners agreed in Paris that exchange rates ought to be stabilized "at about current levels," but clear target zones--and the specific actions that would be taken to maintain currencies within these zones--have not been established.

COMPETITIVE STRATEGY: IMPROVE SKILLS

In addition to balancing the financial scales, restoring America's competitive leadership requires increasing the skill levels of the work force. The Japanese, with a national commitment to excellence in education, are raising an entire population to a standard currently inconceivable in the United States. They have succeeded on the strength of: 1) parents who actively promote learning by cooperating with teachers, supervising extensive homework and not accepting mediocre performance; 2) teachers who are well qualified and highly motivated because of the pay and the respect accorded to them; and 3) a school year that is 30 percent longer than in the United States (240 days versus 180 days average).

Japanese education is justly criticized for its reliance on rote memorization and its failure to foster the creativity that is at the heart of American-style innovation. But their system has been very effective in teaching the "basics."

Meanwhile, we are--for the first time in our history--producing a generation of high school graduates that are less educated than their predecessors. SAT verbal and math scores for U.S. students entering college are twenty to twenty-five points below the levels achieved in 1967. And--in contrast to Japan, where more than 90 percent of students finish high school--a recent study by the Texas Education Agency showed that one-third of the students who started ninth grade did not graduate from high school.

To restore the effectiveness of American schools, we need to provide the necessary financial incentives to attract and retain excellent teachers. We also need a strong national commitment to higher educational standards, especially in math and science. And--most important--we need deeper family involvement in all aspects of education.

An issue of equal concern is the growing shortage of technically qualified college graduates. Although recent figures on U.S. engineering graduates show some improvement, these increases in the number of bachelor degrees are deceptive. Projections of engineering graduates suggest that we approached a peak level in 1986. Some engineering schools are actually reducing their enrollments, and a shortage of engineering Ph.D.s is reducing the supply of qualified teachers.

Universities must work with industry to increase the capacity of our technical education system and alleviate this country's technical manpower shortage. To close this gap, we need to:

1. Encourage more high school students to pursue engineering and science by giving them an understanding of the opportunities open to those with first-class technical educations;

2. Emphasize co-op programs that allow a student to earn the money necessary to pursue a degree and, more importantly, provide the student with valuable experience with industry and with people;

3. Encourage the growing interest of women in engineering; and

4. Develop curricula that support lifelong education
    for self-renewal.

COMPETITIVE STRATEGY: DEVELOP MANUFACTURING PROCESSES

     A third step in restoring America's leadership is
to develop--and use--more advanced manufacturing processes. The very recent nature of the deterioration in
our trade balance is a strong indication that neither
American labor nor American management should be
accused of being the only cause of our trade deficit.
The basic character of America's industry cannot have
changed so radically in the brief span of five or six
years.   But we need to recognize the existence of
severe structural weaknesses in our industrial sector.
And we must face the challenge of enhancing our
manufacturing skills.
     Technological progress has been the driving force
behind most of the productivity gains we have made in
this country--and the United States still leads the
world in the absolute commitment of funds to basic
research.   But we have been weak in executing the
difficult transition from R & D prototype to full-scale
commercial production.   Developing the manufacturing
equipment and processes to make this transition is
often more complex than developing the product itself.
     The Japanese recognized some time ago the
importance of this critical phase of innovation, and
they have committed substantial resources to this area.
The National Science Foundation estimates that Japan
spends better than 60 percent of R & D on manufacturing, compared with 39 percent in the United States.
As a result, the Japanese can take ideas into full-
scale production in only half the time we require--and
often with lower manufacturing costs.
     America's lead in design innovation will do little
good if countries are allowed to acquire our technology
at minimal cost and then, by spending huge sums on
manufacturing R & D, consistently beat us to the
marketplace with our own designs.
     A major weakness in U.S. policy has been the lag
between the advances in technology and the use of
intellectual property law to protect them.   Our competitive advantage in innovation is seriously threatened
by countries that acquire our technology through coun-
terfeiting, patent infringement, and other forms of
piracy and then combine that technology with low-cost
labor to drive the original developers out of the
market.
     The United States has traditionally been the
champion of free trade.   But America cannot remain the
dumping ground of other manufacturing nations--nor can

we continue to be the victim of government subsidies by our competitors. Prevention of dumping--and the payment of a fair price for the use of intellectual property--are essential for assuring the continued development of innovative technologies and products.

At the same time, the government should encourage R & D--and help provide firms with the cash flow necessary to develop advanced manufacturing technologies--by:

First, making R & D tax credits permanent. The temporary nature of the current tax credit makes it difficult to plan long-range projects;

Second, restoring the 25 percent rate that existed prior to the Tax Reform Act of 1986;

And third, expanding eligibility for R & D tax credits to include the whole range of expenses involved in developing and implementing innovative manufacturing processes, machinery, and facilities.

R & D tax credits are preferable to direct government subsidies because they not only provide the necessary funds but, at the same time, they let the marketplace determine where the money goes. Private-sector initiative has been the engine driving our economic growth and no government bureaucracy can pick the "winners" and "losers" more effectively than the free market system.

But while government can help to provide a healthy environment for manufacturing innovation, industry must focus its attention on manufacturing itself. The infusion of a stronger manufacturing culture in boardrooms and executive offices across the country is an essential element in the revival of the U.S. industrial sector.

Many have suggested that high-technology industry could take up the slack caused by the decline of America's other manufacturing industries. But the absolute levels of output and employment in high-technology industries are not enough to offset the impact of declines in other manufacturing areas.

A much greater economic benefit will be realized if we begin to think of high technology not as an industry, but as a powerful set of tools that can revitalize our traditional industries. Using these tools most effectively will require a fundamental change in the way we think about manufacturing. Most strategic planning focuses on financial numbers, basic research, and designing products to meet market demand--not often enough does it focus on manufacturing. This has produced a mentality that looks on manufacturing as a process separate from design and distribution. For the future, we must broaden this perspective, and consider manufacturing as an integrated process.

By "integrated," I mean two things.   First,
manufacturing must be an equal partner with R & D and
marketing, and become an integral part of a total
business strategy.

Second, the manufacturing process itself must be
treated in its totality.   Traditional strategies for
improving manufacturing have concentrated on automating
isolated pieces of the process.   The most efficient use
of resources requires looking at the process as an
integrated system, and then investing in the resources
necessary to optimize that system.

How well we can integrate the manufacturing
process, and incorporate it into the overall business
strategy, will depend on how well we understand the
process itself.   This means taking the process apart,
piece by piece, and reexamining--and challenging--every
task and every procedure.   We must be able to quantify,
measure, and analyze the process--and to do this, we
must have access to complete information.   Nothing
should move, change, or be processed in a factory
without being captured electronically in a data base.

Traditional approaches to analyzing the
manufacturing process have been of limited usefulness,
because the engineers and operations researchers did
not have access to sufficient information and could
address only a part of the problem.   Through continuous
performance and cost improvements in semiconductor
logic and memory, we now have the potential for acces-
sing more and better information through distributed
computing power, communications networks, and artificial intelligence.   These emerging tools, coupled with
the existing tools of modeling and simulation, will
enable us to take a true systems approach to enhancing
the overall process, using hard data.

Low-cost computing power--in the form of
programmable industrial controllers and desk top termi-
nals--has taken computing out of the data processing
room and put it on the factory floor.   With local area
networks, we can link together cell controllers,
computer-aided design terminals, artificial intelligence workstations, and minicomputers.   And these networks provide a way to integrate all the different
sources of manufacturing process information--which
means we can now consider any problem in terms of its
impact on the total process.

Distributed computing power brings a problem, the
information needed to solve it, and the mechanism for
solving it together in the same head.   The person with
the problem now has the potential for interacting with
the system from the individual workstation, in real
time, to solve the problem.

What local area networks can do for an individual
manufacturing plant, worldwide networks can do for an

entire corporation. With telecommunications networks, a semiconductor processing engineer in Dallas can have access to test data stored in Singapore, to compare initial processing parameters with final test results. This allows the processing engineer to assume full responsibility for optimizing total cost and yield for the finished product, instead of optimizing only one stage of the process.

This combination of computers and communications networks offers the potential of achieving revolutionary advances in manufacturing efficiency.

Computer-aided design and computer-aided manufacturing--CAD and CAM--are already helping to increase the efficiency of these individual operations. For example, in semiconductors, some logic designs that used to take up to eighteen months are now completed in a few weeks. But as long as CAD and CAM remain isolated pieces of the manufacturing process, we will only gain small, evolutionary improvements. The power of Computer-Integrated Manufacturing lies in its potential for linking together the entire process.

Human beings will always be the critical part of the manufacturing process, and part of collecting information about the process should involve gathering the empirical data and qualitative knowledge that only humans possess. But until recently, there has been no effective way of systematizing and measuring the process knowledge available in the human mind.

"Artificial Intelligence"--the technology that allows computers to address problems that require human-like reasoning and intelligence--gives us the key to unlock this vital source of information. With its potential for using the information provided by people, Artificial Intelligence opens a new dimension in data processing and will change the way we think about factory automation.

An early application of Artificial Intelligence is expert systems, which involve capturing the knowledge of experts in computer programs that can be used by nonexperts to solve a problem. For example, according to Morgan Whitney, Director of Ford's Robotics Center, Ford was training thousands of engineers and maintenance people to service its manufacturing robots, at a cost of up to $5,000 per person. It still found that "the fundamental inability to keep up with repair techniques is a serious roadblock to the factory of the future."

For a total of $5,000, Ford developed an expert system that allows a technician to diagnose and repair its ASEA robots--without a training course and in a fraction of the time that consulting a manual would have required.

With a thorough understanding, based on both real-time and empirical information about all elements of the manufacturing process, we should now be able to develop and implement a comprehensive, long-term plan for integrating that process and incorporating manufacturing into the total business strategy. Persistent problems can be solved, new opportunities found, and maybe even radical new approaches conceived.

With the widespread availability of these semiconductor and computer capabilities, the barriers to increased efficiency are not technological, but cultural. They are:

First, a tendency to see manufacturing technologies as ends in themselves, and not in their proper role as means to an end. <u>The goal is not increased automation, but increased manufacturing efficiency and reduced overhead.</u>

Second, a tendency to substitute automation for understanding. While automation can help, its use must be based on a total understanding of the manufacturing process into which it is incorporated.

Third, the misconception that upgrading manufacturing operations is too expensive. Technologies such as modular controls and networking make it possible to achieve step-function improvements without scrapping an entire factory and rebuilding from a green field.

Fourth, a lack of determination on the part of top managements. Full implementation might be a five- to ten-year process, and this means a continuity of project management that is beyond what is normal for most U.S. companies.

And fifth, cultural barriers within the workplace. Changing the culture of the workplace requires significant attitude adjustments. People are threatened by new and unfamiliar technology and fear the loss of their jobs. Change creates new pressures on supervisors and requires operators to be more flexible.

Every individual has a vested interest in his or her part of the process. Management can create the best environment for change by involving individual operators in planning and controlling their part of the process, along with the responsibility--and the recognition--for making improvements. Each employee should be seen as a source of ideas, and not just a pair of hands. The same technology that allows us to gather data about the manufacturing process also makes it possible for the operator on the factory floor to control the process, experiment with it, and make it better. When people become involved, they become committed, and commitment produces results.

In today's global markets, America cannot afford a competitive strategy aimed only at Japan. Even if the

Japanese were to implement the structural changes promised by the Nakasone administration, the competitive battle would not be over--it would only shift to other countries. Taiwan, Korea, Singapore, and Hong Kong are already powerful forces in world markets; in time, India and China will be strong competitors. It is time for American ingenuity to come to bear on the problem of competing not only against Japan, but against all international competitors.

A strong undercurrent of the traditional American values is still with us. We have plenty of spirit, of goodness, of patriotism; and the high value placed upon freedom--at home as well as abroad--remains unaltered. But in these changing times, our traditional ideals must be blended with new values: a zeal for winning; a firm belief in fiscal responsibility; a determined effort to tilt our nation's resources toward productivity and investment; a strong commitment to rebuilding an educational system second to none; a revival of the American work ethic; and a firm determination to manage our destiny. Together, the old and the new will form the strong foundation American society will need to successfully meet the challenges that the coming years bring.

6

Changing National
Business Culture

Lawrence Schein

My topic, the changing corporate culture, is so broad
that there is no best method of approaching it. This
means that the lecturer has considerable license and I
propose to exercise that license by taking two cuts at
the topic.
   First, I'd like to talk about some changes in the
national business culture that I believe have profound
implications for the human resources process as well
as for the human resources professional. Then I'd like
to move on to talk about how individual companies
change their own cultures or use their cultures to
adapt to the painful process of change.
   My knowledge and experiences are based on primary
research in which I have conducted hundreds of hours of
interviews with hundreds of executives in nearly 100
manufacturing and service companies across the nation.
These executives have included CEOs, members of their
executive and operating committees, plant managers,
managers and directors of staff functions.

CONTRASTING NATIONAL BUSINESS CULTURES

   Let's talk about changes in the national business
culture. By the national business culture I mean
values, attitudes, beliefs, assumptions and premises
that govern how we do business in the United States,
how we perceive and value work and how we actually
carry out that work in line with the norms that govern
our behavior. These are values, beliefs and assumptions that transcend differences among industries and
between and among specific companies within industries.
For example, the historic adversary relationship
between management and organized labor in this country
is a characteristic of our national business culture.
The Japanese management style, dedication to quality,
and success in implementing a drive toward zero defects
and total quality performance are all very distinctive

to that nation's business culture. The appropriateness of centralized economic planning is a hallmark of the Soviet "business culture."

I think any of you who have been abroad to Europe, or to countries in the Pacific Rim or in South America, have encountered differences in business cultures. I had this experience not by leaving the United States but by visiting some companies that were owned by Japanese interests employing an American work force and some American management. I remember my first such experience back in 1982, of visiting a Japanese-owned semiconductor plant in California. There were a number of things that struck me as quite different.

I've visited a lot of American manufacturing plants. When I was in my twenties I worked in a few of them, and I noticed some differences. One was the importance of time--the importance of punctuality and schedules. Every few feet in the corridors of this plant were clocks mounted on the walls. The manager, who was Japanese, remarked, as he escorted me back to his office, that most of the visitors who came to this facility brought clocks as presents, and they were immediately displayed. There were dozens and dozens of clocks.

The place was also absolutely spotless. We had an interesting interaction with an employee. There was a scrap of paper lying on the floor in the hallway, and an employee was walking toward us. The manager stopped, pointed to the piece of paper, and said to the employee, "Do you not have the public mind? Where is your conscience to permit this?" The employee was startled. Again, the manager said, "Where is your public mind?" pointing to the piece of paper. The employee, a little bit embarrassed, reached over, picked it up, and presumably disposed of it. And that struck me as a difference in management style.

He took me into his office, and he showed me two books on his desk. One was a gigantic loose-leaf book filled with company manuals and procedures. The other was a much smaller book that was his personal managerial diary and record. He opened it up when I asked to see it. It consisted of page after page of illustrations. The illustrations were really more like cartoons. For example, a giraffe's head was floating above the clouds and he translated the Japanese caption that was written underneath the illustration: "A company with lofty sights but does not know where its feet are planted." In another one, he had a frog exercising very vigorously with a barbell deep in a well. He said, "This shows a company that's working very, very hard and very industriously but with absolutely no insight and no vision of the outside world or the outside marketplace." And so the book went, illustration after illustration, aphorism after aphorism.

Also, the manager's age was of interest. He was in his fifties and, I suspect, approaching retirement. He had not been in the United States very long. He had spent six months in what he described as an apprenticeship in an American electronics facility in New Jersey and then had come directly to this particular plant and he had been there only two months. It struck me that if an American company were handling this and saw plant management abroad as a developmental opportunity for a manager, they would have selected a much younger person to be the plant manager. So, there were different ideas and assumptions about managerial career development in this man's firm.

CHANGE IN U.S. BUSINESS CULTURE CAUSES RESTRUCTURING

The thing that changes a national business culture is the same thing that changes a corporate culture, and that is some upheaval, some enormous pressure from the outside. These can include changes in the world marketplace, political conditions, political alliances, demography, and the educational level of the work force. I'd like to focus on a change in our national business culture that has been triggered by several forces coming together simultaneously: lagging productivity, very intense foreign competition in many industrial areas, and a very substantial increase in our trade deficit. These forces have produced a massive restructuring in America--a corporate restructuring that has been described as unprecedented in our history.

Also, I think the experience of the recent recession would play a role here. There has never been as much change, reorganization, disaggregation, disarray, and turmoil on the American business scene as at present. This was said by Warren Bennis at a conference on corporate culture about a year ago. He was referring to a restructuring more intense than he had ever seen in his thirty-five years of studying American business. This restructuring is being carried out in downsizing, in reductions of forces, in the stripping out of whole layers of management, and in unprecedented activity in mergers and acquisitions. The restructuring is only about five or six years old in terms of its major impact, and it goes across industry lines.

JOB LOSS FROM RESTRUCTURING

In 1984, the Conference Board did a study of layoffs and closings of plants and other facilities. The study involved over 500 medium-sized to large companies of which 59 percent reported that they had sustained recent layoffs or plant closings. The fact that 75 percent of the manufacturing companies in the sample had lost employees was not news to us, but that 50 percent of the service firms in the sample had also laid off employees was surprising. This was occurring in a period when companies and the economy were coming out of the recession.

It has been estimated that the _Fortune_ 500 companies have lost three million jobs in the last five years. This is not an unreasonable estimate when you consider that one company alone, General Electric, has lost 100,000 jobs in this period. These jobs are being lost through attrition, layoffs, early retirements and other voluntary separations. The downsizing is not restricted, obviously, to manufacturing companies, heavy industry, or to smokestack industries. High-tech companies, like Wang, Intel, and Apple have all experienced reductions in force in recent months. Even the bastion of employment security, IBM, has increased the number of its employees who are eligible for early retirement.

Blue-collar workers in this country have always known that there was going to be a pinch when the economy tightened up. The clerical workers have begun to learn this over the past few years, too. For example, by the end of this decade both Ford and General Motors will have laid off 25 percent of their clerical forces. And Chrysler, since 1978, has reduced its clerical force by 50 percent.

But what is unprecedented, and what is different, is the impact of the restructuring on middle management and the loss of middle management jobs. For the first time managers are beginning to learn that when the economy tightens up and when companies try to become more competitive, they too, despite all their years of service and all their expectations of career security, face the ax. According to the Bureau of Labor Statistics, between 1979 and 1983 (the period including the recession), 700,000 managers lost their jobs. However, about 75 percent were reemployed in about five or six months. But there's a big difference in where they went back to work. Of the managers who went back to work--that reported being employed in January of 1983--only half went back to work in managerial or professional positions. Those that didn't go back to being managers and professionals wound up working in administrative, clerical, and technical work, which is generally lower-paid work.

It's been estimated that between 1984 and the spring of 1986 an additional 600,000 managers lost their jobs. Reliable numbers for this estimate are hard to come by. When questioned on the figure, the consultant in charge remarked that he had come up with the number by having his staff follow the Wall Street Journal every day. They made lists of the companies that were laying off employees and developed some estimates, and the above number is probably a conservative estimate. The managerial cuts have continued. At the end of December 1986, of the 24,000 employees in AT&T's information services unit that were let go, 30 percent were managers. United Technologies is getting a new CEO and plans are already underway to eliminate 1,250 managerial jobs in that company.

Job losses are also tied in with the merger-acquisition activity. Back in 1975, there were fewer than 1,000 mergers or acquisitions in which an American company was involved. Since then, this type of business activity has just taken off. By the end of 1985 there were over 3,300 mergers in which a U.S. company was involved and one of the impacts is, of course, job loss.

Last year we saw the largest bank takeover in American history and that was when Wells Fargo acquired Crocker National Bank. In May of 1986, when the deal was finalized, 1,600 Crocker managers lost their jobs including nearly the entire Crocker top executive level. Plans were instituted to reduce Crocker managerial positions by another 3,000 jobs over the succeeding eighteen-month period.

RESTRUCTURING IMPACTS JOB RESPONSIBILITY/STRESS LEVEL

Another impact of restructuring is the reduction of job responsibility. The management consulting firm, Lamalie Associates, did a study of executives in firms that had been taken over. They found that within two years half of these executives were gone. The chief reason was not that their jobs were eliminated, but rather that the scope of their authority and responsibility had been reduced. The degree to which they could satisfy their professional egos had been reduced so much that it was no longer desirable to stay on with the company and they left on their own accord.

A by-product of all this restructuring is a "stressed" organization. This may be a new concept. Stress has been raised from an individual to an organizational level. We used to talk about individual stress and I think we all understood what that meant—a worker got burned out or got in trouble with substance abuse. So, companies provided counseling and set up employee assistance programs to deal with alcohol and drug abuse, and family problems. All of these programs were on an individual level. Consider what happens to organizations that have stood for a hundred or two hundred years and then, a takeover, merger, or a buyout is announced, and the organization goes into shock. This is a stressed organization. When there is rumor of a reduction of force or a merger, work begins to slow down and very nearly stops. Productivity, such as it may be, comes to a virtual halt.

I recall speaking to a young woman who had worked for a Wall Street investment house when it was rumored that a takeover was in process. The president of this company announced immediately and publicly that the company was not for sale. However, all the employees knew it was. Work began to slow and it stopped, in effect, for nearly six months until the merger took place. Since the merger, according to my informant, the pace of work has never really picked up. People stopped working; they got on the telephone, they went out on job visits, they did their resumes, they gossiped and they got depressed. She made an interesting point that some of the smartest people who have been produced by our university system work on Wall Street and that new and innovative financial services and products might have been generated were it not for the imposition of the fear and insecurity of an acquisition.

ABROGATION OF EMPLOYMENT CONTRACT

The ante of stress has been raised. The stress point between the employee and manager on the one hand, and the company, on the other, is the change in the employment relationship or "contract." This contract is implicit. It is essentially unwritten, and largely psychological. From the manager's point of view, the contract says that in return for dedication, commitment to organizational objectives and hard work, the company will provide lifetime employment security and career development, with a reasonable opportunity for promotion and reasonable access to increasing levels of remuneration. These contractual conditions are being increasingly abrogated today as companies, in order to stay competitive or to get more competitive, are unable or unwilling to offer lifetime employment.

It is this abrogation that is leading to changes in the national business culture, in the way we think about companies, company loyalty, and work. At the personal level, the impacts are enormous, as I believe

many of you who have been through this kind of
experience already know. One of the most eloquent
descriptions of this that I have heard came from the
head of human resources at a major oil company. He
talked about his having taken an "escalator" when he
went from a university into the company and moved to
the top. He did all the right things, lived in all the
right countries, did his work, followed instructions,
followed directions, and knew he would be rewarded.
Eventually, the perks came and the escalator dumped him
off at the top of the human resources function. Today,
he feels, the escalator is more likely to dump you
right into the street. He asks, "What do I tell a
thirty-five-year-old who comes to me after ten years of
experience in the company and says, 'Why are you doing
this to me? Why are you laying me off after ten years
of excellent performance appraisals, dedication, and
skill?' I lay them off because I can't afford them
anymore." This vice president of human resources continues, "What do I tell a twenty-five-year-old who
comes with an MBA to work in the company? All I can
tell this person is, 'Listen, I can give you a job
since you seem qualified, you seem able to learn.' All
I can give them is a job. I can't guarantee a career.
I can guarantee a job for some slight period of time
and no longer than that."

These developments have caused a considerable
debate within the company. Should the company continue
to function as the corporate citizen it has been or
does it get managed in a very financial way? He says,
of his New York-based firm, "You know we can sell this
building and move to New Jersey or we could rent a
warehouse in Yonkers; we could put in telephones or we
could do business. But that's not what this firm is
all about. That's not the corporate citizen with thousands of employees; twenty thousand annuitants; agreements, arrangements, and contracts with governments;
partnership ventures with other companies and responsibilities all over the place. That is the company
that I want to keep. I don't have the answer. I don't
know what the new contract is except that there is no
new contract."

All of this, by the way, is being exacerbated by
the fact that the baby boom is over and the millions
upon millions of young people produced by that boom
have now been absorbed into the labor force. There is
the enormous pressure of greater and greater numbers,
on fewer and fewer job opportunities, and in particular, on fewer managerial opportunities. This creates
a special problem for the human resources professional. How do you develop a culture to manage decline?
How do you develop a culture to manage diminished
opportunity? Because that is what we are faced with.

We are talking about loss of jobs, loss of careers, and
the loss of an economic frontier. Gordon Smyth, who is
the chief human resources officer at Dupont, put the
question very sharply at a recent meeting. He asks,
"In a period of a contracting economy and of no growth
or slow growth by many industries, how do you maintain
the morale and the dedication to the company objectives
on the part of survivors of reductions in force? How
do you get the commitment to the organization's objectives on the part of young managers who are getting
increasingly cynical about and distrust the motives and
the promises of top management?"

These questions are very telling, particularly
since there doesn't seem to be any end in sight to the
restructuring. It's not as if the first downsizing
will be the last. Many companies are now embarking on
second and third cuts. In a Conference Board survey
that we did last year of Senior Human Resources
Managers in 600 of the largest companies in the country, 60 percent expect merger and acquisition activity
to affect their companies in the next one to three
years.

CHANGED PERCEPTION OF EMPLOYMENT SECURITY/CAREER
DEVELOPMENT

What changes might we see in the national business
culture as a result of the restructuring? I think that
we are going to see more emphasis on promotion and
advancement being based on the criteria of performance
and much less on the criteria of politics, seniority,
or of the old-boy network. The "fat cat" is not dead
but is probably dying. I think we are going to, at
least within the next two decades, place less emphasis
on the importance of vertical promotion and on the
ethic of "onward and upward." More emphasis will be
needed on job rotation, lateral transfers, cross-
training, learning multiple skills, and on having work
being made more challenging.

The Conference Board, in conjunction with the
Kellogg Graduate School of Management at Northwestern
University, is currently doing a study of middle managers' perception of employment security and career
development, given the changes in the business world
that we have mentioned. I have been talking to younger
managers in financial services companies in the New
York area, and the thing that frightens these people
the most is not the loss of a job. It is not a fear of
not making enough money. It is the fear of being bored,
of going into a dead end job, of mastering a complex
task and then having to stay at that task another year.
They want challenge and diversity. A manager at

Citibank said, for example, that one of the things about her firm that she appreciated was the constant opportunity for innovation, for learning, for taking responsibility, and for doing different kinds of things. It seems that there will be, in managerial ranks, more reliance on one's own abilities and professional skills. One's professional discipline may rise in importance, and perhaps, the professional association will begin to replace the corporation   as the repository of loyalty and dedication.   Are we seeing the decline of the   "organization man," of the "company man," and the "workaholic"?   There is some indication that this may be happening.   I think we are going to see more balance in preparation and training.   The young managers that I talk to stress breadth as well as specialization.   Their advice is to get the specialization later with a liberal arts background first.   They urge getting as much diversity and work study experience as possible in the undergraduate years and then thinking about an MBA on down the road.

We are seeing an increasing mistrust of top management, an increasing cynicism about the motives of top management.   Some of the senior managers in human resources look at the new breed of young managers and say, " A lot of these kids are hired guns. They don't have any loyalty to anybody. They come in, they do the job at the highest price and then they clear out."   But the younger managers don't look at it that way.   They see some senior managers as the real hired guns, particularly the people who have been brought in from outside their company, who don't have loyalty to company traditions, or to other employees and who don't have any collegial feelings. These are the senior managers who are seen as much more apt to pull the trigger. And this is interesting, senior managers don't fault their companies for downsizing. They understand that it has to be done. What they don't like is how it is done. They don't like the unfeeling, insensitive, and poorly communicated way in which a great deal of reduction in force is carried out.

I think we're going to see a shift in the management style of American industry. We've been seeing this shift since the 1970s. It's something that Dick Walton at Harvard calls the "transition from a managerial culture of controlling the worker to a managerial culture of getting the worker involved and committed." It's a move from paternalism and authoritarianism to participative management in which decision making is driven down to the lowest possible level and workers become much more involved in controlling decisions that affect their own work.   I don't think that there is a single large American company that has not at least experimented in this area, whether it be with

quality circles, quality teams, employee participation
groups, semi-autonomous work teams, self-managed teams,
gain-sharing plans, socio-technical systems, or other
alternative work designs.

USES OF COMPANY CULTURE

I'd like to turn from the national business
culture to the more parochial cultures of individual
companies and talk about some of the uses of culture
and some of the things that it takes to change a culture. What we're essentially talking about here are
the shared values, and meanings that give individual
firms a distinctiveness from other firms, a different
way of doing things, a different way of valuing work.
Culture can be used as a tool to manage a crisis.
Probably the best national example of this was back in
1982 with the first Tylenol crisis faced by Johnson &
Johnson. This crisis, in which one or more persons
used an analgesic as a murder weapon, killing seven
people, posed a serious threat to the viability of the
company. It could have put an appreciable part of that
company out of business. The product, Tylenol, was in
first place in the analgesic market and was responsible
for a lot of the revenue and an even larger proportion
of Johnson & Johnson's profits.

The company, in dealing with this crisis, fell
back on its corporate culture as embodied by its credo.
The credo was written after World War II by General
Robert Johnson, the son of the founder, and was not
simply a piece of paper. Over the years, Johnson &
Johnson has sponsored credo conferences and credo meetings for their senior managers. At these meetings, the
company says, in effect, that these are the basic
premises, assumptions, and most cherished values of the
company. Managers are challenged to discuss the
implementation of these values in their work.

In the crisis, Johnson & Johnson was up front with
the press, cooperated with the government, asked the
FDA to come in, and cooperated with the pharmaceutical
industry in initiating moves to develop tamper-
resistant packaging. They set up a hot line, a telephone system with hundreds of volunteers to answer
phone calls from the medical and consumer communities.
They cooperated with the police. They sent out literally thousands of telegrams to the nation's physicians
informing them of the nature of the crisis and of the
progress of the investigation. Their spokespersons
appeared on television and gave broad access to the
press. They stopped the advertising of Tylenol, they
stopped the production of Tylenol, and they recalled
and destroyed some thirty-one million capsules. The

whole effort cost $100 million and, of course, Wall Street and the advertising community said, "Kill Tylenol. Don't ever bring it back." And yet, within a few months they had brought Tylenol back onto the marketplace and regained nearly its entire former share. And they did it essentially because Johnson & Johnson realized that, over the years, they had built up, through their corporate culture, an enormous respect for their work and stature as a company on the part of the American consuming public.   When the consuming public learned the facts in the crisis, they didn't blame the company. On the contrary, they found Johnson & Johnson to be a highly reputable firm and indicated strongly they would go back to purchasing Tylenol if the packaging were different and more secure.

In summary, the credo identifies its commitments as follows: "We believe our first responsibility is to the doctors, nurses, and patients, to mothers and all others who use our products and services. . . . We are responsible to our employees, the men and women who work with us throughout the world. . . . We are responsible to the communities in which we live and work.... Our final responsibility is to our stockholders."

I spoke to the CEO, the president, and other members of Johnson & Johnson's Tylenol task force. They all said that they could not have acted otherwise. In effect, they fell back on their corporate credo in coping with the crisis. The public responded with its trust and by bringing Tylenol back to its former preeminence.

Corporate culture can also be used to create an identity and to provide a "glue" to hold a company together. There is a very large financial institution on the West Coast that very recently engaged in a credo development project. As we all know, the banking industry has become progressively deregulated and banking has gone off into a great deal of diversification. There was fear on the part of the corporate leadership of the bank that the several businesses that had been developed were creating their own separate identities and subcultures. Senior management perceived the need for a mechanism that would pull the various parts of this company together and provide for a clearer overall corporate identity.

Inspired by Johnson & Johnson's use of its credo, the bank's management initiated a credo formation project. The top managers first sat down and worked out a series of statements of commitment to their various constituencies, shareholders, employees, and customers. Next, they sent this process down through the organization by having employee focus groups at every organizational level meet to react to senior management's notions of corporate credo. In nearly every case, the troops in the field rejected the thinking of the senior officers. Senior management sat down again and reworked the statements of commitment so that they approached reality more closely. The commitments themselves are not particularly exceptional. But what I think was exceptional was that in the credo-building process the company began to "glue" itself together.

CONCLUSION

Let me finish up by continuing a theme that was developed by Noel Tichy. Cultural changes are not easy or simple. They require a great deal of time, a great deal of trouble, a great deal of patience, and a great deal of pain. Just think, for example, of the impact of the breakup of AT&T on the employees who have been described as "daze-walking." They were in shock when they found out that they were no longer going to have employment security, promotions from within, and consensus management. This was a terribly painful experience that involved three interrelated factors.

You cannot simply change a culture and have it work. You have to also try to change the strategy of the company. You also have to try to change the structure--the social structure--of the company that carries out the changes. In the case of AT&T, there was obviously an externally driven change in strategy. They moved from a monopoly situation to diversified competition. There was a structural change. The Bell system was broken up into a number of independently operating companies. And finally, there was a culture change from a regulatory mentality to an entrepreneurial, competitive mentality. That change, as we can see in the waves of recent and current layoffs at AT&T, is a change that is not proceeding all that easily.

The experience in culture change at AT&T and other firms underscores the principle that changes in values and attitudes alone cannot bring about a desired transition. The new values must be in line with the organizational structure of the company and with its business strategy.

7

Managing Change in Innovative Organizations

Rosabeth Moss Kanter

Yogi Berra said, "The future isn't what it used to be." That's a profound thought. We're in a world where change is so rampant that our views of a future unfolding logically from the past are no longer possible. And to continue the sports analogy, we're really playing very new games in business today. What it takes to win those new games is different from the skills and abilities and organizational systems that we used in the past.

MANAGING RESEMBLES <u>ALICE IN WONDERLAND</u> CROQUET

My favorite metaphor for the game being played in most businesses today is the croquet game in <u>Alice in Wonderland</u>. That's a game in which nothing remains stable for very long. Everything is changing around the players. That describes what it's like to manage today. Here is how it fits:

Alice tries to hit the ball, but the mallet she's using is a flamingo, and just as she's about to hit the ball, the flamingo lifts its head and looks in another direction. This is a good image for technology, for the tools we use. Just as people have mastered the technology, <u>it</u> faces in another direction, as the flamingo did, and there is something new to be learned.

Then Alice focuses on the ball, but the ball is a hedgehog, a living creature with a mind of its own, and just as she's about to hit the hedgehog, it unrolls, gets up and moves to another part of the court. Just like our employees, our customers, our users, they're no longer lying down waiting for us to whack them. They too have minds of their own. They will unroll, get up, and move to another part of the court if they don't feel that they're getting the treatment or the solution they need fast enough. That's been true of employees for years: growing rights-consciousness, career concerns, people less loyal to their employers--

accelerated of course in this era of takeovers. It is
true also of customers, increasingly fickle, less
"brand loyal," willing to shop around and to demand
quality service. It's certainly true of the users of
Management Information System service, as anybody knows
who tries to run a central M.I.S. department and says
to a user, "It's going to take us a couple of years to
get it all reprogrammed for you." The user replies, "If
that's the case, I'm going to do it on my own P.C."
The result is hundreds of different systems, even in a
small organization.

Finally, in that Alice in Wonderland croquet game,
the very structure of the game is in flux. Alice
finally has the mallet ready to hit the ball, but the
wickets are the card soldiers ordered around by the red
queen, and every once in a while, the red queen barks
out another order to have the wickets reposition themselves on the court. The red queen could be the federal government barking out new orders with respect to
regulation or deregulation, and suddenly industry
structure changes or the company reorganizes. There
are other mini-red queens today: Sir James Goldsmith,
Carl Icahn, T. Boone Pickens, and other corporate
raiders who bark out orders to "Buy that stock!" Suddenly there's the company restructuring again, reor-
ganizing to ward off the takeover because they have
been merged or acquired. The wickets increasingly are
even voluntarily repositioning themselves on the court,
as companies go into a nontraditional business and make
very unusual combinations that change the definition of
an industry--like Sears, a retailer, going into
banking.

This is the new game: everything is in flux,
technology is changing at a rapid pace. Customers,
employees, and users are more demanding and willing to
shop around, and the whole structure of the industry is
changing around you. It's very difficult to win a game
like that using conventional methods. That's why innovation is so important. There's really only one way to
win a game like that--by using what I call the "Three
F's," supported by some "P's." The "Three F's" are the
ability to be focused, fast, and flexible, and the
"P's" are a solid base of partnerships. The winning
organization befriends all of those stakeholders in the
business, with minds of their own, who create change--
the customers, the employees, the users, and even the
regulators.

MASTERING CHANGE: THE NEED FOR FOCUS

I will discuss the environment that makes it
possible for people to master change adequately,

through using the "F's" and "P's"--the corporate culture to support winning the new game.

First, in order to win a game like <u>Alice   in Wonderland</u> croquet it is important to be very <u>focused</u>, to concentrate on just a few things rather than many things. If you're trying to play a game where you have to watch things in constant motion, it's very difficult to be on top of all the things you need to do if you're distracted by too many demands, too many requirements, too many unrelated things. That's exactly why this idea of <u>focus</u> in corporate strategy is coming back into vogue. Twenty years ago the conventional wisdom on strategy was to diversify--keep many eggs in many baskets, have a large portfolio of many different businesses or activities, as a hedge against change. But conglomerates that did all sorts of unrelated things didn't necessarily make much money at it. The new wisdom says that too much diversification may not work, not in a rapidly changing environment. Business should be focused on just a few things that they know how to do very well, to build on competence and avoid being distracted by trying to keep up with too many things.

My Harvard colleague, Michael Porter, has new data that demonstrate this very well. He looked at the track record of thirty-three conglomerates in the <u>Fortune</u> 500 with respect to acquiring businesses that were unrelated to their core. Over a thirty-year period, those conglomerates divested an astonishing portion of all the things they bought. They eventually sold about 60 percent of all their unrelated acquisitions. It was interesting to look at who was high and who was low on that list. Lowest on the list were Johnson and Johnson and Procter and Gamble, companies that have a very clear focus in their businesses. They bought the fewest unrelated businesses, and they tended to sell the fewest--under 15 percent of what they bought. At the high end were CBS and RCA that bought a lot of businesses and sold close to <u>90 percent</u> of all of them. Did this effort weaken the companies? Consider what's happened: RCA has disappeared into GE, and CBS has become part of the Tisch empire. On the corporate level, then, having just a few things that the business needs to do well makes a difference in winning a game where everything is changing, and one cannot possibly master everything.

That's just as true inside a corporation. Successful departments cannot possibly do everything. They have a few clear strategic focuses, making it clear that "<u>this</u> is what we're going to concentrate on, and we'll keep those few goals in mind regardless of that flux around us."

MASTERING CHANGE: THE NEED TO MOVE FAST

But, of course, focus does not mean inflexibility.
Organizations also need to change. So the second key
to winning the game of <u>Alice in Wonderland</u> croquet is
the ability to move very fast in today's environment.
Organizations benefit from the "first mover" advantage:
getting in there first before others do, to master an
environment that's constantly changing--or you're
always playing catchup rather than defining the terms
of the game. The companies that are successful in this
new environment are companies that focus on innovation,
on being ahead of the competition, on using all of
their skills to provide a new source of strategic
advantage. Indeed, there are a number of ways in which
information technology is now being used by companies
to get that first mover advantage, to be in there fast
and first, to nail down the customer, and to get the
right location in a retail organization. Information
technology is now a key.

One of the companies that is noteworthy in this
respect is American Airlines. American Airlines used
information technology to get a clear first mover
advantage, the   Sabre System which they placed in
travel agencies to help them make reservations by computer: but naturally, American flights appeared first.
American gained what can be called a "temporary
monopoly," because they used information to be the
first to give a service to their customers. (Later,
this advantage was reduced by legal action.) American
has been first in a number of areas, including frequent
flyer programs.

By being first, you gain the opportunity to get in
there, nail down the customer before the competition
can, and gain the experience that helps you keep
improving.

In contrast, consider Control Data, a much-admired
company, but also a company that a long time ago bet on
the fact that the mainframe computer was here to stay
forever--not a bet that anybody would want to take
today. In essence, we need organizations that do think
ahead and yet are always looking for the shorter-term,
incremental competitive advantages. When a Sabre
System is thrown out as a temporary monopoly, and you
have to give just as much access to other airlines to
get reservations on their flight, you've got another
idea that you're ready to go with. That's what it
takes to win the game today.

In order to do that, those organizations that are
successful are making sure that there is a constant
stream of innovative ideas coming up from their organization. They're trying to make everyone an innovator
and finding ways to develop and build innovations out

of the ideas coming up from the organization. I see strong companies developing a variety of systems and programs to stimulate ideas up from the ranks and even put people in charge of projects to run their ideas.

AT&T's new venture program shows how much potential is there to be tapped in an organization. Several years ago, just before the 1984 divestiture of local telephone companies, AT&T quietly put in place a program by which they would seek employee ideas. They would screen the ideas and choose a few that they would give big funding to for those employees to then run their own projects and, in some cases, run new business ventures in the middle of AT&T. They expected to fund a dozen or so projects, out of maybe fifty to one hundred ideas that would be submitted. They didn't give the program publicity; they just wanted to try it first. So without publicity and with the expectation that maybe there would be fifty or one hundred of these ideas for new businesses, extensions of existing technology, or other innovations submitted, they began the program. Over <u>2,500 ideas</u> were submitted in the first year. Similarly, Eastman Kodak has an Office of Innovation and a whole process for stimulating and encouraging people at every level to be thinking a little bit ahead. Innovation facilitators in every facility help people define a strategy for how they're going to take their ideas further. Those ideas that look like big ideas can get submitted to a venture board and get funding to set up a project or a line of business, a separate subsidiary of Eastman Kodak. Clearly, those companies that want to move fast and gain the advantage of innovation in order to win the new game are getting the ideas flowing and in operation everywhere.

In order to move fast, one also needs an organization in which information is shared broadly across all levels and across all ranks so that people know how to reorient and reposition quickly. In that sense, the M.I.S. function is not only one that needs innovation itself, but it also can be the foundation for innovation throughout the organization, because those organizations that seem best able to take advantage of innovation and move fast are also those organizations with open communication--without barriers or restrictions to the flow of information.

Open communication is expressed in all sorts of ways. One is in terms of the amount of data that is accessible to everybody at every level. For years I've wanted to rename the Management Information Systems Operation the <u>Employee</u> Information Systems Operation. Why only managers? In this new game, we're asking everybody to be able to reorient as soon as they see a new opportunity or learn something new or do something

differently. That information has to get down the
ranks to every level. They need the data as well.

In one leading company, several facilities are run
by employee teams who manage that factory as though it
were their own business. There is one job classifica-
tion, and everybody's on the team, virtually without
managers. It's a highly advanced system. In factories
that are run on that model, costs are half of what they
are in a conventional factory, quality is higher, and
flexibility, which I will consider shortly, is also
much higher. The company is sending technicians from
those factories around the country to train others in
how to incorporate new technology. One of the engines
that drives that system is information. Personal computers down at the plant level enable everybody who
works on a product to find out exactly what's happening
to that product, including its fate in the marketplace.
The manufacturing executives realized that unless production people can know on a daily basis and have good
data to look at, how can they make sound decisions
about what to purchase, how much to produce?

Decisions increasingly are being decentralized in
leading companies. And the thing that makes it work is
information and communication. You can only give that
responsibility to employee teams at the lowest level if
they've got the information behind it to make intelligent decisions. Increasingly, the _information_ is going
to be decentralized even if the information _system_
isn't. Some high innovation companies even take that
communication principle farther; they make sure that
nobody is restricted from finding out anything they
need to know. (Clearly there are some circumstances in
which some information has to be kept proprietary; but
fewer and fewer of those situations exist in major
corporations.)

In a financial services company, one of the most
innovative in the United States as leaders of a whole
new financial services industry, managers simply don't
buy the fact that if you have the wrong status you
can't find out certain kinds of things. They don't
have fixed formal titles, they don't have status
barriers, and they don't even have written organization
charts. All of those traditional things are considered
a barrier to getting the right information into the
hands of the right people. Similarly, a leading bank,
a bank that has bet on information technology as one of
its competitive advantages, was seriously considering
eliminating all titles. Now titles are sacred cows in
a bank, where "vice presidents" of various statuses
abound. Instead, they wanted to have the right people
communicating, getting information into the hands of
the people who are in the best position to make a
decision. They wanted to eliminate status barriers

because one of the purposes of status is that higher-level people get to know things that lower-level people don't know.   Power games that managers play at higher levels of organizations are often games that restrict information.

One telephone company that scored low in innovation, in my research for The Change Masters, had lack of communication as a major barrier to innovation. The field people would accuse headquarters of with-holding certain kinds of data.   In retaliation, they would go out and collect the data so that they could withhold it from headquarters.

Communication is going to be one of the major ways that organizations manage to move fast, take advantage of opportunities, redirect, reorient, and be innova-tive.   In turn, this requires a foundation of commit-ment.   You can't get people to move fast unless they're already committed to the goals and to the organization, and unless they already feel a part of this organiza-tion.   Otherwise, they'll foot-drag, resist, find a million reasons not to act on the new opportunity, not to do anything.   But if you've developed a system in which people are committed to the leadership, believe in them, trust each other, know that they've been involved in the past and know that their needs are taken into account, then when the time comes to direct change, they'll move.   A foundation of commitment in the organization to goals, to the fact that my needs are being taken into account, that I have been listened to, makes it possible to do all these other things that allow organizations to move fast.

"Instant success takes time."   All the things that allow an organization to move very quickly are things that in fact they've been preparing for years by laying a foundation of trust, relationship, knowledge, and information.   This well-built foundation allows an organization to do something that becomes an "instant" overnight hit.

MASTERING CHANGE: THE NEED FOR FLEXIBILITY

The third "F" that it takes to be effective in Alice in Wonderland croquet is flexibility.   Though focused on just a few goals, organizations also need the flexibility to keep bending and redirecting to take advantage of the new opportunities that are arising in an environment of constant change, where technology provides an opportunity a minute.

In order to get that kind of flexibility, a company needs an employee body that's broadly skilled and whose assignments are broad rather than narrow. This is a real change in the conventional wisdom about how to manage.   The conventional wisdom used to argue for narrow specialists, people who have a narrow job in which they know how to do one thing that they do over and over again, to get efficiencies of repetition. In this new environment, any efficiencies of repetition are going to be wiped out by the fact that there's going to be change, and people have to do something new and different. In order to ready people for change, for the fact that it's not going to be the same tomorrow as yesterday, organizations need people who think more broadly, who have more than one skill, who have assignments that have been focusing on getting results rather than just carrying out procedures mechanically and routinely. In some high technology firms, a common joke is that the typical job description for a manager is "do the right thing." How much broader could one get than that?

One of the problems that some telephone companies are having as they try to reorient to the new competitive marketplace is that for years people were measured on working according to the rules, not asking too many questions. Sometimes people didn't even know why they were doing the things they were supposed to be doing. Now a more broadly skilled group of people in touch with more fields and disciplines, who see the big picture, are going to be the success drivers for the organization in the future.

In one case, the typical American way of doing things, the narrow specialized job, bumped up against another country's system. General Motors has a joint venture with Toyota on the West Coast, New United Motors Manufacturing, Inc. One of the concessions GM had to get from the union in order to do this had to do with a number of job classifications. When GM ran this plant, there were thirty-three separate classifications; Toyota for a similar plant had three. Consider the implications of thirty-three different jobs versus three different jobs. Toyota had people who thought more broadly, knew more things to do, and therefore were more flexible when it came time to change.

Educating everybody more broadly is one of the best hedges against a game like Alice in Wonderland croquet. So is teamwork across areas. If you need to redeploy, develop a new project, take advantage of the new technology that's available, or reposition the company in the marketplace, you need people from many different disciplines working together to do that. One of the key lessons about innovations is that innovation always takes teamwork from more than one area. Unless it is very narrowly focused on the concerns of only one function, then innovation always cuts across areas and requires more people to cooperate. Successful companies need an organization structure that's oriented

toward people working across areas, in which people are
comfortable with project teams; care about ultimate
results, rather than just exercising their professional
skills; and know how to operate as task forces across
areas.

One barrier to innovation is a myth I call "cowboy
management." Consider some images used in American
business: performance shoot-outs (as though it's high
noon at the management O.K. corral); development of
deliberate in-house competitions in which people slug
it out; writers who make a virtue of "ready, fire, aim"
as a guide to action; bootlegging the organization's
resources to do a project; and wishing that headquarters back East would go away. These ideas do get some
temporary energy in the organization, but they also get
people focused on wiping out the competition in the
next department rather than getting the overall results
the company needs.

Setting up excessive competitions, where people
focus on killing off other departments and areas to
improve their own position rather than collaborating,
is a very bad way to succeed in Alice in Wonderland
croquet. And one of the problems with performance
shoot-outs is that somebody you need later may die of
the wounds. Collaboration and teamwork is better strategy in organizations today than excessive competition,
hostility across areas, turf-mindedness, and an attitude that we've got to protect our own rather than
collaborating. Even if collaborating means a department may occasionally lose territory because they're
part of a bigger entity, that's the attitude
organizations have to encourage.

MASTERING CHANGE: THE NEED FOR PARTNER ORIENTATION

Behind all of the "3 F's" are the "P's" of
partnerships. No one can win Alice in Wonderland
croquet anymore by treating customers as adversaries,
by treating employees as numbers, by treating suppliers
as temporary conveniences. The best strategy is to
befriend all the groups in the environment on whom the
company depends, so that the whole network gains
advantages together.

The new management wisdom for supplier-customer
relationships is partnership-oriented wisdom. That's
what American Hospital Supply was building when it put
terminals in hospitals so customers could order supplies more easily from American Hospital Supply. That
was a partnership with customers that involved giving
them something that will help them do their business
better while helping AHS do its business better.
Security Pacific Bank could put terminals in car

dealerships for processing loan applications instantly because of a close working relationship with those customers. A partnership means that customers or suppliers are willing to invest in the same kind of technology that you have.

Companies also cannot succeed without partnerships with employees who are willing to bend or even make concessions. Pacific Telesis has a remarkable labor-management partnership, which they call a business partnership with their principal union. The company realized that the only way that it was going to be able to put in a new technology was if the union worked with them on doing it, because it would cost jobs. They reasoned that if they worked as partners in planning the future, they could simultaneously provide job security and move on new technology. This would help them win the new game.

Manufacturers, especially, need partnerships with suppliers. It is impossible to take advantage of systems like just-in-time inventory unless suppliers feel there's a longer-term commitment making investment in the technology worthwhile. That's what it's going to take to make sure we have a competitive system so that we can order across organizational boundaries. In fact, not only do we all have to become partnership-oriented in our attitude as a manager or an entrepreneur, but I think, increasingly, as many of you who work in M.I.S. areas already know, much of the technological development is going to be developed across organizational boundaries. It's going to be developed in joint development projects with your customers or joint development projects with your suppliers or with your users, rather than simply the things we need in isolation. So those are the things I think that it takes to win the new game.

NEED FOR INTEGRATIVE ORGANIZATIONS VS SEGMENTALISM

I have discussed the importance of "Three F's and P's." They require a particular corporate environment. Stimulating innovation and entrepreneurship is a matter of having an organization that I call integrative. It's an organization where people pull together rather than apart, where there might be some overlap across areas: joint projects and joint development activities exist; users and the central group are linked closely; communication and information knit everything together; and shared goals and a focus for the business create unanimity of purpose.

The opposite of this kind of environment that destroys innovation and change, I refer to as

<u>segmentalism</u>.   The organization divides into pieces, into segments, and instructs every piece to operate totally independently, governed instead by performance shoot-outs, hostility, and rivalry.   People are highly specialized just in one segment, so they never get out and find out what's happening in the broader environment.   Everybody focuses on their territory only, without thinking about the wider needs of the organization.

Segmentalism creates a totally rigid, inflexible organization.   It is impossible to take advantage of a new opportunity or mobilize that organization to do anything differently.   People are rigid; they protect their territory; they're inflexible; and they're narrow in their focus.

RULES FOR STIFLING INNOVATION

This picture unfortunately describes too many organizations in America today.   They operate as though they prefer mediocrity and stagnation, as though they were guided by a set of rules for remaining second-rate.   I call them the <u>rules for stifling innovation</u>.

First, be suspicious of any new idea from below because it's new and because it's from below.   After all, if the idea were any good, we at the top would have thought of it already.

Second, insist that people who need your approval to act go through many other levels of the organization first.   That way you can kill them off or discourage them.   Or, if you're lucky, by the time they get back to you, you might be in a different job anyway.

Third, express criticism frequently, withhold praise, and instill job insecurity, because that keeps people on their toes.   How else would they know you have standards?   The "macho" school of management holds that people do their best work when they're terrified.   Of course, it's the opposite of the foundation of teamwork and commitment and recognition of achievements that goes on in high innovation companies.

Fourth, decide to change policies in secret and reorganize unexpectedly and often.   Because if you don't want innovation and change, you don't want people taking initiative.   You want to keep them so off balance that they'll never be able to take initiative. They won't know what's going on, they'll be in the dark all the time.

Fifth, be control conscious. Count everything that can be counted, do it as often as possible.   Because if you don't want innovation and change, you want things managed so tight that you can never take advantage of new opportunities.   There's no slack, there's no extra time to pursue anything new. People are too busy doing the things that are already being measured.

Finally, if you don't want innovation and change, you need attitudes that say that we already know everything that is important to know about this business. We've been doing it for a long time; why don't we just keep on doing what we've always done? This is not the right attitude if you're playing <u>Alice in Wonderland</u> croquet.

CONCLUSION

Companies that want innovation should reverse these rules: be more receptive to new ideas, get those idea flows going. Encourage faster approval, less red tape. Offer more praise and recognition for people, their talent, and achievements. Make sure there are resources that people can use to experiment. Provide advance warning of new plans, spread information so that everybody knows what's going on. Finally, foster the attitude that we're always learning, that we can learn from any level.

All this helps convert change from a threat to change as an opportunity. Change is always a threat when it's done <u>to</u> me, imposed <u>on</u> me, they're making me do it. But it's an opportunity if it's my chance to be a hero, if it's my chance to help us win the game by doing something a little bit innovative.

The real issue in managing change in an innovative organization is to create more opportunity for people themselves to take charge of change.

8

Effective Leadership:
The Key Is Simplicity

F. Kenneth Iverson

Nucor is a manufacturer producing steel and steel products. We operate seven steel mills on four sites. We produced, last year, about 1,700,000 tons of steel. We have a capacity of two million tons, and we're the ninth largest steel producer in the United States. What sets us apart?

NUCOR's TECHNOLOGY AND PROFITS

All of our mills use the latest steel technology. One hundred percent of our steel is continuously cast. In the United States even today, only 50 percent of the steel is continuously cast, whereas in Japan, over 90 percent of the steel is continuously cast. For more than fifteen years, the price of the products we produce, FOB our plants, has been equal or less than the price of these same products produced by foreign steel mills dockside USA. Also, for the last fifteen years we have not laid off a single employee for lack of work.

We operate profitably. Since 1965, when I became president, this company has never had a loss quarter. My predecessor resigned because the company had defaulted on two bank loans, and I got the job because I happened to have the only divisions that were making money at the time. We will have sales this year of about $800 million. In a number of years, our return on stockholders' equity exceeded 20 percent.

NUCOR'S ORGANIZATION STRUCTURE

Certainly, most of the success of this company is due to its organizational structure, and to our employee relations programs and policies. One of the things that we believe in very strongly is that the best companies have the fewest number of management layers. We think that the size of a company is not determined by its sales but more by the number of management layers. The fewer you have, the more effective it is to communicate with employees and the better it is to make rapid and effective decisions.

I attended a lecture one time by Peter Drucker. He brought up the fact that the number of management layers is one of the factors that determines how well a company does. He mentioned that sometimes companies get so many management layers they become unmanageable. Somebody in the audience asked him, "How many management layers is that?" He said, "Nine effective layers." Someone else then said, "Why?" He said, "For two reasons. One is that when you start a memorandum out from the top with a CEO and come all the way down through nine management layers, it can't in any way resemble what it started out as." He said, "Secondly, a young man in his whole career can never really effectively work his way up through nine management layers." In the next sentence he said, "The U.S. Army has nine management layers."

We have four management layers, the first one being the foreman. Above the foreman is the department head, which would be our manager of melting and casting, manager of rolling, division controller, or sales manager. There are about five or six department heads in each division. They report directly to a general manager who is a vice president of the company, and he is the only one who reports to our corporate office.

We're very much a decentralized corporation. These general managers make the day-to-day decisions that determine the success of the company.

The other point that we believe in is to reduce the number of staff. Staff people in marketing, engineering, or purchasing, in many cases, do not help you make better decisions nor do they accelerate decisions. Perhaps we carry it to extremes. With $800 million in sales, our corporate staff consists of a total of seventeen people, including stenographic and clerical help.

DECISION MAKING AT NUCOR

We do have in this company a very strong feeling of loyalty to our employees. We believe that there are two successful ways to manage in relationship to employees. That is: Tell them everything or tell them nothing. Both ways can really be successful, except we happen to believe in telling our employees anything they want to know about the company, unless it happens to be proprietary or it has to be secret for some reason.

We try to impress upon our employees that we're not King Solomon. We use an expression that I really like, and that is, "Good managers make bad decisions." We believe that if you take an average person and put him in a management position, he'll make 50 percent good decisions and 50 percent bad decisions. A good manager makes 60 percent good decisions. That means 40 percent of those decisions could have been better. We continually tell our employees that it is their responsibility to the company to let the managers know when they make those 40 percent decisions that could have been better. Because if they tell us, and if we examine them and agree that a bad decision was made, we will change it. Over the years we've got the employees to the point where they don't hesitate to tell us when they think we've made a poor decision.

The only other point I'd like to make about decision making is, don't keep making the same bad decisions. Unfortunately, over the last twenty years, up until recently, that's what we have done in our steel industry. I have a story to illustrate this. We have people in the Nebraska plant who love to hunt. We have a group that goes moose hunting every year in Canada. A couple of fellows from that plant went up last year. The plane flew them in and when it left them for a week of moose hunting, the pilot said, "Now remember that this plane will only carry one moose." So, they went out on Monday. Tuesday they shot a moose. Since they're not the type who sit around and drink around the campfire and do nothing, they went hunting on Wednesday, Thursday and Friday. On Friday, they shot another moose.

The plane came back on Saturday and the pilot said, "You have two moose. I told you the plane would only carry one." They said, "Awe, that's what the pilot said last year and we put two moose on the plane." The pilot said, "Ok" and he tied one moose to one pontoon and the other moose to the other pontoon and he took off along this lake. He got up about 400 feet and couldn't climb anymore and began to lose altitude. Finally, he crashed into a group of pine trees. As the pilot was pulling himself out from the wreckage he said, "Where are we?" One of the hunters said, "Just about the same place we crashed last year." Don't keep making the same mistakes.

INCENTIVE SYSTEM

All of Nucor employees have a significant part of their compensation based on the success of the company.

The most important incentive system we have is our
production incentive system. We take groups of about
twenty-five to thirty-five people who are doing some
complete task such as making good billet tons, good
roll tons, or good finish tons. We have more than
seventy-five groups of this type in the company. We
establish a bonus that is based on a standard. If the
employee group exceeds that standard in a week, they
receive extra compensation based upon the amount of
increased production over the standard. Very simple.
There is no maximum. It is never changed unless we
make a large capital expenditure that significantly
changes the productivity opportunities for the
employee.

The bonus is paid weekly, and in our steel mills
today it is not unusual for that bonus to run 150
percent of the base pay. All we do during that production period is take the base pay plus overtime, multi-
ply it by 150 percent and give it to the employee. The
average hourly employee at the steel mill here at Utah
averaged about $40,000 last year. They earn every bit
of it.

What's the result of this? The result is that the
average integrated steel company in the United States
last year produced about 350 tons per employee. We
produced about 980. The mill here in Utah produced
over 1,000. Our total employment cost per ton is about
$60 per ton including fringe benefits. The total cost
for the integrated producer was about $135 a ton. It's
no small wonder that when we, or other mini-mills who
have comparable types of programs, get into a product,
the integrated producer moves out of that product.

NO LAYOFFS

We have not laid off or furloughed a single
employee for lack of work for more than fifteen years.
There's a reason why we have that policy. Most of our
plants are located in small towns and rural areas. We
think the rural part of the United States has great
untapped labor resources. People don't want to go to
the cities particularly. They go there in order to
find jobs. Big corporations go there because that is
where the people are. Actually, if you establish manufacturing facilities in rural areas, you find that
there is a flood of people who apply for work. We put a
steel mill in a town of Jewett, Texas, that has 435
people. People said, "Where are you going to get the
300 people to run this plant?" There were only 12,000
people in the whole county. We had more than 2,000
applications from people in Houston and Dallas who
didn't want to live in Houston and Dallas. They wanted

to live in this nice rural community that's about midway between the two.   Under those conditions, of course, we have to accept our social responsibilities, because we are generally the largest employer in the area, and in many cases almost the only employer in the area.   Accordingly, we can't just lay people off, because they have no place to go.   So if we have a slow economic period, everybody works four days a week instead of five. But everybody still has a job.

The other thing I firmly believe in is, you don't get good people if you lay off half of your work force just because one year the economy isn't very good and then you hire them back.   If you do that, you aren't going to get the best people in the area.   That's why we will not lay off people.

EMPLOYEE BENEFITS

We do have a number of unusual benefits for our employees.   For example, we pay $1,500 a year for four years of college or four years of vocational training for every child of every employee in the company.   We have at present about 380 to 420 children of employees enrolled in about 180 different learning institutions in the United States.

In case you should think that we are overly paternalistic, we also have some very tough rules.   If you're late, you lose your bonus for the day.   If you're late more than thirty minutes, or you're absent for any reason including sickness, you lose your bonus for the week.   We do have four forgiveness days.   We have some people who take those forgiveness days in January and February, and we have some people who haven't taken any for five years.   I'll tell you a true story about this that happened a couple of years ago. We had an employee that came into the plant in Darlington, South Carolina, and said to the head of melting and casting, "Bill, Phil Johnson (one of our melters) has had an automobile accident and he's out by the viaduct holding his head." Bill said, "Why didn't you stop and help him?" He said, "And lose my bonus?"

We try very hard to eliminate as much as we can any distinction between management people and anybody else in the company.   We all have the same group insurance program, holidays, and vacations. We all wear the same color hard hat (green). We have no company cars, company airplanes, company boats, executive dining rooms, assigned parking places, hunting lodges, nor fishing lodges, and everyone travels economy class. We think it is very important to destroy that hierarchy of privilege that is so prevalent and pervasive in many corporations in the United States.

MANAGERS MUST BE ADAPTABLE TO CHANGE

While I'm talking about that, let me mention some of the things I think make a good manager. One of the most important things is that a good manager must be adaptable to change. He must, in this day and age, readily accept new technologies as they develop. Because of the mobility required of executives, he also has to be able to rapidly adapt to geographical and cultural differences. Children do this very well, but you find as we get older, it's much more difficult. You have to focus on the fact that "I am going to do it. I am going to make a change."

In 1962, I moved from New Jersey to the little town of Florence, South Carolina. My son was in the fifth grade. I was going to the plant the first day and he said, "Dad, I don't want to go to school." He had already been to one day of school. I said, "Mark, what's the problem?" He said, "Well, the other kids all have a hook with their name on to hang their coats." He said, "I don't have my name on any hook and I'm not sure I can find one. I'm not sure I can find the chair that I'm supposed to sit in." Then the tears started to come and he said, "Besides that, I can't understand what the teacher is saying."

About two weeks later he came home and said, "Dad, do you know what a 'purd' is in a sentence?" I said, "No." He said, "A 'purd' is what you put at the end of a sentence." A month later he came home and said, "Do you know that the Great Lakes are filled with salt water?" I said, "Mark, we lived in Michigan for seven years and you know the Great Lakes aren't filled with salt water." He said, "Well, I've got a teacher in fifth grade that is teaching all the students that the Great Lakes are filled with salt water." I said, "Well why didn't you speak up and say something?" He said, "You don't tell Mrs. Singletary anything." (Just like some managers.) But the real point of my story comes in the Fall when he started into the sixth grade. He said, "Dad, do you know that fellow from Michigan who bought the Chevrolet dealership in Florence?" I said, "Yea, I know him." He said, "He's got a son in my class and gosh, does he talk peculiarly."

COMMUNICATIONS

We run a survey in our corporation about every two years or so. We survey every employee about what they think of the hospitalization program and what they think of profit sharing, and a lot of other questions. It takes about an hour for them to go through all the questions. The one thing that's interesting about it is the fact that with each survey the average hourly employee has said he wanted better and more communication from the foreman.   I think that, in American business, that probably is one of the most important things we don't do well, that is, teach our foremen to communicate with the employee. You don't make an employee a foreman because of his communications skills, you do it because of his technical skills. So you end up with a foreman who may be very good technically, but he really cannot communicate effectively with the employees, who have all kinds of questions.

I am reminded of a story about a personnel man who went out in a plant. This story is to make the point that the training sometimes takes a long time. The personnel man spotted a foreman who had an employee by the arm and he was shaking him. He said, "Listen. We are going to get the yield up in this plant by the end of the week or you're fired." The personnel man went over and said, "That's not the way we do this now days. That's not really good human relations." The foreman said, "Why not? It's management by objective. I told him what the goal was, and I told him the consequences if he failed to reach the goal." The personnel man said, "We have a course in human relations in this company that we'd like to send you to." The foreman said, "I'd love to go." So the foreman went away for two weeks of training in human relations and then came back. The next day, the personnel manager was out in the plant. The foreman had the same guy by the same arm and he said, "Now listen. The yield in this plant is going up by 2 percent by the end of this week or you're fired. How's your mother?" Unfortunately, in many cases, that's exactly the way we do it.

Communication is terribly important.   I really think it ought to get more attention in business schools. But, it's not only communication, it's the quality of that communication. Certainly, a big part of communication is learning how to listen as well as how to say something.

I have a story on the point of quality communications. We have a plant in Fort Payne, Alabama. Before the interstate was put in, I'd fly into Chattanooga and then I'd wind down through Lookout Mountain until I finally got down to Fort Payne, Alabama. One day, I was making that trip and as I rounded the mountain, a woman came around the corner. She went so far out on the road, she almost made me go over the side of the mountain. As she went by, she rolled down her window and said, "Pig!" And I rolled down my window and said, "Cow!" And I went around the corner and ran right into a pig. The quality of communications is vital.

PLANNING

As any executive making a talk of this type, I really want to say a few words about planning. I think that in some companies, it's absolutely ridiculous, unrealistic, and sometimes almost fanciful. For example, consider the objective by a corporate office that says this company or this division is going to grow by 25 percent per year. That's really not planning, it's pearls cast before swine.

The short-term plans in a company, such as the year's budget, the year's production, etc., should always be a bottom-up type plan. In Nucor, we say that we want a 60/40 percent probability. We want 60 percent probability that you can make that plan and 40 percent that you won't. It's not a bit unusual for us to have a division come in with a budget for the year that projects the earnings to be less than they were the year before. It doesn't concern us one bit because there may be some forces at hand. We may be past the top of the construction cycle. There may be some basic reasons why that's going to happen.

Long-range plans are different. Long-range plans, I think, really should be the work of a corporate office putting together all of the projects that are underway, and all of the projects they think might develop in the company. It's a guidebook. It's a guidebook of places the company might go. It also helps prepare you for what might be some unusual crisis in people resources, or in financial resources. It's not a bible. It really helps you avoid difficult areas. I've never met a five-year projection in my life and I never expect to. It's always different than the plan.

It does prepare you for the unexpected as shown by this story. There was a fellow out hunting in North Carolina in the 1920s, a farmer, and he was looking to shoot a bear to feed his family. He hunted all day and didn't get a bear. So about sunset, he dropped down on his knees and he said, "Oh Lord, I am a good Baptist. I go to church and I pray, and you have answered my prayers. I'm out here looking for a bear to shoot to feed my family this winter. Please help me." He got up and he went around a corner of a rock, and there was the biggest bear he had seen in his life standing up on its hind legs with its paws outstretched. He aimed his gun and fired and the gun misfired. He dropped down to his knees and said, "Oh Lord, let this be a Christian bear." The bear dropped down on his knees and said, "Oh Lord, bless this food of which I am about to partake."

PRODUCTIVITY

My remarks would not be complete, of course, without some reference to that super buzz word "productivity." I want to give you just a few brief thoughts.

One is, there is no quick fix. You can't decide you might want to put in quality circles and really have a more productive organization, necessarily. This is evidenced by the fact that about 60 to 70 percent of the firms that put in quality circles have abandoned them. There has to be an overall culture and an overall philosophy in a company to really get the type of productivity that you are interested in. The Japanese system won't work here. There are certain elements of it that we can accommodate and incorporate. But basically, we're a much more heterogeneous society, and workers in this country have different expectations and different goals than Japanese workers. But let there be no mistake. Our workers today have a different attitude than did our grandfathers and even some of our fathers. It was a place that they spent eight hours a day. They built most of their lives around their community, family, or religion.

Workers today expect more out of their jobs. They expect the job to be meaningful; they expect to be able to advance; they expect to participate, particularly in those decisions affecting their work place; and they expect to understand how the company operates, where it's going, and how it expects to get there. If you don't develop programs that satisfy those needs and those interests of workers, you can be assured, in the long run, your company will not be successful.

We do have, in this country, some problems at the moment. It's been blamed on a number of reasons, one of them being the lower cost of labor and more productive labor outside the United States. I had a friend recently who is in the wire rope business. He went to Korea since he was getting beat badly by the Koreans on the price of wire rope, and he wanted to find out why. He took his worker foreman along with him. They were going through a plant in Korea and they found a worker who spoke some English. He said, "Do you like working for this company?" The worker said, "Oh, it's a marvelous company. They give us lunch, and they pay us well." He asked, "How many holidays do you have?" The worker said, "We have fifty-four holidays." He said, "What in the world are the fifty-four holidays?" The worker said, "We get Christmas, New Year's and every Sunday." That won't work in the United States. It's not acceptable to management, and it's not acceptable to the workers.

Well then, what is the answer?   Certainly the
answer is that we have to automate. We have to develop
our processes to the point where the lower labor cost
in the final sales price for competitors outside of
this country is more than offset by the higher costs of
shipping their products into our marketplace.

As I travel around this country, I am concerned by
the attitude of many businessmen who seem to feel they
cannot compete with foreign suppliers. Certainly, it's
not only textiles or automotive, it's steel, farm
implements, etc. They blame it on government subsidies,
lower labor costs, or better technology, and then what
happens? They reduce their capacity, source offshore,
eliminate product lines, and pressure government for
protectionism. My real concern is that they are making
bad decisions. They are making decisions that are bad
for their business and bad for our economy.

We do have, in this country, the people, the
ingenuity, and the skills to compete against foreign
manufacturers in almost every single area. I'm completely convinced of it. What we need is a dedication
to some new management styles. We certainly should not
accept at face value the management practices of the
past, because many of them haven't worked. We need to
try new ideas, we need to make new mistakes, and we
need to be quick to accept new technology. If we do
that we can compete with manufacturing facilities
anywhere in the world.

Appendix to Part I:
Commercializing New Ideas

This appendix describes the organization, processes and experiences for commercializing new ideas. As discussed earlier, John Moore's NSF paper, "U.S. Competitiveness and the National Perspective," cites statistics showing that over 40 percent of our productivity improvements since World War II has been derived from technology innovations. He described NSF programs for increasing education and the applying of new ideas. The paper in this appendix by John Scowcroft describes how the Utah Technology Finance Corporation was created to finance the early stage development of new ideas that have promising commercial applications. Papers by Hal Rosen and T. Peter Thomas discuss the processes and a case example of financing new ideas as they mature into successful products.

A

Utah Technology Finance Corporation

John M. Scowcroft

The UTFC is a public, non-profit corporation authorized
by the Utah State Legislature in 1983. It came about
as a result of efforts of the State Science Council,
who realized there were good ideas in Utah that needed
economic help. It also came about because of workers
in all of the schools, especially the University of
Utah, trying to forecast the needs for jobs creation in
the state of Utah and the foreseeable future. This
work indicated an ongoing need in the state of about
25,000 jobs per year.

There are good and standard economic development
methods that the state and cities, counties, chambers
of commerce, and other economic organizations are using
to create jobs. These are principally marketing,
trying to promote the state as a place to which the
outside business can come and succeed. Hopefully, those
efforts will result in about 5,000 of these 25,000 jobs
that are needed. Something over 15,000 are expected to
come from organizations already in the state that are
already growing, but there is a 5,000-job gap that
needs to be addressed by something else. So it is
thought that perhaps a supplementary approach to development might find success in the state. Present
efforts aren't satisfying the total need.

CREATION OF UTAH TECHNOLOGY FINANCE CORPORATION

The UTFC was authorized in 1983. The Board of
Trustees was given a year by the Legislature to decide
what the UTFC program would be. The Board seized upon
four programs that I'll describe in some detail as we
go on.

A year and a half following was a period of
challenge when the Attorney General of the State raised
the question, "Can a state organization legally do what
this Board of Trustees wants it to do?" There were
arguments that were carried to the Supreme Court in the

State of Utah. It was decided finally just over a year
ago that all of the programs were legal except one.
The UTFC could make research grants, they could make
loans, they could do about anything they wanted to with
the money except invest in companies or invest in
equity funds; equity was out. And since that court
decision, the plans of the Board have begun to be
implemented.

The Finance Corporation is led by a Board of
Trustees appointed by the Governor. Its staff consists
of an Executive Director and an Administrative
Assistant. One thing the Board of Trustees refused to
do was build a bureaucracy. The Board also proclaimed
a sunsetting of the UTFC in 1989. It must show that it
is worthwhile to the extent that the UTFC and its
programs should be continued.

CREATING ENVIRONMENT

There are four committees of the Board, with a
member of the Board as chairman of each. One of those
is the Technology-Based Business Development Committee,
whose charge it is to help create an environment in the
state for technology-based, high-growth businesses. In
high schools, universities, and technical schools there
are programs underway somewhat, but needing development. The Committee has been a major sponsor of
another organization, the Utah Foundation, which in
turn has conducted twenty-eight conference programs
covering a variety of subjects to help small businesses, eighteen seminars or forums, and two venture
fairs to bring good ideas and good investors together.

FUNDING NEW IDEAS

An Innovation Committee also was organized that
established what is called a Small Business Innovation
Program, very similar in words and in style to the
federal Small Business Innovation Research Program,
though financially less ambitious. The federal program, as you know, has two phases. The first phase
funds something up to $50,000 and, if it is successful,
the second phase can fund up to $500,000. Part of the
requirement of the second phase is that the people have
some idea where the third phase is going to come from,
other than government.

The state program is similar, but limited to one
round of $50,000. The first solicitation was about two
years ago. Forty applications were completed and four
projects were funded. One of these four companies now
is dormant. This research was successful, but the

company has not found follow-on funding. Another of
these is moderately successful and growing; but one is
really the star. Lee Scientific, who credits the
Finance Corporation with its first financing, had about
two and one-half full-time employees, no funding, and a
payroll of around $60,000 a year. The Finance
Corporation gave them $50,000 to do their research, and
it worked. Since that time, in eighteen months they
have brought in another $4 million financing, have
grown to a payroll of fifty people and $900,000 a year.

The second solicitation resulted in thirty-eight
applications. This was totally held up by court challenges, but finally there are four of those now in the
process of funding.

The third solicitation was announced in November
1986 and received sixty-seven applications. These are
undergoing evaluation work right now by a team of eight
professionals.

This is a program that can contribute eventually
to generating 1,700 jobs a year, which with a two to
one multiplier would give us 5,000 jobs per year.

LOCAL ECONOMIC DEVELOPMENT

Another effort of the UTFC is conducted by the
City and County Programs Committee. It has two programs, one of which is a Utah partnership where we work
with cities and counties, chambers of commerce, or
anyone with economic development funds who will provide
60 percent or $75,000 to match our 40 percent or
$50,000 to create a $125,000 fund for a local Small
Business Innovation Program. The idea originated with
Davis County who asked the UTFC to manage $200,000 for
the county. This Davis County fund, however, is not
yet confirmed.

We are finding that these programs don't reach the
four corners of the state and we want to spread this
program beyond the Wasatch Front. We're meeting with
the Utah Commissioner of Agriculture and the
Commissioner of the Natural Resources. We are trying
to develop joint programs with them, realizing the
problems with mining and other natural resources and
agriculture. We believe these means will carry the
UTFC program into the four corners of the state.

Another effort of the City-County Committee is
Entrepreneur Support. Small business organizations
need a vast amount of help at times when they cannot
afford it. How can we get this help to them? We are
talking and have talked to accountants, banks, the Bar
Association, and we are talking to management search
organizations. We're looking at help that can come
from the SCORE organization of the Small Business

Administration.   We've talked to the deans of
engineering and business at the major schools.   We're
talking with informal investors.   The Mountain West
Venture group has a very confidential and a very well
developed list of informal investors who will be introduced to idea people.   Besides this, lawyers do estate
and tax planning for these investors.   Trust departments of banks also know them.   Entrepreneur Support
will assemble all of these forces to the organized
benefit of entrepreneurs.

The fourth program of the Finance Corporation
would have invested in venture capital, but the State
Supreme Court disallowed it.   The idea was to put $1
million out for bid, to see who would match it 4-1 and
guarantee to invest in Utah.   There were no takers.
Substantial venture firms were interested in Utah, had
invested in Utah, but didn't want to commit to Utah.
When this difficulty arose some of us who were Trustees
of the Finance Corporation resigned from the Board,
teamed with another group, and organized the Utah
Technology Venture Fund.   The magic of that, however,
is that six of the limited partners are major venture
capital firms in the country.   We have $3.8 million in
cash that we can invest ourselves, but our partners
have resources in the order of $1 billion.   We don't
think any longer that there is a problem of venture
capital in the state.   The only problem is sifting
through our 125 business plans to find the four or five
worthy of funding.

Let me quickly, in conclusion, describe the flow
that we are trying to work for, not only in the Venture
Fund but in the Finance Corporation as well.   This flow
starts with the Finance Corporation through its Small
Business Innovation Program and extends through its
partnerships with local governments and chambers of
commerce, agriculture innovation, and natural resources
innovation.   Success at this stage will pass the funding opportunity on, in more or less this sequence: to
the Federal Small Business Innovation Research program,
informal investors, venture fund investors, and finally
investment banks and commercial banks.   If this flow
succeeds, the UTFC will succeed.

B

Lee Scientific: A Case Study

Hal Rosen

Lee Scientific is an evolution of ideas discussed in previous sessions on developing and using new technology. Let me take you through a little of the process of how Lee Scientific came to be, and how the NSF and other agencies have helped and impacted our existence.

The technology upon which Lee Scientific is based was developed at Brigham Young University by Dr. Milton Lee. For those of you that have roots here at USU, you can feel like you've got a part in it. Milton Lee is the son of Garth Lee who was a longtime chemistry professor here at USU before his death a couple of years ago, so Milton's chemistry beginnings started here in Cache Valley.

APPLYING NEW TECHNOLOGY

The research funding that Dr. Lee received at BYU came from several sources, one of those being the NSF, through the Small Business Innovation Research (SBIR) grants program. The SBIR program was extremely critical. It brought in funding from NSF, the Department of Energy, and some other private foundations. Dr. Lee's group was able to develop the new technology called capillary supercritical fluid chromatography which BYU then patented. The patent was issued in October of 1984.

At the time the patent was issued, Dr. Lee assembled a group, or management team, and founded Lee Scientific to commercialize the technology that had been developed at the university.

I am not a chemist. I am the non-technical founding member of Lee Scientific. I have an accounting and business consulting background. We went through several processes in order to get our company

off the ground. Starting a new company is something
you really can't understand until you've been through
it. Most of the speakers that have talked for the last
two days have come from companies that have 25,000 to
125,000 employees--companies that have been around for
a long time. Lee Scientific has grown from four
employees in November of 1985 to approximately fifty
employees today. That's relatively small in comparison, and yet it's a fairly rapid growth for a small
company.

Lee Scientific manufactures and sells a scientific
instrument called a capillary supercritical fluid chromatograph. Chromatograph instrumentation is involved
in the separation sciences. It is used by analytical
chemists, and is used for separating molecules in a
substance to determine the molecular structure of that
substance. It is used by several companies that have
participated in this seminar. For example, DuPont, Dow
Chemical, Procter and Gamble, Shell Oil, Exxon, and
Kodak are a few of our customers. Most of our customers tend to be large industrial companies, though many
are also universities that are doing basic research.

The process is used in many different
applications: petrochemical, chemical, pharmaceutical,
and research. Government organizations are also very
interested. These include the USDA, EPA, and FDA. To
give you an example of an application, the speaker from
DuPont yesterday showed some polymers. Polymers are
man-made and they are often used as additives or lubri-
cants in substances. Since polymers are man-made, the
molecular structure of the polymer has to be monitored
very carefully. Small changes in the ingredients can
drastically change the final product.

To illustrate, a customer of ours in Texas has
polymers in a lubricant they are using with very expensive aluminum machinery. They need to monitor the
lubricant very carefully since it's expensive. The
aluminum is much more expensive to replace. They try
to maximize the life of the lubricant. They had a
problem; the lubricant was breaking down much too early
and they were ruining a lot of their aluminum
machinery. They were not able to analyze the molecular
structure of when the breakdown took place through
conventional chromatographic techniques. Through our
technique they were able to analyze the breakdown and
determine when it took place, and were then able to
maximize the use of their lubricant, and maximize the
use of their aluminum machinery. So we have created a
niche in the chromatography business by giving answers
to real world problems that have not been answerable to
this point.

FINANCING SOURCES

We tried the venture capital route that has been
talked about.   Often times people refer to them as
"vulture" capitalists.   The name is quite often well
deserved.   I would like to talk more on the other
related devices that are available for starting a new
company and getting it off the ground without having to
go through venture capitalists.   I don't want to down-
grade that form of financing, sometimes it works very
well.   We found another method that we think works
better for us.
    One of the first things that we did was to apply
for SBIR grants.   We were one of the first four companies that were funded by the Utah Technology Finance
Corporation (UTFC).   In August of 1985 we received a
$50,000 grant from the UTFC.   It was critical to our
existence.   It came at a time when three of us had only
received one paycheck in a four-month time frame.   The
money we got from them did not go to our paychecks, but
it did help continue the research and development work
that we had been doing, helped us continue our
shoestring operation, and also helped us raise capital.
    At the same time, we received four other SBIR
grants from the federal government, each for about
$50,000.   One each was received from the Environmental
Protection Agency (EPA), the U.S. Department of
Agriculture (USDA), the National Cancer Institute
(NCI), and the National Institute of Health (NIH).
These gave us tremendous credibility in raising money.
They also helped us bridge the gap of shoestringing and
raising the money that was necessary for commercializing our technology.   The Phase I grants have now
been converted into Phase II grants, and we currently
have contracts on SBIR Phase I and Phase II grants of
over a million dollars.   These grants were the device
that really got our company off the ground.   The grants
helped us have enough credibility to attract a
corporate partner.

FINANCING THROUGH A CORPORATE PARTNERSHIP

When it came to deciding how we were going to
finance our company, we had one venture capital firm
and three corporate partners that were interested in
providing financing.   We chose one of the corporate
partners, Dionex Corporation, located in Sunnyvale,
California.   Dionex is a company that is about eight
years ahead of us and in our industry.   They are on a
rapid growth pattern, financially strong and publicly

held.   It is a company that has been growing at a
compounded rate of 30 percent a year and whose management has been there during that growth period.   They
are dynamic, understand our industry, and also understand what we are undergoing.   They are very involved
in the growth of their own company and have taken the
approach of letting us run our company as we see fit,
yet they provide help when we request it.

Since November of 1985 we have done three rounds
of financing with Dionex Corporation.   Financing has
been in the form of purchase of preferred stock and
they are now our major stockholder.   Corporate partnership provides some real advantages.   One, they have the
patience for a small company.   Business plans look
really nice; in fact, we pulled out our first plan
earlier this week and looked at it.   We looked at what
we thought we were going to be doing, and we missed the
plan by a long way.   We were naive in a lot of things,
but the corporate partner has been through this, too.
They expected it and have been much more patient than a
venture capitalist would have been.   Two, they provide
understanding of the business.   They are in our industry, but we do not compete head to head. We service
different areas of our industry.   They understand the
industry.   They've been there.   They've advertised in
it, tried the trade shows, hired the sales people; they
know how it works and have been able to help us leap
over years of learning experience.

Since they understand so well the business we are
in, they are more understanding and more patient   than
a venture capitalist would be.   They are also excellent
at giving direction, and they've lent a tremendous
credibility to our company.   It is not an easy process
for a new company to sell an instrument that costs
$40,000.   Try to convince Merck, DuPont, or another
large company to buy an instrument that costs $40,000
from a company that doesn't have a history.   Having a
corporate partner that has been in our industry has
lent us the credibility that we needed.   We've also
benefited our corporate partner.   Through our innovation and technology, we've helped   create more respect
for them, so it's a two way street.

An example of's how the corporate partnership can
work happened recently.   We were on the phone with the
employee representing our corporate partner in China
and the eastern rim.   We spent an hour on the phone
with him talking strategy concerning marketing in
China.   One doesn't get that kind of information out on
the street, or you pay an awful lot for it and often
times you can't trust it.   We were able to talk to the
company that markets the same type of instrumentation
and glean and learn from them.   I believe it has saved
us years.   Our corporate partner has been very good

that way, but they've also been very good at letting us run our company, and letting us be innovative and do what we think is best for our company.

PRODUCTIVITY IN EMERGING FIRMS

It was interesting listening to many of the speakers yesterday talk about productivity. Productivity for many of those companies has resulted in a loss of jobs. As they have become more productive they have released employees, often massive numbers of employees. Grant Cannon gave a statistic in his talk about the number of jobs that have been cut by large conglomerates in the United States as they've become more productive. Productivity in this country is really based on small businesses and their ability to meet the needs of a customer in a timely and very efficient manner.

We're just finishing a project that has taken us ten months and cost us approximately $300,000 to complete. Companies that are also in our industry, such as Hewlett-Packard or Varian, would spend three to five years and $5 million on this project. That's the difference between productivity in a large company and a small company. In a small company you don't have the layers of management to deal with. You don't have the bureaucracy. As one employee told me, if you're not good, there's no place to hide. That's the risk of working for a small company and yet a small company is very dynamic. It's very exciting.

It's exciting for a company such as ours that was founded in 1984 and is now the worldwide leader of supercritical fluid chromatography technology. We service about 80 to 85 percent of the worldwide needs for our technology. In real dollars it may not be a great amount compared to the big companies, but it's exciting to work in and it's exciting to assist growth in that technology. I heard a statistic a year ago that there are only twenty-five companies in the state of Utah that exceed $25 million a year in sales. That's not very many. Our company has the opportunity in the next five years to join that group.

SUMMARY

I think the growth of Lee Scientific illustrates the commercialization of new technology as explained at this seminar. It started with a small business innovation grant in a university setting. It was then transferred to the private sector. The innovation, technology, and science were expanded upon through the

help of organizations like the UTFC and the federal
SBIR program. We attracted capital and we created new
jobs. In eighteen months we have grown from four
employees to fifty and we expect that growth to
continue.

We are expanding this technology throughout the
world and meeting the needs of scientists by providing
solutions to problems that need to be answered. In
this way we can have better products, better quality
and productivity, and better scientific advancement.

C

Institutional Venture Partners

T. Peter Thomas

Institutional Venture Partners (IVP) is an old partnership. The original founding partnership was called IVA, and was formed in 1974. That was the year when the total capital that was raised in the venture industry was $50 million. It was a very slow time compared to what we see in today's venture capital industry. Today IVP is managing $153 million under three funds.

I'd like to give you some sense for the venture capital (VC) industry, some of the trends that we see, and then give you a real quick capsule of what it's all about to be working with venture capital.

OBJECTIVES OF VC

What I thought I would highlight for you first is the objective of venture capitalists. We really are out to finance new companies that have rapid growth opportunity. We are typically investing other people's money. We refer to these sources of capital as our "limited partners." The limited partners, as I'll show you later, represent pension funds, insurance companies, and, in some cases, foundations, individuals, etc. Our investments are for equity. We basically either succeed through the company's success or we lose our money if the company fails. We focus on businesses that fundamentally enhance productivity through the standard of life. Those are the businesses that traditionally have markets associated with them. We are clearly trying to finance businesses that have a market. We add value. In this area there are lots of things that I'll talk about shortly, things that we do to help the young venture. We take high risks and some people believe that we have high expectations that go beyond what's fair. I think that if you talk to the majority of entrepreneurs you'll find we're also very fair in our evaluation of what we should get for the

risk we're taking and for the capital that the venture
is receiving. We also have a lot of patience. This is
tied to the fact that it takes a long time to build a
successful business. If you look at the history of how
long it takes to start a company until they can survive
and either be a successful public company or a merger
or acquisition candidate, it's usually five to seven
years. So we have to have the patience to work with a
company during that period of time, otherwise we may as
well not invest day one.

ECONOMIC IMPACT OF VC

     One of the things I think is very interesting is
the value of venture capital. What is it doing for the
United States in general? What can it do for a local
population of states? A 1982 study looked at the money
that was invested by venture capitalists in a total of
seventy-two companies. A total of $209 million was
invested in these companies over a ten-year period.
The study looked at the results of 1982. What they
found was that those companies had combined total sales
of $6 billion and they still had an annual growth of 33
percent per year; they had created 130,000 jobs and
they were doing two other very important things for our
economy: (1) producing tax revenue, and (2) helping
to solve the problem that we have today of
import/export balance.
     What is the value of the venture capitalist to the
entrepreneur? It's a whole list of things. Obviously
from the entrepreneur's standpoint we bring money. And
that's usually the entrepreneur's first interest.   I
believe we also bring other added value. A number of
us have operating experience, we've worked out in
industry, and in some cases we know the industry that
they (entrepreneur) are specifically working with. We
can help them recruit key management, we may have the
specific industry market knowledge, and we serve as a
free sounding board. The president of the company or
his vice presidents can call us in and use us to feel
out ideas on where they go next, be that product,
market, distribution, manufacturing, financing, etc.
We are also used to foster the entrepreneurial climate.
We work with a lot of young companies, so we do our
best to really try and keep the environment positive,
non-bureaucratic, etc. I think the other thing that's
real important, and you would find this out from most
entrepreneurs you talk with, is that we bring some
balance of perspective to the situations that most of
the companies go through. None of us have worked with
a company that really had an easy road to success.

It's a rocky road, it's a lot of hard work, and we tend
to be able to help balance out the ups and downs. We
give the entrepreneur some confidence that we'll get
through this.

AMOUNT AND SOURCES OF VC

     The funds committed to VC during the last ten
years have increased vastly. As I mentioned, in 1975,
there was a total of $50 million committed.   In 1976,
the total venture capital raised was only $10 million.
Because of legislation in 1978, specifically the changes in capital gains tax rates, from a high of 49
percent down to the 20 percent of 1981, there was a
tremendous growth, and infusion of capital into the
venture industry. It peaked in 1983 at $4.5 billion.
The total cumulative capital committed to funds by the
end of 1985 was $19.6 billion. How will the new tax
law of 1986 affect venture capital funds?   It's hard
for us to predict that. I think one of the things you
find if you go back and actually model here how quickly
a change occurs once taxation laws change, you can see
it takes three to five years before a real trend can be
shown. Once it's changed, it's hard to get it moving
back the other direction.
     Where does the committed capital come from?
Pension funds are the largest source of the money today
in venture capital. Legislation occurred, I believe in
1981, that freed up a small percent of the pension
funds to invest in higher risk venture capital. The
other large investors today are insurance companies,
foreigners, individuals (families), corporations, and
endowments/foundations.
     I would like to highlight the stages of financing
that are used in our industry. This is portrayed in
the diagram below covering the stages of financing:

TYPICAL INVESTMENTS	FINANCING STAGES	
$50-500K	Seed	
$1M-5M	Start-up	1st Round
$2M-10M	Expansion	2nd Round
		3rd Round
$2M-10M	Mezzantine	
$10M+	Initial Public Offering Or Acquisition	

We oftentimes refer to a seed investment that is a very small number relative to many of the (VC) funds that exist.   A seed investment is usually an investment that's made without a total management team in place. Often it comes when you have transferred something out of the university, or the lab. You have a few people who think they know what they want to do with it, and they basically would like to get a little further along in proving that the business is a real business.

You'll find that a lot of funds are anxious to invest in the start-up phase.   IVP does some seed financing but we typically don't have the time to provide the assistance to a single individual who wants to start a company. If you've got some of the management team in place and you've got more of the fundamentals in place, we're probably ready to do a start-up financing with you.

DESIRABLE BUSINESS AND PERSONAL CHARACTERISTICS

What do we look for in a new business? We always look at a business plan first, and we try to meet the people that are involved. The primary message we have always had in this business is that we are looking for three things: people, people, and people. And fundamentally you find that the success of most companies eventually ties to the people. If you've got good people, you can sometimes have a mediocre product and still succeed.

We look for focus. You've got minimal resource, and you've got minimal time to get to market. Uniqueness, whether it's a patent, trade secret, or whatever gives you an unfair advantage for a few years is important.   The growth potential of your market is important and the opportunity you have to serve that market. Does the growth curve show that in five to ten years you're potentially a $50 to $100 million company? If the answer is "Yes" and it's going to take $5 to $10 million in capital to get you there, potentially we can see the type of returns we want and it makes sense to us to invest.

It is important that you've got some experience in the business area you're targeting: Are you a management team that has done it before?   Do you have any sense for the business world? Are you more of a scientist than a business professional?   Do you have plans on how you'd like to expand the organization as you get beyond certain stages? Are you willing to work with us as advisors or are we going to be a hindrance to you as an investor?   We like to have very positive relationships with people we invest in. It isn't much fun to work with someone who believes that you are the

enemy, especially over the five- to seven-year period
that you have to live with one another. We'd like to
see real customers. That is a rarity in new start-ups,
but if you do have some, it obviously helps us to
validate the marketplace. We'd like to see attractive
financial characteristics of the business: high gross
margin, high bottom line pre-tax, something that says
eventually you can finance the business from within.

We look for the following personal characteristics
in people: integrity, maturity, energy level, commitment, will-to-win, openness, leadership, sensitivity,
intelligence, financial motivation, and sense of humor.
I think all of them relate to the super person.   I
think the last one is one of the most important. We do
like to see people that have a sense of humor. If you
are going to be a success, you are going to work your
tail off for the next five years and you're going to
have some real rough days and nights and weekends.
Once in a while you need to be able to look back on it
and have a little laugh about the fact that it was hard
but it was actually a lot of fun.

We documented our investment process a few years
ago, and we found the following:

ACTIONS	RELATIVE FLOW
Initial Contact	400
Read Business Plan	300
Initial Meeting	100
Due Diligence	30
- personal references	
- customer contact	
- market research	
- technology review	
Further Meetings	10
Deal Structure & Negotiation	1.5
Syndication (if necessary)	1
Successful Company	1 in 5

These numbers will vary from year to year. We looked
at 400 business plans. A number of those plans tend to
be plans that are coming in over the transom. Our

reference to transom deals refers to the ones that come
to us where we have not had a reference.   They don't
usually get much of a review.   However, whenever deals
are referred to us, or when transom deals look
interesting, we read the business plan.   We usually
only meet with about 25 percent of the deals that we
see, and we work diligently on around 10 percent.
Eventually we get down to investing in less than 1
percent of the plans we see.   Our goal is to have more
than one of five of our deals make it to Initial Public
Offering, but that is the industry average.

CONCLUSION

An   entrepreneurial   environment   is   a   fun
environment.   There are lots of things that are positive.   The bureaucracy is gone, the communication is
better, the responsibility is greater, and visibility
is fantastic.   But you've also got all the negative
things.   There is less support, fewer resources to help
you, much less security, and the pressure is phenome-
nal!   You feel the pressure every day and you know that
you are the reason that you are either going to succeed
or fail.   That's a great positive and a negative.   I
think that founding and growing a company, whether you
do it with venture capital help or through other
sources of funding, is a lot harder work than you can
ever imagine.   It's tremendously difficult, and it's
enormously rewarding.   And if you have a sense of humor
it's also a lot of fun.

Part II

MANAGING HUMAN RESOURCES STRATEGIES

9

Overview

Paul F. Buller and Glenn M. McEvoy

Much has been written in recent years about the importance of an organization's human resources in determining its ultimate success. Indeed, the status of the human resource function has risen dramatically over the past five years. One recent survey revealed that human resource executives have, for the first time, even surpassed marketing executives in terms of average salary. Yet, the frequently heard cliches like "people are our most important assets," and "we value our human resources," still ring hollow in many organizations. Why? In some cases, the answer may be simply that those who utter such slogans just don't mean them. In other cases, the words are sincere, but those charged with putting them into action are at a loss about how to proceed. In still others, implementation is underway, but the company's historical and cultural inertia make change slow and painstaking. Finally, in a relatively small but growing number of firms, people truly are regarded as the key to sustained success.

It is our sense that most organizations today fully recognize the strategic importance of an effective workforce, but are striving to overcome a number of obstacles to making the most of their investment in human resources. The obstacles are many and include: a top management that has typically emphasized capital and technology as assets and employees as liabilities; a personnel department that has historically taken a reactive and isolated rather than a proactive and integrated role in the organization; a labor force whose continual supply was once assured and whose demands on the organization were slight, but now is in short supply and ever more demanding; and a business environment that has moved from placid and forgiving to increasingly more dynamic and demanding. These and other forces have created a tremendous challenge for today's managers--a challenge to acquire, develop, and maintain a highly flexible and effective workforce.

This section of the book describes the experiences of a number of companies with excellent reputations in managing human resources. Executives from these companies identify some of the forces driving the need for a strategic orientation toward human resources. They offer ideas for developing such an orientation, and describe their firms' experiences in applying those ideas.

The first article, a provocative one entitled "Human Resource Profession: Friend or Foe?," is by Andrew S. Grove, CEO of Intel Corporation, and provides a chief executive's perspective of the human resource function. Mr. Grove questions how well the human resource function in general has managed its task to help the firm achieve bottom line success. He challenges human resource professionals to assume more leadership in (1) "laying the track for the management systems" that guide line managers, and (2) "greasing the skids" so that line managers can perform well. His basic message--human resource practitioners must practice what they preach.

In the second article, "Coca-Cola's Changing Perception of HR Function," Douglas A. Saarel, Senior Vice President of Human Resources from the Coca-Cola Company, amplifies Mr. Grove's remarks from a human resource perspective. His major point is that human resource managers must adopt a new orientation about their role in the organization. He argues that the primary purpose of the human resource function is to add value to the firm. To achieve this goal, HR professionals must first learn the business and how to communicate in business terms with line management. Only then can they gain the knowledge and credibility to become true strategic partners in developing human resource systems that add value.

In the third essay, "Travelers Integrates HR and Strategic Planning," Thomas E. Helfrich from the Travelers Companies describes the increasingly challenging environment of today's businesses. This environment is providing the impetus for a greater role of human resources in the strategic decision-making process. Based on his experience, two key elements must exist for the HR function to contribute to the bottom line: (1) top management must have a vision of the future that truly values people, and (2) HR professionals must have a basic understanding of the business and the courage to challenge traditional practices.

Nancy Bancroft, in "Managing Innovation at Digital: A Case Example," builds on the previous articles by identifying some of the key elements for managing a more effective partnership between line managers and human resource staff. Based on Rosabeth Moss Kanter's notion of "integrative action," Ms. Bancroft urges the reader to adopt a holistic approach to addressing organizational problems, considering together task, technological, organizational, and people issues. She describes a systematic process used at DEC for managing innovation, change, and integration.

Next, Rodney J. Falgout, in "Monsanto Upgrades QC Teams to Second Generation Work Teams," describes the evolution of a strategic human resource perspective in one division of his firm. Experiencing limited success with quality circles and other human resource programs, Monsanto recognized that it needed something more to combat the pressures of an increasingly turbulent environment. It developed a business strategy that included specific provisions for managing human resources. This strategy was communicated to all employees and was augmented by other changes in organizational structure, compensation, and training programs. Thus far, this integrated effort has improved productivity dramatically and helped the division to remain competitive.

The next two articles present specific human resource programs with a strategic orientation. Richard S. Sabo, in "Linking Merit Pay with Performance at Lincoln Electric," describes key elements of that firm's widely recognized and very successful compensation system. In her paper entitled "General Foods' Strategic Human Resource Management," Roslyn S. Courtney presents a program employed by General Foods' Information Management Department that integrates business planning, human resource planning, and employee development. Both of these papers describe in some detail how strategic management of specific human resource activities can be achieved.

In the next article, "Measuring HR at National Semiconductor," John Campbell discusses a critical ingredient in the successful management of human resources--namely, the measurement of HR results. Consistent with the previous papers, Mr. Campbell points out that any HR program should be driven by the firm's overall strategy. He emphasizes the need for accountability in all human resource programs and projects, and offers a number of suggestions for managing and evaluating results effectively.

The final article, "Borg-Warner's Change in Organization and Culture," by James R. Deters, provides an apt summary for this section. Mr. Deters describes the development over several years of a strategic human resource perspective in his firm. His experience highlights the following critical ingredients in a strategic orientation, "articulating goals, values, beliefs, and writing them down for all to see, to share, and to challenge." He concludes with an eloquent challenge to human resource professionals to lead the way toward these ends.

In summary, the experiences of these companies highlight the trend toward managing human resources strategically. Companies with a true strategic perspective of human resources have (1) a clear vision and a system of beliefs about the value of people, (2) a business philosophy and mission that acknowledge the strategic importance of people, (3) an organization structure that provides stature and visibility to the human resource function, (4) human resource management systems, programs, and activities that recognize and reward people for performance, and (5) a management approach that encourages appropriate participation of employees at all levels. To achieve an effective strategic partnership with other top executives, human resource executives must first learn the business, and then proactively demonstrate how they can contribute to the company's success. This includes the development of programs and activities that are consistent with business strategy and goals, managing programs in such a way so as to add value, and measuring the results of human resource initiatives. For, as the companies described in this section have demonstrated, an essential ingredient in gaining a sustainable competitive advantage is a strategic orientation to human resources.

10

Human Resource Profession: Friend or Foe?

Andrew S. Grove

Before I start, since I have a subject entitled "Friend or Foe?" I would like to find out who in the audience works in the human resource profession? There are enough of you so that I have some trepidation to proceed. Trepidation comes from the fact that when I first developed this talk with this title and gave it to our human resource manager at Intel he said, "Why do you put the question mark at the end of the title? You know the answer darn well."

It is not that simple. What I am really talking about you might have gleaned out of the introduction. I am an operating manager. I manage a company. I am responsible for a very specific financial and product output. Yet, of course, human resource issues of management are a part of any operating manager's job, whether or not they explicitly recognize it. But conflicts constantly arise regarding the role of the human resource profession in the operation of an output-oriented enterprise such as Intel.

PERFORMANCE DEPENDS ON EXTERNAL/INTERNAL FACTORS

I am going to try to put the subject in some sort of a perspective. Basically when we are running an organization—the organization could be a government organization, a financial organization, or a university—that organization is there to generate some type of a desired output, an output that society considers valuable. How we as operating managers perform depends on two types of factors: external and internal. I will put this in a two-by-two matrix, courtesy of the Harvard consulting group. I will put the external conditions on top and the internal performance of the organization on the bottom.

What are external conditions? External conditions are all those things that corporate managers never recognize in the annual report when they are good, and

always hide behind when the results are bad. For
example when oil prices go up, we cannot help that, so
our performance goes down. When oil prices go down, we
do not mention it. That is the best way to understand
what I am talking about. These are the things that we
cannot really control too much, nevertheless they have
a profound effect on our performance, good or bad. If I
could paraphrase a famous saying, "It is the tide
that raises or lowers all corporate boats at the same
time in the same way."

Our task as managers of organizations is to make
the most of whatever external factors affect the performance of the corporation or the organization. If
the external conditions are unfavorable, and we fall
asleep at the switch internally, we go out of business.
If the external conditions are favorable, we succeed
even with poor performance. When all is going well for
us such that the conditions on the outside are good
and we do a good job, we get a gold medal for our
performance.

Now what is internal performance? That is where
human resource professionals come into play. A lot of
things are involved. First among these is being at the
right place at the right time and with the right product. When it comes to that, quite frankly, it is best
to be both lucky and smart. But the second best is to
be lucky. I will give you an example of this. If I
asked you what company was responsible for the invention of the personal computer, what would you answer?
The personal computer was actually introduced by a
little-known company called Mitz. That was about
three or four years before Apple. They had this
weird-looking box that no one knew how to use. It was a
good product but the world was not there yet. The
software that was needed was not there, and the world's
computer awareness that was needed was not there, so it
died. A year or two after it died, Apple arrived with
a very similar product, but at that point the product
was right for the market. So being at the right place
at the right time with the right product is the best of
all these things.

Second is being in a position to ride the tide. A
perfect example is Chrysler. They were bailed out by
the federal government just before the arrival of
restrictions on Japanese imports. Chrysler was alive
and therefore when the tide came in that lifted all
automotive boats, they were able to ride it.

The third is that we have to constantly and dog-
gedly pursue, sustain, and renew the performance of an
organization. This is the hardest of all types of
corporate activities. Product planning and the like is
easy, particularly when you are lucky and you cannot do
much about external performances. But nagging your

organization into better performance, a bit at a time, is what we are all about. It is hard and most excruciating and, quite frankly, the least romantic and least publicized activity that involves all managers. So, who is involved in it? The middle management is the part of the organization that is involved with pursuing,   sustaining, and renewing the work of   the organization.

Another very important element is to constantly keep an eye on the output. It is very important for us not to be so mesmerized by the daily work and activity that we forget what the organization is supposed to do. The university exists to educate students and not to give papers. Companies like Intel exist to generate profits by serving customers and making electronic products. To do all that, it is very important for us to have management systems.

NATURE OF MANAGEMENT SYSTEMS

What are management systems?   Every company has them.   We have   systems for planning, whether it is a formal system run by a corporate staff with forms and computers, or whether it is done in the old way, by scribbling on the back of envelopes. So, there is a system, implicit or explicit, that all corporations use. All organizations have systems for goal-setting. Some companies have very formal management-by-objectives systems, while in other companies somebody sets goals by barking out orders.

There is always a system of performance assessment, good or bad. It really exists and is well defined, even if you cannot describe it readily. And of course, there is a system of promotion, compensation, and termination. Likewise, there are systems of training and development, formal or informal, ad-hoc or systematic.

There is a communication system wherever you work. It may be well developed and technologically advanced, or it may be that employees find out what is going to happen in the company only when they pick up the daily newspaper. It is whatever system you have.

There is a way in which your organization maintains, promulgates, and hands down the culture of that organization.   Maybe it is through the use of colossal reports, role models that are very well defined, or folklore. There is a system in which we do that.

The main point is that all organizations have management systems. Not all of them describe the systems explicitly. Sometimes management systems are just what is in the air. A most important point is that

very few organizations actually describe their management systems in the way that they operate. For instance, the U.S. government has a voluminous, well-
defined system for performance assessment. Yet you all
read about whistle blowers that are removed from their
jobs for speaking up. Which is the real system? The
one that is described explicitly or the one that dis-
poses of the whistle blowers? Ford Motor in the last
decade had a very well developed career planning system, but when you read Lee Iacocca's book you get quite
a different perspective on how it actually worked.
There is frequently a duality of the system that is
described and the system that actually lives.

It seems that in every organization, both
management and the human resource people have an image
or notion of what their management systems should be
like. What they then describe is what they think it
should be rather than what it is. That is a very
crucial point and a stumbling block in the workings of
organizations.

MANAGERS OWN MANAGEMENT SYSTEMS; HR MAKES SYSTEMS WORK

Who owns these management systems? Not the human
resource professional. Operating managements own them
since it is the management structure that is responsible for developing and delivering the product. They
are the people who own all of these systems that we
have talked about.

On the other hand we pay the human resource people
in our organizations presumably because, in some
fashion, they add value to the workings of the organization, they bring something to the party. In my view
what they bring to the party, when it comes to the
organization's management systems, is to lay the track
for the management systems so that operating management
has a track on which to proceed and perform their work.
It is for them to set up a performance assessment
system, describe it, and teach operating management how
to use it. It is for them to set up a performance
bonus system. It is for them to figure out the appropriate way of systematic communication between
management and lower-ranked employees. It is for them
to grease the machinery of all this so these things
will happen: performance gets assessed, bonuses get
administered fairly, and information proceeds back and
forth between various levels of management.

To draw an analogy that is perhaps more accepted
and more traditional, look at the finance profession.
The profit and loss statements of every organization
are the responsibility of the operating management. I
wish that I could have blamed Intel's 1986 P & L on our

finance organization. I could not. I am responsible
for that as operating manager. But, finance developed
the system by which the game is played. It spells out
how inventories are valued and how reserves are calcu-
lated.   Finance lays the track by which the whole
process of inventory is counted and how all of that
trickles down through the system and gets added up.
This is the analogy. Management owns the P & L and the
management systems of the company. Finance greases the
skids for the P & L to work. The human resource profession greases the skids for the management systems to
function.

HR STAFF PERPETUATES CHARADE; SAY ONE THING, DO ANOTHER

The question is how well has the human resource
profession, in the gross generality of American industry, performed its task.   Intel, probably like all
your companies, has performance assessment categories
such as "superior," "exceeds requirements," "meets
requirements," "marginally meets requirements," and
"does not meet requirements." The best phrase to assess human resource performance is that it "meets
requirements;" not "superior," nor "exceeds requirements," but just "meets requirements."
Why is it not better? I come up with two factors:
the first is what I call the PollyAnna syndrome, and
the second is the problem of the shoemaker's children.
(It has been pointed out to me that I should call the
first one the PollyDonald syndrome, or something like
it, in order to even out the sexist connotation.)
The PollyAnna syndrome basically says that when
we, operating managers, say one thing and do another,
the human resource profession acts like nothing is
amiss. Operating managers have a hard time looking at
reality and seeing reality. The reason is that it,
reality, is full of worms. Reality is generally not
pretty. We say we have a meritocracy, but it is hard
to implement and enforce meritocracy. Assume you have
a 4 percent budget increase and I tell you to apply
meritocracy. You cannot give poor Joe nothing so you
have to give him something. You cannot give him 1
percent or 2 percent so by the time you rationalize
your plan for poor Joe, who really should not get
anything, you give him 3.5 percent. This means that
you will not have enough money to give the people who
really make things work in your company any better than
6 percent. So, meritocracy has turned out to range
from 3.5 percent to 6 percent. That is not meritocracy. Then you start arguing that Joe has been
here for twenty-five years and all of a sudden the
whole idea of meritocracy has gone out the window.

We tell it like it should be, not like it is. The
human resource profession should be our conscience, but
often they turn the other way and participate with us
in perpetuating this charade. Some of the worst euphe-
misms come from the human resource profession.   For
example, I have described Intel's categories of performance assessment. You have no idea how many battles I
have had with our human resource people who are aghast
by the phrase "meets requirements." They say that is
not good.   It does not make your people feel good
about themselves. But, I cannot help the mathematical
fact that half of our people are below average! But if
I say that, I get crucified, and the first nail comes
from the Human Resource Department.   It is very difficult for operating managers to face the facts and the
reality. We need help; help to prod us to face reality, not to talk us out of it.
    I will give you an example. I write a weekly
column for a newspaper. People write to me about work
problems, kind of Ann Landers style, and I try to give
them advice. This is a recent question:

> I have been working at a bank as a temporary,
> part-time new accounts clerk. This position
> offers no benefits and minimal pay. Although in
> theory I am a part-time employee, I have routinely
> been working forty hours per week. Also, since I
> had worked at this bank in the past as a teller I
> am often asked to work in that capacity at no
> extra pay. In fact I often work next to brand new
> tellers who make more money than I do. I have
> talked to the personnel director about this
> situation but he offered no help, because the
> bank's policy did not allow the use of temporaries
> in teller positions.

    So, it is okay for a bank to use a temporary in
the teller position when you cannot pay the teller
salary because of the policy that does not permit the
use of a temporary in the teller position. It's Catch-
22 coming from the personnel profession.
    I think the simple rule that I would like to leave
with you and that we all ought to try to live by is
that, whether you are on the human resource side of the
fence or the operating management, if it is too embar-
rassing to say, then do not do it. If you are doing
it, then steel yourself to be able to say it exactly
the way you do it.
    The second phenomenon is the problem of
shoemaker's children.   Basically, the human resource
organizations have typically been the worst in following management systems. When it comes to performance
assessment, they have a tendency to be non-meritocra-

tic. When it comes to training, typically the human resource people train themselves less rigorously than the other professionals in an organization.

A particular weakness of the human resource people is careful execution of the details.   I have faced innumerable human resource people who complain to me that they do not have time to do their work because of their paper work. The paper work that they are complaining about is the performance review and the related compensation work for the people in their organization.   They do not have time to do their work because they have to attend to the paper work.   First of all, the system of the paper work was generated by them. Second, that piece of paper, which is an administrative trivia to the person who complains, is the annual performance assessment of an employee in the organization.   It is the change in his/her compensation as a result of an annual performance assessment. If that is not their work, what is their work?

Typically I found that the quality of execution of this kind of thing is much worse than the way we process sales orders, production scheduling orders, and the like. These deal with inanimate objects. By contrast, the paper work the human resource people complain about, the administrative trivia, concerns the lives of fellow employees.

So, this is another factor. You can all make up your own reasons why it has come to be that way, but it is a very unsatisfactory situation when the organization that is expected to be the promulgator of the human resource systems, and the conscience of all the rest of operating management, does not provide a proper example.

BEWARE OF FADDISH HR APPROACHES

I have one last thing that I would like to discuss concerning other ways that the human resource profession can be more helpful and less of a hindrance. <u>Try to resist the lure of fads</u>.   There is a $25 billion training and development industry in the United States. (When I first heard that number, it was almost as big as the semiconductor industry worldwide. It is now bigger than the semiconductor industry.) It is a huge industry. Now these people have products to sell, and just like the retail stores at Christmas, they must come up with new products. They cannot sell the same thing over and over. So each year we have some new acronyms, such as MBO one year, MBWA the next, and then corporate cultures. Now, the human resource departments buy the products of this $25 billion industry, and bring them home. But it doesn't work that way.

Pursuing, sustaining, and renewing is grubby work, day
in and day out; these hard tasks include assessing
performance, doling out compensation, training people,
setting objectives, and all these other things. The
new fads are really distractions that, like the diet of
the day, offer a simple answer when there is nothing
but hard work that can truly give you results.

HR CAN HELP RESTORE U.S. COMPETITIVENESS

In my view, which is colored by the fact that I am
in a badly beleaguered industry that is in a recession, the manufacturing industries of this country are
in deep trouble. We have lost the strong competitive
position that we had worldwide in the decades following
World War II. We have become just one of the leading
industrial countries rather than the leading industrial
country. For us to hang onto the standard of living
that we are accustomed to, we need a renaissance of
performance, of corporations, organizations, public
organizations, the people who manage these organizations, and the people who work in them. The human
resource profession has the potential for a major "added value" in all this. As I mentioned, the added
value comes in laying the track for the management
systems and in greasing the skids so they can perform
well.
I am not asking for the human resource profession
to do this task on their own. I am asking for this to
be done in partnership with operating management. The
key thought that I would like to leave with you, the
key that is missing very often and should never be
missing, is to bring the utmost intellectual integrity
to the job. Simply put, say it like it is, and if it
does not feel good to say it like it is, then change
what is, rather than what you say.

11

Coca-Cola's Changing Perception of HR Function

Douglas A. Saarel

As I was preparing my remarks I was reminded of a similar occasion some years ago. I had settled myself into a comfortable chair to pull my thoughts together in undisturbed quiet. My small son, Ted, came up to me and wanted to play. Quickly recognizing that my preparation might be at an end, I looked at the table to my left and noticed a copy of <u>Life Magazine</u> with a brightly colored picture of the earth on its cover. I quickly ripped off the cover and tore it into small pieces. Then I suggested to Ted, "This is a puzzle. Go to the kitchen table and see if you can piece it together." Ted toddled off happily. However, within a few minutes he was back and said, "Daddy, I put the puzzle together." At this point I was amazed, thinking I might have a prodigy on my hands. Sure enough, when I looked into the kitchen, there on the table was the globe put together with an abundance of scotch tape. I said, "Ted, how did you do it?" He replied, "When I was walking into the kitchen, I tripped on the edge of the rug and dropped some pieces of the puzzle. A lot of them turned over and I saw that they made a person. So when I stuck the pieces of the person together and turned it over, the pieces of the world were put together too."

Let's stop a minute! Why do we smile at this story? Why do we react as we are now? Is it because there's some general truth expressed by this story, although some may argue that the point goes a bit far? What is that truth?

We all agree that people are important. We happen to be talking about ourselves. "Our human resources are our company's most valuable resources!" How often have we heard that phrase? How often have we said it ourselves? How often have we assented to it? It must be true!

If so, all of our organizations should reflect that value in their policies, practices, commitment of resources, strategies, plans, and quarterly reports to

the human resources press. Our constituencies must be
closely following our human resources progress. Our
schools of higher learning must be teaching courses
about leveraging our human assets, acquisition strategies of human resources, and return on human capital.
Prestigious business schools must be concentrating on
human resources rather than financial or marketing
management.

DESIRED PERCEPTION OF HR FUNCTION

But, I'm going a bit too far. Then, why the
discrepancy between what we preach and what we practice? Perhaps it's because we believe in our heads
that human resources, our people, are our most valuable
resources, and that we who represent that part of the
business add value. But we are not convinced in our
gut that this holds true. Maybe we have let ourselves
be influenced by well-intentioned but unknowing others
who suffer from the same perception.

Some time ago I was speaking in Savannah before a
regional group of human resources managers from colleges and universities in one section of our country.
The keynote speaker was a local university president
who, thinking he was complimentary, suggested that
personnel and human resources people were very important because of the friendly handshakes and fine pats
on the back they gave to newcomers in an organization.
He went on to say how these most important actions made
the new environment warm not only to new members of an
organization, but also to all employees. The remainder
of his soliloquy was a variation on this theme. Needless to say, although slightly exaggerated, this perception is not too far off from the way human resources
is being viewed by the well-meaning but unknowing
majority of businessmen and women today.

Perhaps it's this perception by others, which we
begin to share ourselves, that prevents us from getting
a credible emotional hook into the reality that the
effective leadership of our human resources and the
participation of us who influence that part of the
business do add measurable value, a value that we will
define more clearly later.

Some of you may be thinking that perception
doesn't matter. I disagree with those who think this
way. Camus once said words to the effect, "There is no
truth, only perception," and the way we are perceived
is the way we are treated. Let's consider some of our
own perceptions concerning what we do and the part that
human resources (both the function and our people) play
in our enterprise.

There are few human resource executives or general managers that I know, or have known, who don't perceive that human resources is a staff service, that they are internal consultants, that they are specialists in compensation, benefits, labor relations, organization development, or that they are human resource generalists. Few do not perceive that their role is to provide professional services or support to their operations clients. In fact, it is in style to refer to us as human resources practitioners, or that we represent a profession.

Although we need to know our technology, to me, if we perceive any or all of this as our prime accountability, we are deluding ourselves that we are making the contribution or leveraging the contribution that our human resources should be making to our business. Because, if we believe these standard utterances as to what we are, it is not conceptually different from carrying the proverbial watermelon to the company picnic. If I could banish only one word from our collective vocabulary, it would be the word "services." Those who provide services to the business are not perceived as part of the mainstream of the business. I think we need to make a creative and conceptual leap in terms of how we believe in our guts about ourselves, and by extension, how our businesses feel about us and the value that both our function of human resources and the human resources themselves can add to an enterprise.

CHANGING THE PERCEPTION OF HR FUNCTION

For this change of perception to occur, I believe we need to do three things: first, we need to be businessmen and women, and only then view ourselves as human resources professionals. We need to know our business, not just our products and our customers, but our financial strategy and structure, our marketing models and practices, our distribution system and its many intricacies. We need to know them as well as those individuals who are responsible for them. Second, we need to know how to add value to each of those systems in a variety of ways through more efficient application and leveraging of our organizations and our human assets. Third, we need to learn how to articulate what all this means in credible business language, which is also our own.

Further, we do ourselves, our organization, and our people a disservice when we speak in the jargon we often do, process consultants, organization dynamics, and interventions, to name a few. Yet, I think we have to be careful we don't go too far! Since ours is a

credible part of the business, some language that is
unique to us makes sense, just as marketing talks about
its market segmentation, manufacturing its through-put,
and finance its burden and overhead.

Also, I think we should be careful that we do not
borrow too much language from other disciplines of the
business, because if we do, we become stepchildren and
not credible in our own right. I was guilty of this
"language borrowing" a short time ago when I spoke of
leveraging our human assets. People are far more than
assets in a capital sense, although a partial analogy
is useful on some occasions. So let's develop and use
our own simple business language that connotes adding
value through human resources and articulate that part
of the business we represent in ways our other business
partners fully understand.

I really believe that this lack of business
knowledge and the inability to talk in terms of the
business severely limits us--in the way we think about
ourselves, our people, operations management, and the
business. We lose real confidence to initiate business
actions. So we lose ourselves in a forest of jargon,
safe services, and personnel work. Our human resources, our most valuable resources, are managed and
administered to, not led.

CONSIDERATIONS IN CHANGING PERCEPTION OF THE HR
FUNCTION

How can we change this? First, let's change how
we think about ourselves, and then our relation and
that of our people to the business.

Do those of you who have owned or have driven a
car made prior to 1960 recall how it appeared under the
hood?   It was quite simple.   There were six wires
attached to six spark plugs on one end and a distribu-
tor cap on the other. There were points, a carburetor,
and a battery and not too much more besides the engine.
And if you were reasonably competent mechanically, you
could fix many problems that arose from difficulties
under the hood, as well as operate the car.

What do our more modern automobiles look like
under the hood today? Quite different! There are
diodes and sending units and air conditioning conden-
sers as well as fuel injection and a potpourri of other
complex apparatus. Can the person who operates this
modern car also take care of problems that arise under
the hood? Generally, no! It takes one or more highly
skilled specialists.

So it is with business! The businesses of
yesterday and the requirements and abilities of those
who operated them and who could take care of problems,

by analogy, under the hood are no longer valid. Today, and tomorrow especially, it takes and will take a highly integrated team of specialists as well as the operator to run a modern business enterprise success- fully. No one would seriously suggest that, if some- thing goes wrong under the hood of a modern automobile and the car stops, then the team, working together to get it going again, is not equally as essential to the smooth running of the automobile or, by analogy, the business, as is the operator. And we in human resour- ces are responsible to assure that the engines of our business continue to run smoothly and efficiently in the direction the drivers take it.

Another way of looking at our human resources which clearly emphasizes the impact of perception oc- curs at budget time. Typically, the subjects of per- sonnel and labor are indicated as costs. Further, in accounting terminology, they are considered as "burden" or "overhead." If one takes a moment to consider the subtlety of these indications and to personalize them, we, among others, are considered as burdens or over- head. This is not a stimulating connotation and cer- tainly not one that provides motivation to our "most valuable resource" to either add value to the business or to make a major contribution to the operating sys- tem. As a contrast, cash, equipment, and other non- human resources that are perceived to drive the business are termed assets by our financial accounting process. As such, they are investments in the business upon which returns are expected to be generated. It appears to me that the perception of an asset differs significantly from that of a burden. And this is a key point. How can we seriously expect our human resources to be perceived as adding value to the business when we preach about them as assets but in real practice apply them as burdens?

Once we make the perceptual leap that human resources are assets upon which investments should be made and returns generated, our treatment of them in practice changes dramatically. These assets can appre- ciate and add value as skills, knowledge, and improved behaviors are added to them. An example of this is the way training can be viewed. Too often training for training's sake is justified as a worthwhile activity with little expectation of measurable results. How- ever, if training is viewed as an investment upon which a return is to be generated, our expectations change dramatically. For instance, if we are teaching selling skills and a sales manager considers sending a few of his or her sales representatives to the program, it will be expected that the investment of time spent off the job in the program will generate additional skills, knowledge, and behavior that will be manifested by

increased sales when the sales representatives return
to their territories. If such increased results are
not forthcoming on the job, a further investment by the
sales manager is doubtful. The training program will
fail.

A final consideration for both operating
management and their informal staff partners to recognize is that the investment in human resources is most
likely the largest investment made by their organizations.   In the Coca-Cola Company during 1985, the
average direct cash compensation worldwide was approximately $25,000. If one adds an additional $10,000 for
benefits and employee taxes, a total expenditure of
$35,000 is made per individual. If one considers that
the productive life of the average human asset is
thirty years and that an annual cash investment increase of eight percent is made for the thirty years
per employee, the future value of that investment is
over $4 million.   On a present value basis, that
investment is approximately $1 million.

If one then adds an annual investment of 100
percent of salary (a conservative number) for support
of that human resource (space, telephone, electricity,
supplies, etc.) the total present value investment made
when an average employee is hired by the Coca-Cola
Company is approximately $1.7 million. The point to be
made is that if management considers that they are
making an investment of this magnitude each time they
hire an employee, they will behave differently and
expect a generous return on their major investment.
When this change of perception is translated into practice, the human resource will in truth be "our most
valuable resource."

12

Travelers Integrates HR and Strategic Planning

Thomas E. Helfrich

I will be talking about the integration of human resources with strategic business planning, from a corporate perspective. The first part of my remarks will deal with why it is increasing in importance, and the second half will cover how we, at The Travelers, are addressing the important linkage between HR and our businesses.

IMPORTANCE OF INTEGRATING HR/STRATEGIC PLANNING

Conceptually, the integration of human resources into strategic business planning is logical, not terribly complicated on paper, and is a subject that has been given much lip service by human resource and operating leaders over the past five to ten years. The fact that this seminar is devoting so much time to this subject is an indication that the transition from the theoretical to the practical is much easier said than done.
Let's briefly reflect on some information that will make this discussion more relevant and place the shifting role of HR into clearer perspective. One cannot seriously discuss this shift and the importance of HR and business planning integration, however, without a review of the enormous social, attitudinal, and business changes that have occurred in little more than a decade and affect each of us, all employees, and our businesses. Think, for a minute, about some of the value shifts that have occurred and consider the human resource and strategic business implications:

Increasing divorce rates and single parenting

Dual careers

Drugs in the work place

Women's rights

Increasing concern for the environment

Health and safety laws

AIDS

Self-awareness

Paternity leaves

Sexual harassment

All of these impose a significant agenda of social
change and affect not only the internal operation of
our companies, but what goods and services our companies produce and how they are marketed. Coupled with
these changes is the fact that people in today's society are better educated to the possibilities of life,
less in awe of authority, and better able to choose
where and when they work. They have grown up sharply
less tolerant of the way work has been organized and
led.
     At the same time, our businesses have undergone
enormous cost pressures, intense increases in competition, merger threats/opportunities have abounded, and
layoffs, plant closings, and special retirement packages have become an everyday occurrence at even the
most successful and respected companies. Against these
changes let's see what lies ahead. First let's look at
these demographics:

1.   Between now and 1995, the absolute number of
     sixteen- to twenty-four-year-olds will decline.

2.   For the first time in U.S. history, there are more
     people over the age of sixty-five in the population
     than teenagers--and the trend will continue.

3.   Women will account for two thirds of the labor
     force growth during the 1980s and 1990s.

4.   From now until 1995, the minority labor force will
     grow at almost twice the white rate.

5.   High school dropouts will increase--to over
     1,000,000 each year. One out of four ninth graders
     will not graduate.

6.   Two-worker households will increase to 86 percent
     of all households by 1990. That is up from less
     than 50 percent in the early 1950s.

7. Today, 23 million adults are functionally illiter-
    ate and the numbers are expected to worsen. Among
    17-year-olds, 40 percent cannot draw inferences
    from written materials and 66 percent cannot solve
    math problems with several steps.

    Next, let's recognize our changing work place:

1. During the 1970s and so far in the 1980s, of the 20
    million new jobs that have been created--5 percent
    were in manufacturing while 90 percent were in the
    service and information industries.

2. Within the next ten years, as many as 15 million
    manufacturing jobs will be restructured. At the
    same time, an equal number of service jobs will
    become obsolete.

3. The nature of our work over the next ten years will
    certainly continue to change, not so much that jobs
    will become high tech, but technology will alter
    how they are performed. Because of this, high-
    level skills will be required and those seeking
    entry-level jobs will face more intense competition
    for the fewer unskilled or semiskilled jobs.

4. The 1980s and 1990s will see an unusually large
    concentration of baby boomers competing for fewer
    high-level jobs--creating a large number of job
    satisfaction issues.

    While most of us are familiar with these social,
attitudinal, and demographic facts, the key questions
are what impact have they had (or are they having) and
how are our businesses planning for the future? Let me
give you an example: IBM, AT&T, and GE, to name only
three, have, in the past, spent significant effort in
the area of manpower planning. Several years ago the
reality hit that in the year 2000, each of them, independently, will have the staffing requirements to hire
every electrical and mechanical engineer that graduates. What has resulted are plans (and existing programs) to make elementary and junior high school kids,
who are interested in math and science, aware of them
as wonderful places to work. Consider the recent television advertising of Dow Chemical. The focus is not
on the product--it is on Dow as a great place to work--
and we know who watches TV--kids.
    Therefore, I am convinced that either by design or
necessity truly successful companies will have their
human resource management and business planning functions interwoven. Leaders will expect, and demand,
that their human resource people become more "line"

oriented--that there are individuals who understand the business side of the equation, but can also walk the employee equity tightrope.

The balance of my remarks will focus on what we are doing at The Travelers--how the integration occurs, what role HR plays, and finally, an assessment of what's gone well or not so well.

INTEGRATION AT TRAVELERS

We, like most of your companies, have one-year and three-year business or strategic plans, but have only recently added a specific piece on what human resource needs and implications exist in the achievement of the business strategy. In each business, the operating HR people work closely with the business planners. We, at the corporate level, on the other hand, develop a company-wide HR perspective, making certain we are integrating with the businesses so that as the individual business and corporate plans are developed, the HR requirements and business strategies will be integrated.

What I've just described is not unique or complicated and can easily represent what I mentioned at the outset as lip service. The process sounds OK, but where is the substance, the checks and balances, and the assurance it does not become another bookcase ornament?

In my estimation, there are two key elements that must exist for HR to, in reality, be a key and essential part of the business--transforming it from a function often associated with being too reactive or too theoretical, to one that significantly contributes to bottom-line profitability.

First, there is the need for top leadership with a vision of the future; individuals who can provide the organizational direction, "people specs," and business framework for the enterprise. Second, there is the need for HR professionals who have the breadth and courage to not only rethink the way we design our jobs and handle work relationships, but can then be pragmatic in their approach in order to achieve operating buy-in.

How we began the transition, and the key to our success to this point, lies with a CEO who knows precisely where he is driving the company, knows the type of individual he needs to attract and retain in leadership roles, and recognizes the imperative to relook every aspect of our approach to HR--recruitment, development, compensation, effectively dealing with top and bottom performers, and you know the rest of the list as well as me. It is the understanding that as we have shifted to playing in the financial services game, our competitors are tougher, faster moving, and more diversified and, therefore, the people and organizations we need must be able to anticipate and respond more quickly.

On the other hand, our top HR individual is adept at wearing both a business and HR hat. He, and those of us who work for him, not only are expected to challenge conventional wisdom, but must stand ready with recommendations that are directly linked to the three things that the CEO has stated are essential in successfully differentiating us from our competition:

1. The quality of service we provide both internally and externally.

2. Our ability to use the full potential of the revolution in information processing and telecommunications technology.

3. Our ability to more effectively utilize people and achieve total human resource excellence.

Therefore, we have initiated an organization and management review process that focuses on both the current state of affairs and the formalization of actions that will help drive the attainment of business priorities that have been defined in the business plans. The Organization and Management Review complements the business plan by specifically addressing the current health of our organizations and people, as well as developing specific action plans to deal with identified longer-term objectives.

The major difference that I've seen among the many plans and human resource reviews that are regularly held in most businesses, is that the successful ones have the commitment of the organization's leadership and the leaders are tenacious in making certain that specific action plans are developed and executed.

The short-term painful decisions are made if, in the long run, there is value added to the business strategy—going after the very best talent, broadening job experiences—not simply job rotation—but assignments where there is real responsibility—and giving individuals enough rope early in their careers to demonstrate what they can or cannot do.

It is not enough to simply do things right—i's dotted and t's crossed—but it is essential that we do the right things. In essence, we are talking about the transformation from managing the business to leading the execution of a business strategy. It is an operating management driven process. HR plays a critical role in the development of the process, and the execution, but there must be operating ownership.

ASSESSMENT OF TRAVELERS' INTEGRATION

Now, let's shift to an obviously subjective, but hopefully realistic, report card.
On the plus side:

1. The company is clearly viewing its business and how it is and will deal with its employees in a more strategic manner. Executives are required to plan and defend how organization or individual development actions tie into the business strategy.

2. There is commitment from the top officers to make the process work--to make personally difficult decisions by wearing the bigger company hat.

3. There is integration of a discipline that both identifies "measurable" action items and then tracks progress.

4. The development and implementation of an interactive data base is under way to maintain accurate organization and personnel information.

5. A focused look at both the demographic and technological implications of the next decade.

Let me assure you that while we have taken an important step, there are miles yet to go and its success will not truly be known for some time--it will only be at the point when this process becomes the way we do business.
To date, there has not been total exposure or total commitment below the top levels of the company. My expectation is that the transition from understanding the concept to making it happen will be a difficult one. From a total company perspective the next level "middle" managers play an essential role in the execution of the strategy--recruiting the future leadership, working closely with them on their career development, and, to a large extent, shaping their future role--either inside or outside of the company. The ball is going to rest in their court and we in HR play a pivotal role in the next step.
What all of this points out to me is that we in the HR field have an opportunity, really a requirement, to really make a difference in our businesses.
With the changes we are experiencing in our industries, how well we participate in the molding of the

strategies and their execution will have a true impact--either positively or negatively.

- What courses of action should our corporations follow in achieving their objectives?

- What changes in organization structure, management processes, and people are required?

- What types of action programs are required in the pursuit of these strategies?

These are real questions that we in HR must address head on. The question is whether we will become active players, developing the game plan and carrying it out, or whether we elect to sit on the sidelines waiting for the operating managers to send us in with the play.

13

Managing Innovation at Digital: A Case Example

Nancy Bancroft

In reviewing Ms. Kanter's book, <u>The Change Masters</u>, I was struck by several observations she had made that reinforced my own experience in directing the process used in making a systemwide change in an organization not too long ago.

To set the stage, the case example is that of a purchasing department within Digital Equipment Corporation. My group was asked if we would be able to help them significantly improve their performance. We were in the early stages of developing a methodology to improve organizational effectiveness while implementing new technology. The success of this case has led us to our current position of providing consulting to Digital's customers. I will review the steps of the methodology we employ in this presentation, this case example, and some learnings about managing innovation.

CORNERSTONE TO SUCCESS: INTEGRATIVE ACTION

Ms. Kanter defines innovation as the process of bringing any new problem-solving idea into use--the generation, acceptance, and implementation of new ideas, processes, products, or services. She notes that organizational innovation is sorely needed today.

One of the keys to innovation Ms. Kanter identifies is integrative action, or treating the problem as a whole. This point has served as the cornerstone of our approach from the beginning. We strongly believe that in order to succeed, one must pay attention to elements in addition to technology to produce the results needed. What is frequently ignored are those other elements in the success formula such as the business goals, the organizational structure and facilitating mechanisms, the people--their personal goals, aspirations, skills, and desires, and the work flow itself.

In fact, we find that people frequently see technology as the whole answer, when it is truly the tip of the iceberg. If we do not pay attention to the organizational and human issues attendant to planning for and implementing that technology, we run the risk of disaster. The many examples we read about of mis-used, underutilized, or even sabotaged systems can in most cases be traced back to the lack of effective processes for planning. In most cases the user's per-spective has been omitted.   I cannot tell you the number of times I have asked the MIS people if they have asked their users what they want. The answer is they already know what they want so there is no use in asking them.

We are not proposing anything new here. As in many innovations, we are putting together, in a new way, pieces that have been proven to be successful. In our experience, however, they have not been consistently applied to the field of technology.

SOME KEY CONCEPTS

We start with the notion that improving productivity is the result of accomplishing the same mission with fewer people (resources) or an expanded mission with the same amount of resources. The impor-tant word here is mission. If we are not clear as to what a group is organized to do, we cannot effectively help them to accomplish that mission. We may, for example, define and prioritize the critical success factors before we start to work on bringing in automa-tion. We certainly don't want to automate tasks that ought not to be done in the first place.

We encourage as much participation as possible. In many projects, as in the purchasing department, this means a task force from the organization. This is important because these are the people who really know what is going on, and because they develop commitment to the change (people do not resist their own ideas), and because of the tremendous organizational and indi-vidual learning that takes place in this type of activity.

We look at the problem set as a whole. We use Leavitt's diamond model as shown in Figure 13.1.   It includes four elements of an organization that we iden-tify and plan for: the people themselves; the tasks they do; how those tasks are organized, the communica-tion paths, the decision-making processes, etc.; and, finally, the technology used. Each one impacts the other and the effective organization achieves the best fit of the four in their pursuit of their goals.

Figure 13.1   Leavitt's Diamond Model

KEY CONCEPTS

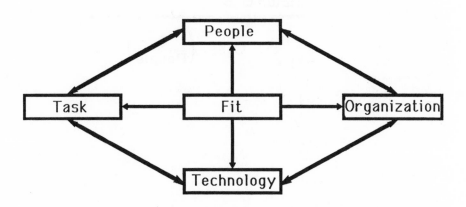

CONSULTING PROCESS

The process that Customer Management Consulting uses is based upon the Beckhard formula for change, as shown in Figure 13.2. This says that effective change is a function of a dissatisfaction with the status quo, a vision of the future, and some clear steps to get there. These elements must be greater than the cost of that change, both in financial and human terms.

We devised a five-step process that uses each of these factors. Prior to starting any project, we have a series of meetings to define the scope of the project, to insure the commitment of upper management, and to develop understanding of the process itself.

The first step is Data Collection. In this step, we first customize and then use a series of surveys and questionnaires to gather data about the current state of the organization. We need to know what is the level of dissatisfaction regarding the status quo. In this step, we gather information about the four elements in Leavitt's diamond shown above: the people, their tasks, the organization, and the current technology. We ask about the environment surrounding the organization, what gets in the way of people doing their jobs, and what suggestions they have about the future.

Figure 13.2.   Beckhard Formula for Change

$$\text{EFFECTIVE CHANGE} = \frac{\text{Dissatisfied with Status Quo} \times \text{Vision} \times \text{Steps}}{\text{Cost} \left(\frac{\$}{\text{People}}\right)}$$

The second step is Data Analysis. Here we make sense of the usually enormous pile of data that has been collected. While using statistical packages for data tabulation, we also place a great deal of emphasis upon the intuitive nature of analysis. We work to separate symptoms from true causes of problems.

An organization can't change unless it is aware of the pain it is feeling. In many cases, people are surprised at the clarity with which we are able to express their concerns, in others, they discover situations about which they had no prior knowledge.

Third, we find it is important to have people consciously spend time thinking about the future. In order to create the future the way you want it to be, one has to be systematic about the exercise. Of course there are barriers and constraints, but unless people are encouraged to think freely about what will make their organization the best, the most productive, the highest performing, they tend only to focus on those barriers.

The driving force in the purchasing department that went through this process was that although they felt that they were doing a good job, they were sure that there were ways in which they could substantially improve on their performance. They developed a vision of an organization that would manage the purchasing process in such a way that they were able to insure the best purchasing decision based on criteria of timeliness, quality, and price. They wanted to do this in the most automated way possible--automating the process from the requisitioner to the vendor. Furthermore, they wanted to develop an organization full of highly motivated people performing at their highest potential.

Step four is to develop a list of technological and organizational alternatives--steps to get there. The team in the purchasing department worked hard at designing alternatives and at finding the best combination of them to achieve the future goals they had set for themselves.

Finally, in step five the team puts the whole thing together in an action plan along with evaluation criteria. They make sure that the benefits outweigh the costs and sell the package to their management. The purchasing team had met several times along the way with their steering committee to insure that there would be no surprises with what they would eventually recommend. The implementation plan was therefore accepted readily.

IMMEDIATE/LONG-TERM RESULTS

During the course of the project a number of issues surfaced that could be resolved immediately: redesigned flextime system, instituted departmentwide feedback system, reassigned work responsibilities, improved department morale, increased employee productivity involvement. These primarily fell into the category of communication and involvement. From the beginning stages the project served to draw the employees together. They began to feel as though they were part of something important.

With respect to the long-term results of the project, the team was able to implement exactly the type of future they had envisioned. These results include installation of a VAX, new purchasing system (requisitioner to vendor), dedicated MIS support, career path planning (exempt and non-exempt), performance metrics, ongoing productivity involvement, steady head count, improved service, greater corporate impact, and role model. They installed a system that allowed them to automate the entire purchasing process from requisitioner to selected vendors. They gained dedicated MIS support because of the importance of this system to the corporation and to the purchasing community at large. The team placed great importance upon redesigning the jobs so as to provide career growth for their people.

Performance metrics were designed for the department. This has continued to be an area of focus. The department today still struggles with how to measure the quality of a buyer's contribution to the corporation. Their effectiveness depends upon a number of factors such as the type of commodity they purchase, whether they can combine multiple contracts, and the length and sensitivity of the bidding process.

Further long-term results are that the department members have maintained an emphasis upon productivity, personally and organizationally. The head count had remained steady until recently when the department assumed responsibilities for another purchasing group. By combining these departments, they added seven people. However, they were able to do the increased work load with only four of those people so they did not replace three people who moved on to other jobs. In fact, they found that several of their best people had been offered other jobs, but had elected to remain in the department because they found the environment so stimulating. This is critical since in measuring the value-added portion of this project, one of the important factors was lack of turnover. It takes a lot of training and time to develop a buyer into a key contributor to the corporate bottom line.

The manager designed a customer satisfaction survey as a result of the project and found that the level of satisfaction and of service rose. The department could identify ways in which it had contributed to corporate profitability because it had developed the ability to spend more time in researching vendors and in negotiating larger contracts. Finally, the department found that it had become a role model for other internal purchasing groups and, through certain publications, for other corporations as well.

LESSONS LEARNED

This case serves as an example of managing innovation at two levels. First is the invention of this process that Digital Equipment's Customer Management Consulting group uses, and the second is the implementation of a new way of doing work in the purchasing department. From an analysis of both of them, several conclusions can be drawn.

First, it is imperative that an individual be the sponsor for the innovation. Without one person who takes the change on as a personal mission, the innovation will not succeed. In cases where that sponsor has left, another can take on the project only if the same sense of personal investment is present.

Second, as Ms. Kanter states in her book, that person needs to obtain the power to do what he or she has set out to do. For the manager of the purchasing department, this was relatively easy although power is not given forever. My experience is that once something is fixed, it has to be KEPT fixed. The manager had to periodically make sure that upper management was happy with the project's progress and results. Even two years after the initial planning period, this communication was found to be important.

In my own case, the situation was different. I was trying to get an organization, Digital, which is highly successful in encouraging innovation, to accept an approach that was countercultural. There was little resistance to the idea of innovation, but there was great confusion as to why anyone would want to pay attention to the organizational factors involved in technical change. That was five years ago. Today the idea is very well accepted. My strategy was to conduct guerilla warfare. I just kept on doing what I was doing and saying the right things. I consistently managed to keep around enough believers to keep my group in existence until the majority of people began to understand that the company had to do some things differently and that my approach to selling the total solution was necessary.

Today there are a number of organizations who have jumped on the bandwagon and are combining the use of organization development methodologies with the introduction of technology.

To conclude, I'd like to emphasize the need for planning for technological change and in the process to pay attention to the organizational and human factors. There are ways to do this that increasing numbers of MIS people are adopting.

14

Monsanto Upgrades QC Teams to Second Generation Work Teams

Rodney J. Falgout

I'm delighted to be in Utah, a state whose history is rich in the struggle for personal freedom, to talk about American industry's struggle to give its employees the freedom to excel. We're all familiar with the need to help employees, but how do we translate that philosophy into reality? How do we change a culture that took decades to develop, and how do we change it quickly enough to have a positive impact on profit and employee morale? I'd like to share with you how we did that at the Fibers Division of Monsanto Chemical Company, an operating unit of Monsanto.

In his book, <u>Managing in Turbulent Times</u>, Peter Drucker writes, "In turbulent times the first task of management is to make sure of the institution's capacity for survival, to make sure of its structural strength and soundness, of its capacity to survive a blow, to adapt to sudden change and to avail itself of new opportunities."

Today, I'm going to address the latter part, management's task--indeed responsibility--to avail itself of new opportunities, to maintain or establish a competitive manufacturing base.

QUALITY CIRCLES PROVIDED FOUNDATION

By the late 1970s and early 1980s, Monsanto's fibers plants were beset by the same problems common to much of American manufacturing--eroding profits, foreign competition, rising costs, and seeming inability to do much about any of it. We responded at the time in much the same way as the rest of American industry. We divested or eliminated businesses that no longer fit our long-range business direction. The remaining businesses were restructured resulting in sizable reductions in the work force.

We felt a need to involve our employees in this transition and in 1979 we successfully launched the first of a series of successful quality circles. Like other companies, Monsanto found the quality circle approach worked to break down barriers between managers and wage employees. And, like many other companies, Monsanto eventually realized that the very characteristics that made quality circles attractive--being voluntary, operating parallel to the cultural structure, and functioning temporarily--also limited their value. Quality circles, while part of the solution, weren't the total solution.

That led us about two years ago to decide to take a harder look at how the philosophy that spawned quality circles could be expanded to become a much broader, much more integrated part of our business.

First, let me set the stage. At that time, 1985, the pressures on our fibers business were pretty severe. We knew we had to do something to relieve them. We also knew that whatever we did wouldn't work unless we had the wholehearted support of _every_ employee in our organization, not just wage employees this time and not just those who chose to volunteer to participate, but everyone. Our challenge was to develop a strategy that would competitively position our business for the 1990s. This challenge assembled a consolidated effort that developed a strategy or direction for manufacturing referred to as the "plant of the '90s."

INTEGRATING HRM INTO THE BUSINESS PLAN

Our human resources offer us a significant opportunity to improve our competitive position. Today I will share with you our direction for managing our human resources in the 1990s. I will:

- give you a perspective of how we began;

- review the evolution of major events;

- tell you about the planning process we used to develop the human resource management plan for the 1990s;

- define the issues that need resolution;

- review our desired benefits; and

- conclude with results to date.

Competitive challenges caused us to reassess our basic and traditional ways of managing our resources. That reassessment pointed to the need for change. One major change needed is in the management of our human resources. Today we have many human resource programs operational in the fibers business, but they are programs. They lack strategic direction and need to be integrated into our business plans. We recognized the need to develop a plan that focuses on integrating our human resource management process into our business direction.

Our plan required the need for culture change. Culture is a word used frequently today. Its meaning varies as it is applied in different situations. We define culture as what we do, and how we do it.

Before discussing the culture change needed for the plant of the '90s, I would like to take a few moments to review major events that occurred in the evolution of our directional plan. Employee involvement began in 1979 in fibers when the direction toward increased productivity and employee involvement was begun with the implementation of the quality circle concept in all of our locations.

In 1984 the Fibers Division emphasis was on plans to strengthen our businesses with aggressive three-year goals to improve productivity and return on capital. In 1985 the Fibers Division went through major restructuring. It was a very traumatic year when decisions were made to exit certain businesses. Two early retirement programs offered during the year provided us with opportunities to accelerate productivity plans and goals set in the previous year by not replacing people associated with ongoing businesses.

The transition from the quality circle concept to total involvement began in our plants. This transition began the integration of our human resources to support our business direction.

Let's review the Planning Process for HR Management used in the development of the human resource management plan defining the direction needed to bring about the culture change for the plant of the '90s. Our goal is to achieve a competitive advantage in the 1990s. To accomplish that goal we recognized the need to reassess the traditional methods used in managing and operating our plants.

A team commissioned by myself, made up of personnel superintendents from our locations, was given the objective to develop a plan to change the culture for the plant of the '90s. The plan defines the direction we will take to change our culture. It consists of two major components: a mission statement and a strategy to achieve our mission.

Our strategy focuses on key areas that will have a major impact in accomplishing that mission. There is a need for our employees to understand our business direction. We found that communicating to employees the direction we want to take them is important to making the changes we want to make now and in the future.

A mission statement was developed to accomplish that objective. It will serve as a consistent foundation to move us into the 1990s.

SELF-MANAGED WORK TEAMS

Let's review the key areas that will have major impact in changing our culture. The first is organization. As we look to the plant of the '90s, we see an organization design that places responsibility, authority, and decision making lower in the organization through self-managed work teams focused on product line results. (A self-managed work team is a natural work group of seven to twenty employees focused on achieving defined goals within defined parameters without a first-line foreman.)

Jobs and organization structure will need to change to accomplish this direction. These changes will evolve as we move toward the plant of the '90s. We've already made organizational changes. For example, a business unit in our South Carolina facility, composed of about 250 employees, operates with no first-line foremen. They operate with self-managed work teams.

At our Texas facility, we now have our maintenance organization administratively reporting to our production units to create a team effort to accomplish product line results. Job responsibilities have also changed.

At some of our locations, production operators are performing maintenance work and maintenance technicians are performing production work. Production operators are assuming more responsibility in analyzing product quality, giving them more control over the entire process utilizing statistical process control.

WORK ENVIRONMENT IN THE 1990S

As we look at work environment in the plant of the '90s, we see it as being a lot more flexible than it is today. There will be more freedom for employees to do what is needed to accomplish defined business objectives and goals. Each unit will have specific key business goals defined that are simple, understood, and measurable. Self-managed work teams may establish their own work practices to enhance their unit results within certain defined parameters.

As we move to the plant of the '90s, we see the evolution of the self-managed teams, and compensation systems that focus on performance, skills, knowledge, competitive rates, and profitability. Benefits will be designed to minimize the difference between wage and salaried employees. Other forms of recognition will be developed to be compatible with the achievements of the work teams, such as achievement awards and bonuses.

A major area that needs attention as we move toward the plant of the '90s is communications. We feel communications should promote ownership and part-nership concepts in our locations. Employees need to understand the goals and objectives of our business. They need feedback on how we're doing against those goals and objectives. They should be aware of how customers feel about our products--the products they're producing. They need to understand how the plants are evaluated as well as the competitive environment we face in our business. And they should definitely know our direction, our long-range plans for the future, and our communications should be timely and direct.

The management style in the plant of the '90s will see the boss/subordinate role replaced by the facili-tator role. The new style will promote employee involvement in accomplishing results at all levels in the organization and promote partnership/ownership. The style will increase responsibility and accountability at the lowest levels of the organization.

Training will be vital to help us make the changes needed for the plant of the '90s. Areas for emphasis are (1) the development of skills and knowledge to implement and maintain advanced systems in the plant of the '90s, (2) developing team skills in problem-solving and conflict resolution, and (3) the retraining of people for new job skills will be essential as we move forward. Integration of job responsibilities will become commonplace. We must also develop leaders ver-sus supervisors by training our managers in a different style as they work with people.

As we developed our plan, these issues were identified:

- Corporate benefits

- Compensation practices

- Job security

- Accounting/audit considerations

- Enrollment

- Legal aspects

- Fibers division coordination

- Changing management style

Some of these issues have been resolved and plans are being developed on others.

DESIRED BENEFITS/RESULTS ACHIEVED

We feel key benefits, such as the following, will result from our direction:

1. Reduced costs by operating with fewer people;

2. Improved communications through fewer levels of supervision;

3. A multi-skilled work force which will result in the need for fewer resources;

4. Improved commitment to results. We're seeing that happen already; and

5. An improved utilization of our human resource skills, knowledge, and experience base.

Our results have been extremely successful to date. Not only have we achieved a 50 percent productivity improvement that seemed out of reach a few years ago, but we have replaced the adversarial relationship between management and wage employees with a feeling that we're all on the same team working toward the same results.

At some of our plants, it has meant the difference between mere survival and outright financial success, and those results have been achieved with a minimal amount of disruption, physical or emotional, and with a fraction of the supervisory foremen we worked with before. We were able to reduce 35 percent of our first-line foremen in the last eighteen months with further improvements expected in the near future.

Why have we been successful?

First, a foundation was laid with quality circles. That earlier success allowed implementation of the self-managed work teams to be seen as a gradual transition, a natural evolution.

Second, the change took place at the same time as other cultural shifts--major reductions in a work force that had been relatively stable and major changes in a marketplace that suddenly was more demanding.

Third, the changes were institutionalized throughout the business. Every level of employee, every product line, every plant, and every part of the business was involved.

That doesn't mean there haven't been problems and obstacles that needed to be overcome. There was scattered employee resistance to suddenly being thrust into a decision-making role after the security of being given explicit instructions. And there was resistance from managers to sharing information and responsibility with other employees.

But on the whole, our direction has been extremely successful. Supervisors say morale and productivity are up, employees are taking more active roles in making decisions and solving problems, and our sites are producing more and better quality products with fewer workers. As one spinning operator at our Greenwood, South Carolina, plant put it, "I want to be challenged. I want to find out what I'm capable of achieving. Without the opportunity to test myself I'll never know what I'm really able to do."

Our plant of the '90s program has given employees the opportunity for greater self-fulfillment. It has given the company the opportunity to remain competitive. That's an unbeatable and successful combination plus a testimony to the power of trust and a greater stake in our business. With it, we all win.

15

Linking Merit Pay with Performance at Lincoln Electric

Richard S. Sabo

The Lincoln Electric Company of Cleveland, Ohio, is the world's largest manufacturer of arc welding products and a producer of industrial electric motors.

The Lincoln Electric case is a classic case, as proven by the Harvard Business School. Out of 35,000 cases published worldwide, the Lincoln Electric Company case is the one chosen most often by anyone who purchases a Harvard business case. Today I am going to describe some of the principles that make Lincoln unique, with special emphasis on "linking performance with pay."

TRULY LINKING PERFORMANCE WITH PAY

Every one of our approximately 2,300 workers participates in the company's incentive plan, and all but two have been sharing in the year-end bonus for fifty-three consecutive years. Now, how do you go about fairly and honestly determining how every person in the company should receive over 50 percent of their annual income in one year-end bonus check? Well, it's definitely not easy. It takes a great deal of time and effort, because you must be as fair and as honest as you possibly can in determining each person's share.

The first thing that we do is eliminate the president and chairman from our year-end bonus calculations. We pay those people on the basis of performance by giving them a percentage of sales. In other words, if sales decline, the top executives will take the first pay cut.

Once the top two executives have been removed, we will evaluate each job in the company to determine the fair base rate for that job. This is done by committee, and it is our objective to have the hourly rate for that type of job to be average for the Cleveland area. That keeps us competitive with other industries in our area.

We have another committee that will do the same thing for all salaried positions in the company. The only salaried positions in the company are exempt salary positions. In other words, we will pay no overtime to a salaried employee. Now, we take the hourly rate that has been established, and for our roughly 1,200 production workers, we will apply a piecework rate for each individual who is working on production. In essence, we will apply piecework in any area of the company operation where we know how to apply piecework. This definitely links performance with pay, because there is no minimum guaranteed base rate.

After we pay the piecework rate to the employee, we then have roughly 800 jobs in the company that are hourly rate. Those hourly rates have a range, so that an individual needs to perform to his or her highest capability in order to get to the top of the range for that job. So that is a second method of paying for performance.

The salaried people have exactly the same arrangement. Each job in the company has a salary range. An individual may or may not reach the top of the salary range. As you heard earlier, once you have given a person the salary, it is very difficult to take it back. So, we are reluctant to give the salary until we are certain that the individual has earned it. Now, all of what we do is well and good, but it still does not guarantee that people will perform at their maximum capability.

EMPLOYEE RECOGNITION

Our next step is to merit rate individuals every six months. You may have heard earlier that merit rating appraisals are rarely used by executives. Nothing could be further from the truth at Lincoln Electric. Our chief executive officer reviews the 2,300 merit ratings every six months, and he looks at every one of them. So, the merit rating is very vital and very important. Essentially, we rate the people on four cards.

To measure output, you simply ask how much work did the individual do, and it doesn't matter whether that individual is a pieceworker, an hourly worker or a salaried person. You measure their output. Second, we measure the quality of the work. How well did they do the work they were assigned? Third, we measure their dependability, which to us is defined as the ability to work without supervision. We eliminate supervision wherever possible. The average ratio of foreman to worker is 1:100. We do not feel that it is the

responsibility of the foreman to make people work.   If
you hire mature, skilled individuals who have a desire
to perform, it is not necessary to make them work; they
will do the job.   It's management's role to create an
environment where they can do their best.

Then we will rate the person's ideas and cooperation.   This is the rating card that keeps people from
stabbing each other in the back to gain a greater
income.   The average rating for an employee is 100
points, or 25 points per card.   This rating takes place
every six months, and we will take the two ratings for
a bonus year and develop an average for that person.

The rating may be compared to a grade that you
received in school; however, the big difference is that
our rating determines what percentage of the year-end
bonus each individual will share.   So, if you could
visualize an individual who earned $10,000 during the
year and received a merit rating of 100, they would
receive a $10,000 bonus at year-end, if we were, in
fact, paying a 100 percent bonus.

Now, for the last fifty-three consecutive years,
we have paid an average bonus of 95.5 percent.   The
merit rating would change the individual's income,
depending on what the rating could be.   The ratings
generally will fall between 90 and 110, but there is no
upper limit, nor is there a lower limit, so that people
could receive merit ratings of 140, 150, or 200 or
more.

But, the bonus payment is paid to everyone in the
company except, as I mentioned earlier, the president
and chairman.

OTHER BENEFITS TO LINKING PAY TO PERFORMANCE

Now, how successful has our system been?   It has
allowed Lincoln Electric to maintain its position as
the world's largest manufacturer of arc welding products over the entire fifty-three-year period that we
have been operating under the principles of incentive
management.

Second, our turnover rates are very low.   We have
a monthly turnover rate of less than .3 of 1 percent.
The average age of our work force is just over forty
years of age, and the average number of years of
experience in our machine division is 16.75 years.

This experience has resulted in very high quality
products.   We offer the only five-year warranty in the
electric motor field, and in the welder field we offer
a three-year warranty.   Our rejections rates are minimal.   The cost of rejections in our Motor Division has
been less than .3 of 1 percent of sales.   Add to this,
the fact that we have never had a strike in our

company's history, and you can see why Lincoln's market share continues to increase.

The company has no debt, and we have been able to pay a dividend every year for fifty-three consecutive years. This is in spite of our industry experiencing a severe recession over the last four or five years, because the manufacturing base in the United States has certainly declined.

In closing, I do want to call your attention to our Incentive Management Seminars that are periodically scheduled in the Cleveland, Ohio, headquarters. There is no charge. We sponsor these as a service to industry.

16

General Foods' Strategic Human Resource Management

Roslyn S. Courtney

I want to review an approach used for integrating business/human resource/ and employee planning for General Foods' Information Management Department. We call it the Professional Development Program or PDP. This approach, or elements of it, can be applied to any functional or professional organization with skill requirements. It is a way to understand and manage a skill-base over time. I refer to strategic human resource <u>management</u> rather than human resource planning because of the execution emphasis of our program.

The PDP laid out a structured approach to all employee career and development planning and tied this employee effort to overall department and human resource plans. It required management to define and widely communicate organizational strategies and goals, as well as skill expectations for the next three years. The focus was both longer-term and immediate. Our strategic planning timeframe was three years, but in addition we had immediate expectations of employees to engage in formal training and on-the-job development. The approach heightened employee awareness and understanding of department human resource planning and called for employee input and involvement in the planning and review process.

My comments will highlight the elements of the PDP, relate these elements to a general planning model, explain why we created supporting processes and activities, and review the benefits and requirements for implementing a successful program.

IMPORTANCE OF COMMUNICATION

You will notice an important thread that runs throughout this discussion--a very strong emphasis on management's efforts to communicate to the organization at large. While many human resource plans are discussed only within the ranks of top management, these plans and objectives were broadly communicated. We moved away from the notion that plans should be locked up and kept secret, that the general direction and state of the organization was the kind of information available only to the privileged few at the top.   Our assumption was that a broad understanding among our managers and employees was necessary to motivate and change an organization of about 450 professionals.   The objective was an ambitious one and covered a large and diverse group of employees. Management was asking for involvement, follow-through, and action.

PLANNING MODEL

The model illustrating our planning processes started with the business and human resource plans and was then expanded to include employee planning.   The employee planning and assessment part enriched the quality of information available to management and actively involved employees in the process.
By 1982, the strategic plan for the Information Services Department was a thorough, sophisticated statement of goals, strategies, programs, and activities.   Management throughout the company was experienced in strategy formulation and evaluation, and this management team was no exception.
The human resource plan was not so well developed as the business plan.   We had detailed succession plans, issues focused on near-term gaps in staffing, and an action plan for addressing the issues.   But, there were no descriptions of the ideal organization, the mix and allocation of skills necessary, and the kinds of programs required to execute the strategies in the strategic plan.   And, there was no way to assess the organizational skill-base, or the current placement of skills.   We needed to develop our human resource planning if we were going to more aggressively manage and change our capabilities.

APPROACH

In April of 1982, I chaired a task force composed of key managers from each major unit of our Information Services Department.   Our charge was to recommend how we could insure the availability of skills and the proper placement of resources in the next three years.
This group quickly realized that we lacked two parts of the evaluation equation.   We needed an ideal skill profile and an assessment of our current skill-base.   Once we had these two pieces, we could compare

them and identify the gaps. The conceptual model was very simple: Ideal versus current, gap analysis leading to action plans.

We introduced the prototype program in 1983. Managers were asked to explain the program and distribute materials to their subordinates. We found--and it is not surprising--that the quality of execution depended on how much support we had from each manager. Some took it very seriously and others did not. The learning from this prototype experience called for changes in process and format for the introduction the following year. We wanted a wider management involvement and support for the program and a greater and more uniform understanding among employees.

ELEMENTS OF PDP

Let's go back to the planning model to look at the elements of the PDP.

The management component included an ideal human resource profile, an assessment of the current organization, and specific action plans. The ideal profile identified thirty-four skills that management considered critical for the department's success. Examples of these skills are developing subordinates/resources, project planning and control, and business analysis. The number of skills was limited and the list was re-examined every year. In addition, the skill requirements were defined for all the jobs in the department.

A second part of the ideal profile was an organization plan. Management was required to project the number and kinds of positions needed in the next three years and the skill requirements of these positions. This projection would ideally be completed with the strategic plan. Expected skill proficiencies were ranked on a scale of one to four. A one indicates full competency to work independently in an area. A four would indicate no skill proficiency.

The assessment part of the plan was completed after we collected the employee data. By aggregating the assessments prepared by each employee and manager, we could look at the current organization in a number of ways--by grade level, kind of position, by unit or total department-- and then compare the current to our projected ideals.

This comparison generated our action plans, which included training and development programs designed to build priority skills. Management could identify critical areas from the strategic plan and from the gap analysis. Development moves and job rotations were also part of the action plans.

The assessment part of the human resource plan can be accomplished in a number of ways. We chose to involve employees and to require a rather precise evaluation that used our 1-4 rating scale. The employee component consisted of four pieces:

1. Each employee completed a <u>staff resume</u> that summarizes important work experiences, including assignments in other companies. The resume becomes a starting point for reviewing a person's strengths and experiences.

2. The employee and manager both complete a <u>skill assessment</u> and then meet to discuss these independent assessments.

3. The assessments are the basis for a <u>development plan</u>. The plan identifies formal training and on-the-job experiences to increase skill proficiencies. On-the-job development is emphasized, requiring the manager to think about work activities, task force assignments, and special projects that enhance an employee's skill base.

4. An important piece of employee input is a statement of <u>career and development interests</u>. Here the employee directly expresses interest in specific jobs, assignments, or project areas. This information is used in management reviews to make sure the employee perspective is accurately registered.

SUPPORTING ACTIVITIES

A person is motivated to work on a PDP if he/she believes the information is used and will be the basis of management decisions. Our prototype experience raised issues of how to increase overall support and quality of input. As a result, we initiated several supporting processes and activities.

First of all, we started an annual human resource review that follows the PDP process. Each unit director and his staff reviews summarized strengths and weaknesses, needed new assignments, plans for promotion, career issues and opportunities for each professional employee. The organization is aware of the review and the way PDP information is used during the review discussions.

Second, a top management personnel committee meets every three or four weeks to deal with all recruiting and staffing decisions and to work out the details of the human resource reviews, and the PDP data are used to plan individual moves and job rotations. Interestingly, the committee and its agenda have replaced the previous one-on-one negotiations that were required to move a person from one unit to another. A major management goal was to provide challenging and diverse assignments for professionals and use people most effectively.

A third activity is our roll-out communication program. Each year, employees and managers are reminded of the PDP requirements and new expectations or changes. The vice president and the unit directors present the PDP and results from the previous year to each major unit. This direct form of communication insures that every employee gets the same message and there is no dilution or distortion of the message through management levels. The roll-out process is another way to keep top management very visible and involved in leading the program.

Still another supporting activity centers around collecting employee feedback. Employees are asked to critique the PDP to highlight strengths and weaknesses. We have found focus group discussions to be an excellent way of getting employee issues in front of management for resolution.

BENEFITS

While the PDP was originally designed to strengthen our human resource planning process, we have realized broader organization development benefits because of the communication aspects of the program, the employee involvement, and the emphasis on implementation and management of human resources. The key benefits are:

1. We integrated department (business) and employee planning, as well as business and human resource planning.

2. In that process, management was required to think through future requirements and skill expectation.

3. We now have very current, usable information to support human resource decisions.

4. We established uniform expectations for career planning and development planning. Our people welcomed a structured approach that required each manager and subordinate to follow an assessment and planning process.

5. We have a basis for setting our training priorities. We can set development and training priorities based on skill gaps.

6. Finally, employee participation and morale are higher. Employees are more involved and feel more in control.

CRITICAL SUCCESS FACTORS

From this experience, what conclusions can we draw? What are the key requirements for implementing a successful program?

1. Integration with the ongoing business/activities of the organization.

2. Management involvement, <u>ownership</u>, and visibility. The human resource plan must be management's tool and generate management's programs.

3. Training--to develop an understanding among managers and employees.

4. Currency of data--human resource data become obsolete quickly.

5. Staff support--someone must be assigned to champion the program and manage the follow-up.

6. Demonstration of results--people have to see progress/changes to keep interested and motivated.

7. Keeping a program dynamic and relevant--change is evolutionary and priorities shift over time.

SUMMARY

We at GF are very pleased with the PDP. It was expanded to cover our Marketing Research Department and has been a model for other areas of the company. The approach or elements of it can be tailored to any functional group and could be designed to fit the specific goals and operating style of the organization. We have found that yearly changes and refinements have improved the program and make it a useful and ongoing management tool.

17

Measuring HR at National Semiconductor

John Campbell

My objective is to present some ideas and strategies that we have found to be successful at National Semiconductor for measuring human resources. I shall also discuss some general philosophy about what organizations can do to measure their human resource efforts.

Before I begin, let me give you my general idea of what that phrase, "measuring human resources" means. It deals with the following ideas:

- What will we measure and why?

- How will we measure it?

- What will we do with the data?

As HR professionals, particularly in recent years, we have been patting ourselves on the backs when we read in the business literature how far HR has come recently as a profession. In many organizations, HR has increased substantially in prestige, credibility, and organizational power. One sign of HR's newfound popularity is the growing number of people trying to get into the field.

In describing HR's <u>perceived</u> status, let me share with you a couple of quotations from a 1981 article in the <u>Harvard Business Review</u>:

> Personnel work has seldom been attractive to fast-moving, younger general managers who see the field as out of the mainstream of the business. They also see personnel as a staff function that is strictly advisory, lacks authority and power, and deals with small-scale troublesome problems. A personnel job is seldom an attractive position for a manager who wants to run something independently. Because of personnel's conflict-ridden, pressured, contradictory nature, the decisions

personnel managers make are touchy and cumbersome. Because they involve other managers, they are not only time-consuming but also often frustrating.

For these reasons, few outstanding managers move into personnel, and those in it often have problems getting out. The detail, time required to gain expertise, low status in the organization, and lack of clear-cut authority can swallow up and overwhelm all but the very best in the field.

HRM BECOMES MEASURABLE

Does any of this hit home? Although written over fifty years ago, does any of this hold true now? This reminds me of the time when my four-year-old son, Mark, asked me what I did for a living. In trying to keep it in simple terms, I said, "I do personnel work--I hire people, fire people, write letters, talk on the phone, read reports." He had a puzzled look on his face still, so I gave him some more duties. "I figure out how much people get paid, I help people find jobs, etc." He still looked puzzled and finally as he started to walk away he said, "I wish you were a policeman like Trevor's daddy." He understands what a policeman does. When there's a problem, the policeman fixes it. It was simple because it is measurable. Come to think of it, that is about what I do most of the time.

Anyway, companies that lead out in HRM practices have done some specific things to take that lead:

1. They develop a long-term program to develop general managers with HR skills and experience.

2. They treat the HR department as a functional operation with strong authority and responsibility. This attracts the top HR talent.

3. They move outstanding managers into HR functions for two- to four-year periods. After five to seven years a top management group has been developed, who have in-depth experience in HR.

It seems to me that one of the major ways (besides these three) that HR has come out of the dark ages (so to speak) is that now we are subjecting ourselves to the same kinds of management measures as in the other disciplines--like production, engineering, finance, etc.

POTENTIAL MEASURES OF HRM

The central theme behind all of this focus on measurement is that measurement brings accountability (some of you have probably wondered when I was going to get to my main point; well, here we are; it's accountability). For just a minute, think of what you are held accountable for. What are you measured against? Does your boss or do you yourself have specific HR measures you are held to?
Is it:

Turnover?

Headcount targets?

Aging of open personnel requisitions?

Number of lawsuits not filed?

Company productivity?

Product quality?

Performance ratings' distributions?

Number of grievances?

Training classes conducted?

The list is endless. I have developed the following laundry-type list of potential HR measures that can be used to track an HR organization.

EMPLOYMENT/STAFFING

Aging of Open Requisitions

Cost/hire

Acceptance Ratio: Offers/Acceptances

Recruiter Effectiveness

Revenue Ratio: Revenue/Headcount

Expense Ratio: Expenses/Headcount

Income Ratio: Pre-tax Income/Headcount

COMPENSATION/BENEFITS

Compensation-Revenue Ratio: Comp./Revenue

Compensation-Expense Ratio: Comp./Expense

Compensation-Income Ratio: Comp./Pre-tax Income

Benefits Costs/Revenue

Benefits Costs/Expenses

Benefits Costs/Pre-tax Income

Health Care Costs/Headcount

Merit Review Distribution

Grievance Costs

EMPLOYEE RELATIONS

Absence Rate

Turnover Rate: Total Terms/Headcount

                 Voluntary Terms/Headcount

                 Involuntary Terms/Headcount

Number Lawsuits/Headcount

Number Unemployment Appeals/Headcount

Unemployment Tax Rate (where experience-based)

HR DEVELOPMENT

Training Effectiveness: pre-test, post-test, control
group productivity rates of "trained" groups

Training Costs/Expenses

People trained, courses held, level of participation

Use of Diagnostic Measures: participative mgt., management style, supervisory skills

Top Management Involvement/Support

OVERALL HR

HR Expenses/Total Org. Expenses

HR Headcount/Total Org. Headcount

I have separated out by the various HR functions the measures that I see being used by companies, talked about in the literature, or used in national surveys. This list is not intended to be all-inclusive, but to present the idea that there really are a large number of potentially useful ways to monitor your HR efforts. As you know, many of these measures now have standardized formulas. This is helpful as organizations want to compare their own rates or ratios with other organizations. I'm not trying to advertise their program, but the ASPA personnel association has joined forces with the Saratoga Institute in California to help employees collect and input measured data, analyze it, and have it compared with other U.S. organizations.
    As we look at some of these, ask yourself:

- Which ones are more meaningful for my organization?

- Which ones would be most appropriate for us to track?

- Are there some we'll track and then show to the rest of the   organization?

- Are there others we'll track but keep to ourselves? (We don't want to expose all our weaknesses!)

DETERMINING APPROPRIATE HR MEASURES

    How do you determine the appropriate measures? The first thing to do is look at the overall strategic plan for the organization.   From that overall organizational perspective you can establish an HR mission plan/statement. From there we begin to prioritize all of the HR projects and programs.   At National Semiconductor, all HR programs are classified into one of these four categories:

1. Legally required

2. Very desirable

3. Moderately desirable

4. Marginal but desirable

Once the priorities are established, the actual measures can be established and data collection can begin. Although not listed in the group here, the one item that comes up again and again as HR's top priority is its productivity. Let's talk about that for just a minute. We all know intuitively that productivity, particularly in today's world economy, is probably the chief concern of any profit-oriented organization. HR historically has been concerned about productivity, but only indirectly. We do training, we hire qualified people, we compensate fairly--all in the hope of improving overall productivity.

How much should HR share the responsibility for productivity with the operations people? How much should my pay raise, bonus, or stock options be tied to "line" groups' successes?

At National Semiconductor we believe that HR is responsible and should be held accountable for productivity--more so now than ever before. Like most organizations, HR's role at National is in a state of change as shown here:

MOVING HR

	FROM:	TO:
ROLE:	- Controller - Police	- Consultant - Facilitator - Strategic partner
POWER BASE:	- Threat (lawsuits, Unionization, etc.)	- Expert - Referent
OBJECTIVES:	- Programs - Systems - Activities	- Performance - Productivity - Quality

HR staff have the overall goal to help managers do what is really important to them:

1.   Identify what people do.

2.   Measure it.

3.   Let people know how they are performing.

We follow a philosophy that says "the way to get power is to give away power." Let me share some examples of how we do that. Line managers at National used to complain about the exempt performance appraisal system. We said, "You design the system." From all of their input a system was implemented that they now own and like--because they designed it.

Another example: salary equity among employees has always been a concern for operations managers.   To fix what they perceived to be a salary equity problem within their group required a trip down to the compensation staff.   After hearing the problem, the HR compensation staff did the reviews and analysis and then came back to the manager with some options.   Now a plan is in place that makes the line manager the salary equity administrator for their group.   Once a year at the focal review for merit increases, the manager receives an equity line plus an equity budget that they now administer to improve the group's salary equity in our PAY FOR PERFORMANCE SYSTEM.   This is not particularly unique or unusual, but it demonstrates a principle: our HR managers are held accountable not only for the specific performance objectives in their own functional area (employment, compensation, employee relations, etc.), but they are also held accountable for how well they are making line managers use effective HR principles and techniques.   My job right now is to transform all managers of exempt employees into good salary administrators--to give part of my compensation power away.

TRACKING HR MEASURES

We track some specific HR measures at National. For better or for worse, these are the things that we care about.   You need to really think about what data you are going to collect and look at.   There is a natural tendency to focus on the data you track and that data alone in making HR decisions.   It's easy to exclude what may be other relevant data.   Be sure you track the right things, that you don't include unnecessary "stuff" or exclude items you'll wish later that you had tracked.   Here are some measures that we monitor:

- Headcount--by various categories

- Additions per four-week period

- Terminations

- Turnover

- Performance review distributions (We are trying to see how our merit system is working.)

We have standardized the above measures and the reporting so we can analyze the data as follows:

- Companywide

- Each manufacturing facility (Utah, California, Texas, Connecticut)

- Our semiconductor group versus our information systems group

- Each manufacturing unit (MOS I, MOS II, MOS III)

As in any MBO type system, National expects each manager to have a set of specific, concrete objectives that are negotiated with his or her supervisor and become the measure of his or her performance. Accountability for performance toward those objectives is the key. Here are the specific procedures and methods we use to keep our managers focused (accountable) for their objectives:

A. A corporatewide five-year plan is developed and updated regularly.

B. From the five-year plan, one-year objectives are established for all groups and employees.

C. After the first six months, all managers are reviewed as to how well they performed on their objectives.

D. After the second six months they are also reviewed. That review becomes part of the actual performance review that determines their merit pay increase.

E. Within the six-month periods, there are quarterly reviews at the three-month mark.

F. There are written reports turned in at the end of each four-week period.

G. Finally, the genius of it all is the weekly one-on-one accountability meeting. These meetings between a supervisor and each subordinate professional are held weekly or less frequently depending on the need. They last at least one hour and provide for mutual teaching and information exchange. The subordinate sets the general agenda for the meeting that should be held at or near subordinate's work area. The agenda covers performance, data on productivity, trouble/potential problems, important developments since the last one-on-one meeting, and

future plans.   Notes should be kept for establishing agreement.

       I know this sounds like a lot--but it works.
I'm not a believer in a lot of unnecessary meetings
or reports, but this strong emphasis on accountability, exchange of information, and participation
keeps the entire organization focused on results
and performance. I've heard managers complain about
these one-on-ones, saying, "I don't have time for
meetings--I've got work to do."   These "meetings"
are the essence of what managers should be doing.

       One very revealing survey to conduct is to find
out how much of your managers' time is spent managing
versus the time they spend doing individual contributor
type work.   I would suggest that where you find a
disproportionate amount of individual contributor work
being done, those managers aren't earning those extra
dollars you are paying them to be managers.   These
accountability meetings and reviews and particularly
the one-on-ones are where the real management and/or
leadership in our organization does or does not take
place.   To quote Mr. Andrew Grove:

       What is the leverage of the one-on-one?   Let's say
       you have a one-on-one with your subordinate every
       two weeks, and it lasts one and a half hours.
       Ninety minutes of your time can enhance the
       quality of your subordinate's work for two weeks,
       or for some 80+ hours, and also upgrade your
       understanding of what he is doing.   Clearly, one-
       on-ones can exert enormous leverage.   This happens
       through the development of a common base of infor-
       mation and similar ways of doing and handling
       things between the supervisor and the subordinate.
       And this is the only way in which efficient and
       effective delegation can take place.   At the same
       time, the subordinate teaches the supervisor, and
       what is learned is absolutely essential if the
       supervisor is to make good decisions.

CONCLUSION

       So there you have it--we've talked about
appropriate HR measures, what we collect, how we
measure our HR staff and their responsibilities. We've
covered how we hold people's feet to the fire--to
establish real accountability.   We haven't covered the
myriad of formulas and other specific HR measures out
there, but I've tried to portray the culture of HR

accountability and measurement I have come to know "and love."

Referring back to that initial <u>Harvard Business Review</u> article I quoted at the start, it is measures and accountability techniques like this that will continue to improve our status and image until our "HRM strategy, when successfully carried out, becomes a uniquely dynamic competitive weapon. It is more important than ever to recruit and develop a high-quality group of employees, for companies with a head start are hard to catch. Their good people attract others like them, while conventional organizations have to accept what is left."

"HR can act as a catalyst and an operating mechanism to accelerate the building of an effective work force. Where this is accomplished, people are energized and committed and become the most powerful, fundamental corporate competitive resource of all."

18

Borg-Warner's Change
in Organization and Culture

James R. Deters

I really couldn't think of an agenda more important
than one addressing "change." John Naisbitt said it
all in <u>Megatrends</u> when he wrote: "The current level of
change involved is so fundamental yet so subtle that we
tend not to see it, or if we see it, we dismiss it as
overly simplistic...yet we do so at great risk to our
companies, our individual careers, our economy as a
whole."

I must admit to disagreement with some of Mr.
Naisbitt's points. But not that one, and not with his
conclusion that it's a "fantastic time to be alive."

I'm new to the human resources function. I'm no
expert in the area. Also, I don't want to imply anywhere in my comments today that Borg-Warner is a perfect role model. We're struggling with those same
changes facing others and we certainly don't have the
answers.

I hope to share with you some of our struggles and
our experiences. Maybe my being a newcomer to the
human resources function might be an advantage--some of
my opinions are not burdened by past "knowledge."

In picking a specific topic for "sharing company
experiences," I debated about topics such as quality
circles, the employment security policy we implemented
two years ago, plant closing guidelines, communication
programs, employee assistance, profit sharing and all
salaried plans.

Finally, I decided to share a very fundamental
Borg-Warner strategic issue with you, one that's been
evolving for some time, but one that's developed momentum in the 1980s. This issue deals with organization
clarity, with change, and with corporate culture.

HISTORICAL ROOTS

To get the proper cultural backdrop for my story
it would help to have a pretty clear view of the

strong, at times almost fanatical, historical roots of
individual business autonomy in Borg-Warner. Now, if
this were a movie, right about here my voice would
trail off, the picture would fade out and back to the
spring of 1928--back to a jovial dinner meeting in
Chicago. Five quite successful automotive supplier
businessmen are dinner guests of two investment bankers. The economy is booming, their business is
booming, and they're clearly proud of their individual
achievements. But they recognize that single part auto
suppliers are vulnerable. They know that some alliance
might make sense. Well, none of these individualistic
businessmen had ever met each other before, but it was
a good dinner, they liked each other, and so
negotiations to form a new group began the next day.

The new company named Borg-Warner Corporation was
born on June 5, 1928. And just months before the great
market crash of 1929 the confederation was essentially
complete. But from the beginning, none of the company
heads were willing to submerge the identities of their
own companies. Instead of agreeing to an outright
merger, they established Borg-Warner as a holding company. Each participant maintained its own identity and
operating method. The "central office," as it was
referred to, merely furnished the financial glue. By
the way, the name "central office" persisted until the
early 1970s when it was finally accepted as "corporate
headquarters."

Well, the attitude, the culture, the very
organizational fiber of Borg-Warner was formed right
there at its birth. It was passed down over five or
six management generations. Organization as a loose
"confederation of tribal fiefdoms," as it has been
called, was tested and reinforced by many very
successful years. After World War II the autonomy
concept was supported even more by diversification--
first within the manufacturing area and more recently
in the services area.

Much of this broadening of business interests came
from acquisition--acquisition of businesses and
acquisition of management. Our approach to acquisition
has always been friendly rather than hostile. So in a
way the management in each acquisition generally had an
option to join or not join the "alliance." Just like
those early diners in Chicago, they also savored the
idea of sharing financial strength without giving up
their autonomy. Again the tradition grew and streng-
thened. The Borg-Warner cultural steel was being
tempered.

By the late 1970s Borg-Warner consisted of the
following seven quite diverse business areas totaling

over $4 billion in consolidated and unconsolidated revenues:

Air Conditioning

Chemicals (plastics)

Automotive

Industrial

Financial Services

Protective Services

Affiliates

Services accounted for almost one-third of Borg-Warner income and the goal was (and is) 50/50 services and manufacturing. As the 1970s unfolded, it became increasingly obvious that the organization was fuzzy--some were sure we were, or ought to be moving toward more centralization; others thought it should be more decentralization.   There were concerns: Were we a holding company or an operating company? Should we be a holding company or an operating company? What's the role of corporate headquarters and of corporate staff? Was there, or should there be, an overarching set of Borg-Warner values or beliefs?

BELIEFS

For some time in the 1970s our chairman, Jim Bere, had been thinking about issuing a statement of beliefs. he said recently:

> I didn't have the nerve to do it in the mid 70's. Some in the company felt it was stupid because some actions in the company couldn't measure up to a standard of ethics. Some felt it would detract from the business of running the company.

Well, by 1981, Bere was certain that regardless of any past mistakes, a norm was essential. In May 1982, all Borg-Warner senior managers (about 100) and their spouses were gathered at a chairman's meeting. It was only the third time such a gathering had ever occurred to discuss long-range issues. Every topic of the meeting pointed to and climaxed in a discussion and publication of "The Beliefs of Borg-Warner."

176 James R. Deters

Actually, the development of this beliefs statement was really the fruit of ten months of intensive work. It involved soliciting views from management throughout the company regarding what the current beliefs were--or ought to be. Outside help was used including a business philosopher at a well-known university. He helped us find an historical spring-board for each of the beliefs. At the close of that meeting here's what Bere said to the group:

> For some time, Borg-Warner has been a company in search of identity. I am not sure that during most of the half century of our history we ever really ourselves questioned what we stood for, what needs we fulfill, or where we were going. But before we can know surely where we are going, we must first know where we are.
> So we are beginning a long re-examination. Why now? There are three reasons: (1) Because right now we are at a turning point in a long and massive transition in this company; (2) because in a sea of violent change, guidelines are vital to avoid being swamped; and (3) because profits and values are inseparable.

I'm not going to discuss the beliefs in detail. Here they are in summary form. To reach beyond the minimal we believe in the following:

1. Dignity of the individual

2. Responsibility to the common good

3. Endless quest for excellence

4. Continuous renewal

5. Commonwealth of Borg-Warner and its people

PRACTICE VS BELIEFS

And for anyone wanting to study a little more about the beliefs development, a Harvard Business School case study has been written on the subject.
So how are we doing in institutionalizing the beliefs? Well, fair--just fair. If we judge by quantity of media, i.e., brochures, posters, etc., progress would be excellent. But, you know, genuine acceptance and practice is quite another thing. I would say we've gone through the media stage, and it's increasingly clear to our managers that we're very serious about implementing the beliefs. I'm convinced that those

Borg-Warner managers who do not integrate these beliefs into their actions will be rejected over time. We're building beliefs discussion forums into all internal training programs and identifying them in the criteria for selecting high-potential managers--our senior managers of the future.

But something else is happening that impresses me a lot more! And that's the increasing challenges coming from employees at all levels, challenges about our practices versus our beliefs (and being in human resources I have heard a bunch of them). I guess it's my controller orientation but having these standards to measure against--however uncalibrated they may be--is a darn good start to telling you what needs fixing. We must be willing to deal with questions of "How can you do _this_ and say you believe _that_?" But we're not pollyannish about the difficulty of how such principles can be meaningfully shared in the large diversified and decentralized organization I've described. In my opinion, at least a generation of management or ten years will be required to complete this process. If Moses were alive he'd still be working on the Ten Commandments. So we'll not be discouraged. We're convinced that the sharing of common values allows operating personnel to still make difficult decisions without looking to headquarters, and in reality, this sharing will give power to a broader base of management. In a very real sense, sharing beliefs, to us, means sharing power.

ORGANIZATIONAL CLARITY: CENTRALIZATION VS
DECENTRALIZATION

I'd like to share another related change we've been grappling with. It more specifically addresses the issue of organization centralization and decentralization.

By the early 1980s it was also abundantly clear that the Borg-Warner "commonwealth" needed more than shared values. It also had a real mental tug-of-war about organization. The ebb and flow of centralization and decentralization over the years and the firm roots of autonomy had created a pretty unclear picture of the relative roles of corporate headquarters and our various businesses. To quote from a speech I made to our policy and planning committee in 1982:

> Over the years all of us have been strongly committed to decentralization in Borg-Warner. We know in our bones it's the right way to manage such diversified businesses. But we've had a fuzzy perspective of what decentralization really

means. That is, what decisions should be and must
be made by corporate management, and what limited
involvements in our businesses should corporate
management be concerned with to monitor its
responsibility to its constituencies. A rightful
zeal about corporate bureaucracy and decentralization has made us overly reluctant at times to give
direction and to establish principles, to write
things down or somehow clearly communicate them.

Well, for over a year we held small meetings
between the CEO and other members of top management. I
initiated the session, acted as scribe, and as best I
could, referee. Terms like "holding company" and
"operating company" were examined intensely to reduce
semantic differences. A task force of group line officers was asked to digest all these discussions into a
"position paper." In summary, that position paper set
out the following operating principles:

1. It affirmed a strong commitment to individual
    autonomy of our business units so long as there are
    not material exceptions to agreed performance
    targets.

2. It affirmed that, while Borg-Warner is multi-
    industry in structure, a commonality of values as
    expressed by the corporate beliefs must exist
    substantively across the company.

3. Four additional areas of corporate-level concern
    and involvement were identified: strategic direc-
    tion, management quality and succession, financial
    controls, and management compensation.

Within this framework, group heads were to conduct
their business very much like CEOs for freestanding
businesses. Supporting staff functions were to be
primarily in the group and not at corporate headquarters. Corporate staff functions would be limited to
law, finance, human resources, communications, and
corporate development (planning, acquisitions, and
investments). To reinforce these concepts, we began
organizing the separate businesses as subsidiaries.
The corporate policy and planning committee would act
as a board for each subsidiary. (The policy and planning committee consists of the CEO, President,
Executive Vice-Presidents and the Vice-Presidents of
the staffs.)
These concepts have been in place now for about a
year and a half. I can't say there's complete agreement or even understanding. It's still evolving and
will be for some time. We are now stepping back and

re-examining the structure and process. We want refinements but we also want to keep the change process under way.

Revising our logo last year was a subtle part of that change process. It was done probably more to send a message inside the company than outside, a message that we're looking to the future and, as always, it's different.

CONCLUSION

The two and one-half years I've spent immersed in the human resources function have been a real blessing and I've loved every second of it. But let me conclude with two points I've tried to share with you through these short stories.

First, I think organization clarity is pretty important, not the false clarity that comes with rigid, detailed organization charts that draw little boxes around jobs--and around people. And, not the clarity itself, but the seeking of it through open, continuous and evolutionary dialogue, the process of communicating. It's really a constant process of change that examines and re-examines the structure as reflecting the reality of the operating style.

Second, a true transformational change must be based on strong and thoroughly communicated strategies and values. The task is, perhaps, to establish a climate where change can be absorbed in an orderly way rather than forced through policy, procedure, and directives. Persistent patience seems to be the tactic here. Older management generations are not going to institutionalize the change nearly as effectively as the new wave of younger managers. Doesn't that suggest the need to take a new look at management development through "education" versus "training," and to emphasize affecting attitudes as well as skills?

Institutionalizing open two-way communication rather than a one-way media-oriented approach also seems essential. To me, articulating goals, values, and beliefs and writing them down for all to see, to share, and to challenge is, possibly, a critical ingredient in a strategic orientation.

These two points have been the principles behind the experiences in developing beliefs and organization clarity. If Borg-Warner and other companies are good places to work, I suspect it's because of what they do and because of what they are trying to do.

I want to make one final point. It seems to me that the human resource function is just crying out for leadership. You know there are profound changes under way. The traditional management model of establishing

order, of exercising control, and of achieving efficiency in the work force just isn't working. This also
means profound opportunity in the human resources function. Frankly, if I were in college today I'd give
very serious consideration to the area as a career.
It's possibly the hot spot for the next decade or two.
Those already in Human Resources have enormous
opportunity to impact their institution if they:

1. Think strategically, moving from a technical to a
    cultural orientation.

2. Provide their management with a strategic framework
    in simple terms.

3. Avoid quick fix recipes such as trying to open a
    package of theory "Y," mix with Maslow's hierarchy,
    season with job enrichment and participation, and
    serve with quality circles. Also, I've seen a lot
    of "one minute managers" searching for excellence
    in a copycat list of activities. I've never seen a
    "one minute leader"!

4. Convince management and themselves to quit thinking
    of human resources as assets only in the financial
    sense.

    To paraphrase John Naisbitt--for human resources
people that see change as opportunity, it really is a
"fantastic time to be alive."

Part III

MANAGING INFORMATION SYSTEMS AND AUTOMATION

19

Overview

Charles M. Lutz

Modern information systems in business are in a state of flux as they continue to evolve as a major productivity tool. Rather than being a singular monolithic entity, modern information systems are a collage of disjointed and often competing subsystems. This conflicting and competitive nature of information subsystems detracts from the overall productivity of the information system and the firm, but it is a direct result of the evolution of modern information systems.

At least four different layers of information systems and six different subsystems have evolved (see Figure 19.1). The lowest level and the first to evolve was transaction processing which mirrored the physical system of the firm. This is the "traditional" data processing function that was used primarily to speed up tedious and time-consuming tasks, such as inventory control and payroll. Such functions still constitute a high percentage of data processing programs. The transaction processing level deals with structured problems, known parameters, mostly internal sources of data, and a high level of detail. This level emerged as a separate entity, highly controlled, and staffed by data processing experts using sophisticated hardware and software. This level still constitutes the bulk of information systems in most businesses.

The second layer to evolve was built on the transaction processing level and was designed to provide management with summary data for control and decision making. This layer supports the operational and tactical levels of management and was originally conceived as the "management information system." The MIS level deals with less structured problems, some unknown parameters, and a detailed summary of mostly internal information. Unfortunately, when this level emerged in the 1960s, the software was not available to deliver the promised results. The manager still had to rely on the data processing staff for writing programs and producing the needed information. It was and is

Figure 19.1 Business Information Systems

User	Problem & Time Frame	IS Level	Software	Integration
Top Management (Strategy)	-Unstructured -Long-range (2-5 years)	Strategic Information Systems	Artificial Intelligence	Telecommunications & Automated Office Systems
Middle Management (Tactical)	-Semistructured -Short-range (1-2 years)	Decision Support Systems	Decision Support Systems	Telecommunications & Automated Office Systems
Line Management (Operational)	-Structured -Current (Quarterly)	Management Information System	Third/Fourth Generation Languages	Telecommunications & Automated Office Systems
Support Staff	-Highly Structured -Immediate (Daily)	Transaction Processing	Third Generation Languages	Telecommunications & Automated Office Systems

not uncommon for a backlog of months, even years, to exist. Reports were inflexible, often not in a readily usable format for decision making, and usually out of date.

The third layer really began to emerge with smaller, more flexible hardware and more sophisticated software. This layer, the decision support system, also was designed to support the operational and tactical levels of management. This level is coming into its own with microcomputers and distributed data processing. Decision support systems are usually user controlled and deal with semi-structured problems and both internal and external sources of information. Decision support systems are very software dependent for their success and are improving as users are better able to define their requirements and to quantify the variables affecting the decision.

The fourth layer supports the strategic management function and is just beginning to emerge. This level deals with unstructured problems, long-range future projections, and primarily external sources of information. Strategic information systems demand a high level of expertise to determine strategic goals for the firm. Artificial intelligence, particularly in the fields of expert systems and natural language systems, is being applied to the strategic level with varying degrees of success.

A note of caution is appropriate; the levels discussed above are neither universally recognized nor agreed upon. Moreover, technology and emerging software cut across layers with little regard for our organizational charts or figures. For example, robotics is generally regarded as part of artificial intelligence but is being introduced at the physical level of the firm (e.g., GM's manufacturing automation protocol, or MAP) and is mirrored at the two lowest levels of information systems. We are truly dealing with a collage of interwoven subsystems. The six subsystems of modern information systems include the four levels already discussed (transaction processing systems, management information systems, decision support systems, and strategic information systems) and telecommunications systems and automated office systems. The last two subsystems are integrative in nature and cut across all four IS levels.

Telecommunications, to include its subset data communications, is the integrating force that has the potential to tie all four IS levels and the other subsystems together. Local area networks and wide area networks can link the detail of the transaction level to the decision support system's summarization and extract facility to provide the decision maker with the most accurate, timely, and complete information.

Automated office systems, sometimes called the electronic office or the office of the future, also have the potential to be an integrating force within the firm. Office systems exist at every managerial, professional, and staff level. Through integrative software, the automated office workstation can merge and summarize data in the form of spreadsheets and graphics, with text, and produce a desk-top printed report in a few hours. The integrative power of automated office systems and telecommunications is just beginning to be exploited.

MANAGING INFORMATION SYSTEMS

In Part I, Dr. Rosabeth Moss Kanter summarized the impact of the information system's evolution in one word, change. Managing information systems demands a generalist who understands the business as well as the technology and who can deal effectively with people as well as with machines. In our lead article, Des Cunningham of Gandalf Corporation discusses how a firm can exploit change and increase productivity through information technology.

There are at least three reasons to invest in information technology. The first is to attain a competitive advantage in the marketplace. Second, the

competition may be the driving force and we are playing
catch-up, a matter of survival. Or, last, it may be
the right thing to do, it is effective. A major problem is quantifying the discrete costs and benefits of
such a productivity investment. According to
Cunningham, there are several requirements that should
govern any investment in technology. These include, a
comprehensive information systems plan; integration of
software, hardware, and communication technologies;
information sharing through a corporate data base; and
applications solutions that interface with each other.

Cunningham then discusses how Gandalf implemented
and managed their information system, an integrated
Manufacturing Resources Planning (MRP) system. It is a
complete system having most of the components shown in
Figure 19.1. The key is the telecommunications network. Gandalf's integrated network includes: public
packet switching networks, private leased lines, and
private satellite communications. Telecommunications
networks must provide security, reliability, flexibility, and adaptability. The inclusion of the telecommunications network as part of the information system
is essential if productivity is to be increased.

Wayne L. Hanna of McDonnell Douglas discusses
Manufacturing Automation Protocol (MAP) and Technical &
Office Protocol (TOP) as ways of increasing productivity by dealing with today's pressures while preparing
for tomorrow. Factory automation has been stalemated
due to industry's failure to communicate. Hanna states
that this failure has created "islands of automation"
within our factories, developed piecemeal and incompatible within the corporation as well as with the
environment.

Productivity demands that these islands be
integrated and integration requires standards. The
International Standards Organization has developed the
Open Systems Interconnect (OSI) model for standardizing
data communications networks. The OSI model is based
on seven layers of protocols from the physical (or
lowest) layer to the application layer. The model
functionally defines the characteristics of each layer
for compatibility in exchanging data on different computer processors. Unfortunately, international protocols have only been established for the three lowest
layers. Protocols in the higher layers are often
proprietary and usually incompatible within the
industry.

The concept of MAP/TOP was an industry effort to
achieve the basic objective of interoperability and
data compatibility. MAP's success depends on wide
industry support. GM took the lead in factory MAP and
Boeing took the lead in office multi-vendor networking.
Hanna describes the progress on forging international

standards to date. To assist in forging these standards, a consortium of seventeen vendors proposed the
formation of the "Corporation for Open Systems," a
private company with membership available to both user
companies and vendors. Support for standards has been
driven by the need to improve productivity within our
factories and improve our competitiveness with the
world.

However, as Robert Rubin points out in his
article, improved productivity demands quality in
information systems. Quality is subjective. As John
Guaspari concludes in his delightful modern fable about
quality, "I Know It When I See It." For the more
precise, Robert Rubin offers metrics, or ways of
measuring quality, for information systems. He
observes that all metrics reduce to the same measurement in the eyes of the client: have you helped more
than you cost?

The computer is an excellent tool for improving
productivity. Rubin discusses using information
technology for simplifying the organization, for
increasing communication, and for controlling the
resources of the firm. Quality is defined by the
customer, the line management of the firm.

Duncan B. Sutherland takes a different approach to
managing quality in information systems. He believes
that the key lies not in technology but in the very
nature of the office. Increasing the productivity of
individual information workers does not necessarily
result in an overall increase in productivity. Managers know, even if they will not admit it, that synergy
is a myth. In real life, not only does 1 + 1 NOT equal
3, most of the time it doesn't even equal 2! Pure
mechanization has not had the same effect in the office
that it did in agriculture or on the factory floor.

Sutherland states that it is knowledge, not
information, that's critical in sustaining a competitive edge. And knowledge resides in the human
resource, not technology. Therefore, Sutherland proposes using technology to accomplish three objectives
for the firm. First, maximize the intellectual contribution of individual knowledge workers. Second, minimize the number of people a company employs. And last,
eliminate unproductive capital investments in
technology. To be successful, management must find new
ways of organizing knowledge workers and technology to
sustain an intellectual advantage.

MANAGING OFFICE AUTOMATION

Ronald P. Carzoli discusses the second integrating
subsystem, the electronic office. Office automation

has progressed through three distinct stages. Phase I, which began in the 1970s, emphasized efficiency in the traditional office activities. Phase II, which began in the early 1980s, emphasized effectiveness by using technology to change and expand the traditional office activities. Phase III, which is just beginning, uses technology in office automation to fundamentally change the way we conduct business.

These fundamental changes are reflected in flatter organizations, enriched and enhanced jobs, improved relations with employees and customers, and increased productivity at all levels. According to Carzoli, several lessons were learned during the early phases of office automation. First, you fit the technology to the organization rather than trying to change the organization to fit the technology. Second, the best way to improve office productivity is to prevent interruptions. And finally, the human resource is still most important for any increase in productivity. He concludes that as the electronic office continues to evolve, two things are vital: that we manage technology to become increasingly competitive; and, that we recognize the impact that this new technology has on our human resources.

Rodney D. Becker of Control Data continues the theme of the importance of the human resource in the electronic office, or the office of the future. He states that the place where people, technology, and work environments merge is in the design of the job. Thus, as technology evolves, the fundamental nature of the job must also change or the office of the future will look like the factory of the past. Becker argues the management must take responsibility for job redesign, facility design, and appropriate organizational design for office automation to realize the productivity goals set for it by the firm.

Job redesign may also be the vehicle to solve other emerging issues such as health and safety concerns. Stress and other potential health hazards can be reduced by designing task variety into jobs. Task variety can also make a job more challenging, more rewarding, and, ultimately, the employee more productive. At CDC, it was realized that without the synergistic effect of combining of telecommunications, office technology, facilities, and the human resources, improvement in productivity was virtually impossible. Becker summarizes by saying that in integrating technology into solving business problems, we must not forget the organizational dynamics and, most importantly, the human factor.

MANAGING INNOVATION

Dagnija D. Lacis moves us into another information subsystem with her discussion of decision support systems based on fourth generation languages. Fourth generation languages can dramatically speed development of complex application programs and provide better information to decision makers. She defines fourth generation languages as software packages designed to facilitate development of application software and which can be mastered by non-technical staff. Systems analysts can use fourth generation tools to develop prototypes to increase user participation in the development of realistic system requirements. Prototyping allows end users to try working models or applications systems before the development of the final product.

Lacis also describes the benefits of using fourth generation tools. First is the reduced application development and maintenance time and costs. With some organizations spending over 80 percent of their programming effort maintaining old programs, this benefit should significantly decrease costs. Second is the better communications between user and information systems departments. The most difficult job of any systems analyst is correctly defining the needs, implicit and implied, of the user. Third, information systems can be adapted more easily to the dynamics of the business environment. And, finally, the corporate investment in application software will be better protected.

Richard L. Chappell of Arthur Andersen discusses the highest level of information systems, strategic information systems. He suggests that the key to a successful strategic information system is gaining a sustainable competitive advantage. Information technology is being used as a strategic weapon by more and more organizations. Business is being confronted with competitors who are aggressively investing in a proliferation of new information technologies.

Arthur Andersen, working with Professor Michael Porter of Harvard, believes that information technology can assist in developing that sustainable competitive advantage by altering the industry structure and shifting the competitive balance in your favor, by improving your relative position within the existing business lines, and by creating entirely new business opportunities. Information technology is unfreezing the structure of many industries and altering relationships within the firm and the environment.

Chappell also discusses emerging trends in information technology and makes recommendations for the future. The emerging trends are dictated by the decline in the cost of hardware and the concomitant increase in power. A result is a multi-tiered network comprised of both centralized and distributed information processing. The second trend is the continued proliferation of electronic work stations, projected to be one for every white-collar worker by 1990. The primary recommendation to deal with these trends is to establish a chief information officer within the firm to act as a catalyst to bring about the effective use of information technology.

Michael A. Brewer of Unisys gives us further insight on how to manage creative information programs. Change represents the ultimate challenge in that it never fails to extract greater creativity from us. The key question is how to manage change, with an iron fist or with kid gloves? This question is central in information systems management since information systems are often central to and the arbiter of change. Brewer suggests a little of both, a firm hand in a kid glove. This takes a certain amount of detachment. For example, the innovative manager has to be firm and objective enough to overlook the elegance and creativity of a design proposal if it does not solve a problem. Even creativity must be businesslike.

The pace of technological change has considerably shortened information technology life cycles. For example, a mainframe computer has a product life cycle of twenty-four to thirty-six months. A workstation's product life cycle is only twelve to twenty-four months. Brewer points out that to sustain an existing product, it must be performance upgraded annually and technologically upgraded every two years. The market window for a product is very short and elusive. Product development tends to be asynchronous, a high-powered effort driven by changing market demands. A six- to nine-month lead over the competition may equate to half a product lifetime.

The information systems manager then must begin with an agreed-to, thorough business plan. The plan should be open and written down, "owned" by the staff, flexible, and continually updated. Planning for information systems must include financial aspects, to include what-if scenarios. As Brewer states, the key is to welcome change and manage it, with a firm hand in a kid glove.

In the final article in this section, Michael Connors of IBM projects our discussion into the future. He first discusses three drivers of change: a transitory social-economic structure, information technology, and communication integration. As already noted, the

social-economic fabric of the United States, and of
much of the world, is changing dramatically with an
increasing proportion of the workforce being knowledge
workers. Further, the rate of technological improvement, as measured by the price performance ratio, is in
the range of 20 to 25 percent annually and is projected
to remain that high until the end of this century.
This means the costs for increased capabilities will
continue to decrease. Finally, media integration
(data, text, image, and voice) has already begun.
Text-to-voice is already here with voice synthesizers.
The complicated part is voice-to-text. Connors
predicts voice-to-text will become a reality in the
1990s.

These drivers of change present information
systems managers with several categories of business
opportunities. Opportunities include: office information systems, mechanical design automation, and artificial intelligence. Office information systems will
continue to be a major integrating factor with an
electronic workstation on every desk. Increases in the
quality and quantity of work produced should translate
into increased productivity. Mechanical design automation, especially three-dimensional modeling, can
assist engineers in such vital areas as tolerance
design and help reduce manufacturing cycle time.
Building automated prototypes rather than machined
prototypes can be a key ingredient in maintaining a
competitive edge in business. Finally, artificial
intelligence or knowledge based systems have the potential for giving a larger portion of our population an
increased degree of computer and technological
literacy. AI can address the unstructured contextual
problems dealt with by the strategic information
system. The drivers of change create new uses, new
users, and most importantly, new business challenges
and opportunities.

CONCLUSION

A modern business information system, modeled
after Figure 19.1, can effectively reduce the separate
"islands of information" and provide timely, accurate,
complete, and relevant information to the decision
makers. Better decision making increases efficiency,
doing things right, and effectiveness, doing the right
things. The end result is an organization that uses
all its resources more effectively and thus is more
productive. A modern, integrated business information
system not only can help solve today's problems, but
offers the potential for providing the creative and

innovative resource needed to deal with tomorrow's
opportunities.

20

Exploitation of Information Technology

Des Cunningham

It's a pleasure to be here to share my ideas about
productivity with you. Gandalf, the company I co-
founded sixteen years ago, is in the productivity business. We supply a large range of network services to
end-users including data communications equipment,
voice interconnection, network design and planning,
applications solutions, and customer training. There
are 1,400 people employed worldwide by Gandalf and its
subsidiaries. In our fiscal year, ended July 1986,
annual revenue reached $108 million Canadian. This was
a 25 percent growth of which I am especially proud
considering the tough year in general for the data
communications and computer industries in the United
States. This year we will accomplish another 25 percent growth to at least $130 million CAN (or $100
million U.S.). We have had profit increases in each of
the last eight quarters--so we're doing something
right.

Today, I'm going to discuss productivity
measurement in general; list several requirements for
investment in information technology to ensure it
promotes productivity; and, explain how we've applied
these concepts within Gandalf to run our own business.

First of all, let me say that we believe in
information technology's impact on productivity and we
actively use it. I'd like to say we're a perfect
example. However, I'll be honest with you and tell you
that we're only working toward achieving our overall
plan.

PRODUCTIVITY MEASUREMENT

What is this idea called productivity anyway?
It's used as a magical word in some quarters to justify
million dollar computer systems as well as $2,000 per-
sonal computers. In my business, productivity is a
term that has tended to relate primarily to the

manufacturing floor. That's where it can be measured
and justified most accurately. For instance, we are
automating and mechanizing our assembly facilities by
upgrading machines and investing in robotics, in-
circuit testers, inline mechanical transport systems,
and are considering automatically guided vehicles. We
are just now implementing in our Canadian plant a
factory information system, using bar codes, designed
by one of our subsidiaries, which allows us to more
efficiently control the status of work in progress.
These investments increase productivity which leads to
decreased costs and increased profit margins. But if
it is done well it can lead also to better response to
our customers and hence increase revenues as well.

For these types of investments, productivity
benefits are relatively simple to measure. Inputs and
outputs can be quantified. In other departments,
certain clerical functions can also be measured.
Junior buyers in purchasing can be measured on the
number of requisitions expedited in a day. The number
of keystrokes per minute and the error count can be
used to measure a word processing operator. The number
of calls handled by the switchboard operator in an hour
can be tracked, and so on.

However, the same types of simple measurement of
quantity are not applicable to white-collar or knowledge workers. Justifying a laptop personal computer,
a local area network, voice mail or a laser printer is
not as simple a process. Will a manager who has a PC
with a 386 processor make better quality decisions or
just generate more and perhaps poorer quality decisions
at a faster rate due to the overload of data that is
now available? There are hard costs involved in these
investment decisions, and at the time, they are compared to "soft" benefits. Executives, managers, and
professionals are today's knowledge workers. They
collect, analyze, interpret, synthesize, amplify, and
distribute information to perform an unstructured job.
Measuring the impact of information technology on their
productivity is a difficult but not always impossible
task. For example, an accountant analyzing slow receiv-
ables establishes some correlation between our shipping
and billing process and customers delaying payment.
Corrected shipping procedures rectifies the problem and
improves cash flow. Local processing of orders in
sales offices leads to a better response to customers
and encourages additional business.

Quality engineers analyzing fault reports identify
and correct problems that have been costing the company
many thousands of dollars per month. However, given the
high expense of this employee group and the increasing

percentage of the labor force they are predicted to become, quality measurement is an issue that needs continued study. We invest millions of dollars in technology for these employees without always being able to measure the real benefits or payback except in general or global terms.

REQUIREMENTS GOVERNING INVESTMENT IN INFORMATION TECHNOLOGY

Why do we undertake and continue this investment? In my company, we have several reasons. First, to attain some form of competitive advantage. The competitive advantage may be as a result of lowering costs, delivering faster, increasing product or company differentiation in the marketplace, or creating new products, services or markets as a result of marrying information with innovation. Second, we do it because the competition has already done it. We may be playing catch-up to try to cash in on the benefits our competitors have realized. Or, it may be a matter of survival in that particular product or marketplace.

Third, it may seem like the right thing to do. Call it vision, intuition or gut feeling. If we don't try new ideas we will never know. After the trial is over, results can be measured and effects examined and we may find out that the idea bombed and created havoc or, now and again, that it was hugely successful. It may have been successful for reasons not even originally considered. I believe we should always attempt to measure and justify the expected costs and benefits of a productivity investment decision because experience shows that such investigation teaches us something new about the process of interaction between individuals, departments and systems. However, it's not always possible or as clear cut as I'd like and it's often necessary to take the risks and plunge in head first.

In Gandalf we are committed as a company policy to investing in information tools and technology that will improve the productivity of all our staff. We have dramatically increased the use of information technology so that employees have access to tools that will assist them to more effectively perform their jobs and provide better service to our customers. I believe that for an organization to succeed in using technology to improve productivity it is vital for the top managers to accept and make clear that this commitment is a policy and not just a series of opportunistic adventures.

Generally, there are several requirements that govern investment in technology. These include the need for:

1. An overall information technology management plan

2. Integration

3. Information sharing

4. Information accessibility

5. Application solutions

Corporate, departmental, and individual requirements must be satisfied by means of an OVERALL INFORMATION TECHNOLOGY PLAN to establish that the objectives of information management are allied to the business purposes and goals of the organization. Planning studies must show that integration of systems and data sources is feasible regardless of physical location. Multi-vendor hardware, software and communication technologies can be INTEGRATED to supply a strategic corporate solution. New technology must interface with the old to protect previous investments. No one vendor can effectively satisfy a corporation's total information technology needs. There is also a significant risk involved in depending on one vendor to supply all solutions--you get locked in. The need for integration of networks and sources will only continue to grow in importance in the future since this activity is the one that leads to the creation of new knowledge that will be a real competitive asset.

Information needs to be SHARED across the corporation instead of each department having its own independent data source. Not to make information available across departmental boundaries is a certain negative when it comes to productivity. Managers of different groups should ACCESS the same data and be able to manipulate it as needed for coordinated decision-making purposes. Easy to use tools must be available so that users are able to query, manipulate and perform "what if" analysis as desired. Information must be accurate and timely. Users must be able to get at the data regardless of physical location or storage mechanism. APPLICATION SOLUTIONS are relevant to the hardware, software and communications aspects of the technology. However, total solutions and a grand strategy are required which enable individual application solutions to recognize and interface with each other. For example, order entry, scheduling and shipping releases, billing, sales commissions, customer histories and market analysis are all separate but related application solutions using much of the same data.

Recognizing that customers who require information systems for productivity reasons will also be looking for complete solutions to certain applications, Gandalf has invested in three business areas that satisfy our criteria for business expansion and I would like to mention these briefly. First of all, we have a division that manufactures and supplies factory information systems that, for a low cost investment, enable fairly complex manufacturing operations with many workstations to input data relating to work undertaken and to extract data regarding work yet to be done at the workstation. This technique involves bar code reading and we have adapted a product normally used for the apparel industry for use in our own electronic manufacturing facilities whereby workstation data can be entered directly into our Manufacturing Resources Planning (MRP) system. This greatly helps in controlling work-in-progress and all levels of inventory management.

A further area of application solution is the Voicestation system which we sell to improve productivity in high-level and busy office environments where it is advantageous to make use of voice mail and voice-annotated text in order to enhance communications between busy executives. Users of this system can readily identify productivity gains arising from better communication.

Finally, we sell dispatching systems to taxi cab companies who virtually run their businesses from our system. Taxi cabs are equipped with small video displays and communication pads that receive and transmit data through a standard radio set to a central dispatching computer. The taxi cab owners estimate that it is possible, over a one- to two-year period, to double the size of their business by these means, and clearly the productivity of the cab itself and the drivers are greatly enhanced while at the same time giving the public a better taxi service.

GANDALF'S IMPLEMENTATION OF INFO MANAGEMENT
REQUIREMENTS

How has Gandalf addressed these requirements internally? Currently, over two-thirds of our 1,400 employees have accounts on the corporate network which includes eight mini or mainframe computers, the largest of which is a DEC 20/60 at this time. Users access any one of a number of application programs that are available to support specific departmental requirements, corporate requirements, or general office needs. Some of these applications are tied together while others

are not. A project is under way now to replace the
stand-alone applications that should logically be
integrated so that additional benefits will be
realized.

Three years ago, we invested in the implementation
of an integrated MRP system. The on-line system is
used by our three manufacturing facilities to schedule,
manage, and control all activities associated with the
production of over 400 end-products and thousands of
component parts. Since the facilities are located in
Canada, the United States, and England, timely coordination is essential to ensure on-time external and
internal deliveries of quality products and parts. Our
major vendors dial-in to our system to track purchase
order requirements for their specific products. This
has resulted in significantly improved on-time delivery
by those vendors. This is beneficial from our viewpoint because we can carry less inventory, expediting
effort is reduced, and the parts are on the floor when
they're needed which minimizes rescheduling of
production jobs.

Effort during the coming year will be directed
toward the integration of our computer-aided engineering and computer-aided design systems to the MRP system
to result in a faster transfer of accurate and quality
information between systems. We need to tie together
these "islands of information" much more effectively
than today. The connecting manual entries that are
presently required due to the lack of integration
result in delays and the introduction of errors.
Faster movement of information is needed to support the
large number of new products and enhancements released
every month.

The users in the sales organizations access the
central computers for on-line order entry and order
status checking, determination of product availability,
sales forecasting and historical customer information.
Orders are entered directly on-line from over thirty
sales office locations throughout Canada, the United
States, England, Europe, and Australia. During the
next few months, we will be providing automated support
to our sales reps for territory and account management,
proposal preparation, lead followup and sales call
reporting. We hope to reduce administrative paperwork,
improve sales presentations and provide quicker
responses to our customer requests.

In a manner similar to our vendors, major
customers will be able to enter orders from their own
site without having to contact a sales rep over the
telephone to place the order. Distributed computers
will be installed in Canada, the United States, and
England to support automated service call handling and
dispatching for customers requesting field service

support for problems, installations, or routine maintenance. These changes will enhance our already excellent reputation for service and we hope it will further differentiate Gandalf from the competition. Internally, they will improve the productivity of our service technicians and minimize our field inventory investment.

Competitive advantage can also be achieved by investing in expert system technology. This is an excellent example of an investment that will improve internal productivity by capturing knowledge from a limited number of subject matter experts for use by generalists. Network configurations and problem diagnosis are good application examples. However, it's almost impossible to justify the cost with concrete figures or savings. Intuitively, I know it will be extremely beneficial once the systems are developed and operational and our staff has learned to use them.

The finance users access systems for the processing of accounts payable and receivable; general ledger reporting; standard cost, variance and gross margin analysis; fixed asset management; and employee payroll information. As in most organizations, finance was the earliest user of electronic data processing output. The systems they are using today are similar to previous systems. There is one significant difference, however. The information that is provided today is more timely and comprehensive allowing real-time decisions and the establishment of priorities for management actions of all kinds.

The hardware and software engineers in product development maintain all product and parts information on our corporate system ranging from part specifications, quality standards, and manufacturing tolerances, to customer documentation, software quality procedures and engineering changes. This ensures that only approved parts are ordered and subsequently accepted on the receiving dock, that the correct configuration of system is manufactured and delivered against customer orders, and that modification levels of hardware and software are known by all concerned, particularly including the staff responsible for installation on the customer site.

In the office automation area, Gandalf has been a longtime proponent and user of electronic mail systems. In an average week, our users generate a total of 10,000 mail messages (from 1,400 total employees). This represents a significant amount of communications activity. Electronic mail and document transfer is especially important within Gandalf due to the numerous time zones in which the dispersed facilities are located. Communication is facilitated that may otherwise not occur; or occur with phone tag delays; or,

heaven forbid, be pursued with typewritten memos sent
through the traditional mail system. Although we use a
private mail system, there are gateways to public mail
networks to facilitate communication with customers and
vendors.

Support for word processing and other areas of
office automation are now addressed primarily through
the use of personal computers. My sole criterion for
selecting word processing machines has always been that
they must communicate through our communication network. Recognizing the importance of information sharing, we've established corporate standards for word
processing, spreadsheets, communications and data base
management so that sharing among users is facilitated.
We've started to use in-house Desk Top Publishing which
promises cost savings and productivity improvements.
We're using it for the creation of product manuals and
documentation, training materials, customer proposals
and internal newsletters.

The most essential aspect of the technology which
makes these information systems beneficial to us is our
PACXNET communications network. We have an extensive
telecommunications network that links every sales and
field service branch office, manufacturing facility,
technology center and departmental location with the
corporate computing center. The network is essential
to the effective sharing of information by employees
regardless of location. Our goal is to make our worldwide decentralization transparent to our users. PC-to-
PC communication is easily supported as well as access
to external new services and data bases. Without the
ability to communicate, many of the productivity benefits of information technology would be nonexistent.
The computer would be only a personal productivity tool
rather than the corporate strategic tool it can be
today.

Telecommunications has strategic implications for
every organization given the global economy in which we
exist. The ability to communicate depends on the existence of a reliable, flexible, adaptable and cost-
effective telecommunications network. Airlines, the
banking industry, governments, universities and manufacturers recognize the importance of networks to their
business. At Gandalf, we believe it's critical to
communicate with our customers, vendors, sales force,
distributors and remote office personnel in a timely
and effective manner. Our network is the key. Gandalf's
network utilizes: public packet switched networks;
private leased lines; and, private satellite
communications.

The remote sales offices use the public X.25
network to access corporate computer resources or other
offices. Leased lines are installed between multiple

offices in the same city or country. The U.S. facility located outside Chicago is linked with Ottawa via satellite.   In October 1986, Gandalf became the first corporate crossborder satellite customer of Telesat Canada.   Expanded use of various applications will be promoted using our network such as audio and video teleconferencing, electronic blackboards and improved integration of facsimile and telex.

Requirements that are essential for all aspects of our communications network include: security; reliability; flexibility; and, adaptability.

SECURITY features are a must for the network as well as the computers and software.   The majority of computer crime is committed by someone within an organization rather than outside the company.   Information is a corporate asset and must be safeguarded. Equipment and service RELIABILITY is essential due to the impact information technology has on our business. Service must be available under even the most adverse conditions when Murphy's Law is most likely to be tested.   The technology must be FLEXIBLE to meet the needs of different user groups within the organization to minimize double investment.   Accommodation of multiple solutions must be feasible without a significant amount of extra time or cost.   Finally, the technology must be ADAPTABLE to today's fast-paced changes.   The ability to upgrade without totally obsoleting existing technology is important.   Upgradeability to new revisions of hardware and software must be supported.

The ideal network is a strategic management tool providing flexibility and the freedom to choose from the best computing options available--now and in the future.   This philosophy is the backbone of Gandalf's product and service strategy.   We have actively implemented this concept in our own telecommunications network that supports our business.   The inclusion of a reliable, flexible and adaptable network as part of an information technology plan is essential if an increase in productivity is going to be realized.

21

McDonnell Douglas View
of MAP/TOP Standardization

Wayne L. Hanna

Worldwide competition resulting in thousands of lost
jobs has forced the United States to reevaluate its
manufacturing capabilities. This reevaluation has
shown that:

1. Most companies are required to use a wide range of
    products from various vendors in designing,
    manufacturing and supporting their own products.

2. It is accepted today that no single vendor can
    satisfy all the requirements necessary to support
    an integrated factory.

3. Vendors use their own "proprietary" technology in
    developing the communication protocols used by
    factory processors and device controllers. It is
    the exception rather than the rule that any two
    vendors' products can "talk" with each other.

4. Studies indicate that interdevice communication
    accounts for 30 to 50 percent of the cost of new
    automation. Custom device programming, major
    rewrites of application programs, multiple, spe-
    cially tailored wiring systems and high cost main-
    tenance are causing major capital expenditures to
    deal with the situation.

5. Within companies, departments and divisions have
    developed automation capabilities over the years in
    a piecemeal manner independent of each other. It
    is the exception rather than the rule that the
    applications from any two departments or divisions
    can talk with each other.

6. Incompatible applications along with vendors'
    proprietary protocols have resulted in "islands of
    automation" within our factories.

A major factor in achieving greater factory productivity is the integration of these islands of automation in order to achieve easy, timely and cost effective sharing of essential product data as it moves from design through manufacturing, quality assurance and into product support. The degree to which a company successfully meets this objective may well determine its competitive position within the marketplace.

THE NEED FOR STANDARDIZATION

A key ingredient to integrating the islands of automation is to obtain interoperability among the vendors' products used to support our factories. The question is: How can this be achieved?
We know from past experience that:

- You cannot go to a single vendor (IBM, DEC, etc.) and obtain a multi-vendor solution.

- One vendor will not build its products using a proprietary solution from another vendor.

- An individual company cannot afford to have products built according to its own internally generated requirements.

Companies for years have attempted to mandate a single vendor solution to automation by using internal procurement restriction. A survey of industry will show that this method has failed.
Also, rarely will one vendor jeopardize market share by agreeing to build its products using the proprietary technology of another vendor thereby implying the inferiority of its own technology.
The cost to build or have built products that meet the uniquely specified automation requirements of a single company has been shown to be ineffective. Even studies within the U.S. government are recommending moving away from this approach in order to control costs.
It is within this framework of past experience that the industry move to standardization has dramatically accelerated within the past two to three years. Standardization is the common ground on which users and vendors can unite.

THE STANDARDS PROCESS

Three basic questions arise when addressing the subject of obtaining interoperability among vendors' products through industry standardization:

1. Does there exist an industry structure in which the appropriate standards can be adopted in a timely manner?

2. Will vendors build products according to these industry standards?

3. Will the vendor products comply with the industry standards in such a manner as to achieve multi-vendor interoperability?

STANDARDS ORGANIZATION

By 1983 the International Standards Organization (ISO), a collection of standards organizations, had developed the Open Systems Interconnect (OSI) reference model to such a point that it was emerging as the basis for standardizing networks. The American National Standards Institute (ANSI) represents the United States within the ISO structure. While there is a great deal of jockeying among companies, nations, and organizations within ISO, it has created a structure in which industry can be brought together for the purpose of adopting national and international standards.

The OSI reference model is a seven-layer model that functionally defines the characteristics of each layer in order for applications on different computer processors to meaningfully exchange data with each other. The exact specifications are (or will be) supplied as the standards organizations adopt specific protocol(s) standards for each of the seven layers. ISO and the OSI reference model are a major achievement for industry. If we cannot discuss the functions of communications within a common structure using a common language, how can we hope to achieve interoperability among multi-vendor products?

Several standards have been adopted within the layers. However, many layers still lack standards. While there has been a great deal of work expended, the process of moving a specification to a fully accepted industry standard is historically a very slow process.

STANDARD PRODUCTS

In terms of products being built against the existing standards, the message has been very mixed.

1. In the 1983 time frame, products were beginning to appear against IEEE 802.3 and IEEE 802.2 for layers one and two due to major efforts by companies such as DEC and XEROX. Products against these standards are generally referred to as Ethernet.

2. Today, the number of vendors offering Ethernet type products has rapidly increased. However, many of these products do not comply with the standards thereby defeating the basic purpose of providing interoperable, multi-vendor products.

3. There was no clear indication of emerging products against IEEE 802.4 and IEEE 802.5, two more standards for layers one and two. IBM was noncommittally supporting 802.5 but was rejecting 802.3 (Ethernet). DEC was supporting Ethernet but rejecting 802.4 and 802.5.

4. It was clear that vendors were aligning themselves against each other by supporting different standards. By doing so, vendors could claim to customers that they were supporting standards, however, the basic objective of achieving interoperability was again being defeated.

5. Prior to 1984, there was no visible activity relating to compliance testing.

MAP

It was in this basic industry climate that the concept of MAP (Manufacturing Automation Protocol) took hold, a climate characterized by the following conditions:

1. A United States industrial complex facing a ten-year decline in productivity resulting in loss of market share along with thousands of jobs.

2. Factories splintered with multi-vendor, proprietary products and incompatible applications.

3. Standards slowly beginning to emerge from a splintered standards organization controlled primarily by vendors and utilities.

4. Vendors noncommittally admitting to the need for industry standards but failing to meet the basic objective of interoperability.

The missing link for creating a highly integrated factory environment was the user community involvement. A unified industry solution cannot be obtained until the user community is able to clearly and forcefully state its requirements.

General Motors (GM) formed the basis for providing the missing link through its internal effort to specify its requirements for factory automation. This effort resulted in the specifications for the manufacturing automation protocol thus giving birth to the acronym MAP. In summary:

- GM has been working on MAP since 1980.

- MAP is seen as the vehicle for integrating the "islands of automation."

- MAP's basic purpose is to directly attack the high cost of new automation related to communications.

- MAP is based on open and public standards, not closed, proprietary standards like IBM's SNA or DEC's DECnet.

- MAP fully supports the activity of the International Standards Organization.

MAP's success requires wide industry support. It was within this context that McDonnell Douglas Corporation agreed to work with GM to determine industry's interest and willingness to support the MAP objectives. In March 1984, the first MAP User Group meeting was held in St. Louis for the purpose of defining a MAP implementation strategy.

MAP STRATEGY

The basic MAP implementation strategy is based on the following:

1. Within the user community, generate a set of communication specifications using GM's manufacturing automation protocol as a baseline.

2. Using the OSI reference model, assist in the definition of industry standards where they are lacking and provide user feedback/pressure to accelerate the acceptance of industry standards.

3. Provide marketplace feedback/pressure to encourage computer/device manufacturers to develop a set of non-proprietary communication products using industry standards.

4. Ensure product compliance and interoperability with the existing standards by working within industry to establish compliance testing organizations.

     Significant activity has occurred since the first MAP user group meeting. This activity included:

- Two multi-vendor network demonstrations were held at the National Computer Conference (NCC '84) in Las Vegas in July 1984. They were:

The Factory MAP Demo sponsored by GM

Allan Bradley	Hewlett-Packard
Concord Data Systems	IBM
DEC	Motorola
Gould	

The Office Demo sponsored by Boeing

Advanced Computer Communications	Honeywell
Charles River Data Systems	ICL
DEC	INTEL
Hewlett-Packard	NCR

- Several User Group meetings were held in the United States during March 1984 through January 1986. Attendance ranged from 100 to 500 executives representing from 50 to 200 firms.

- The European MAP User Group was organized beginning in March 1985. Several meetings have been held.

- Japan is in the process of organizing a MAP User Group.

- A worldwide MAP Federation has been organized to develop a single MAP user specification and coordinate international standards activity to ensure that user requirements are being met.

- In January 1985 the Industrial Technology Institute (ITI) was the first organization to commit to perform vendor compliance testing for MAP products.

- The MAP Steering Committee was organized and has grown to include:

    Ten voting user companies

Boeing	General Motors
Deere & Company	Inland Steel
DuPont	McDonnell Douglas
Eastman Kodak	Proctor & Gamble
Ford	U.S. Air Force

    Ex officio members

    American National Standards Institute
    Canadian User Group Representative
    European User Group Representative
    IEEE 802 Chairman
    Industrial Technology Institute
    Japanese Representative
    National Bureau of Standards
    National Electrical Manufacturers Association
    Society of Manufacturing Engineers (SME)
    PROWAY Standards Committee Chairman

    MAP Executive Committee

    TOP Executive Committee

- SME was selected to provide full-time secretarial/ administrative support for MAP.

- The MAP User Group, along with the MAP Steering Committee, was expanded to include the Technical & Office Protocol (TOP) in order to cover total company communication requirements.

- A seventeen-vendor MAP/TOP demonstration was given at Autofact in Detroit in November 1985.

- Commitment made to organize a private company to be called the Corporation for Open Systems (COS). This company will be funded by private industry and will coordinate on a full-time basis the implementation of standards including compliance testing.

- Over twenty vendors have committed to build products supporting MAP.

- Many user companies have implemented or are in the process of implementing MAP pilot projects.

COMPLIANCE

Compliance testing is the "glue" necessary for the successful product implementation of the standards. It must accomplish the following objective: Ensure vendor's products comply with the specific standard in such a way as to provide interoperability.

If interoperability among vendor's products is not achieved, then the total exercise of generating standards and developing products will be wasted. We, as an industry, must not lose sight of this basic goal.

Compliance testing requires the following:

- A commitment by industry to "mandate" compliance testing

- A commitment by industry to pay the associated cost of compliance testing

- The development of the required hardware/software testing tools

- The certification that these tools will accomplish the basic objective

- The creation of the specific organization(s)   to perform the product compliance testing

The major activity associated with the development of compliance testing tools to date has been done by the National Bureau of Standards (NBS). The tools developed by NBS for some of the standards have shown great promise. However, to complete the job will require a significant increase of their budget. The policy of the current administration is to hold or reduce NBS's budget with the understanding that this type of activity should be accomplished within and paid for by private industry.

Two proposals were made in 1985 to address the area of developing compliance testing tools:

1. NBS proposed that a private company called "Open Systems Institute" (OSI) be organized using NBS expertise and staff. Funding would be provided by private industry; i.e., each company paying $100,000 per year.

2. A consortium of seventeen vendors proposed that a private company called the "Corporation for Open Systems" (COS) be formed and funded by private

industry. Three categories of membership available to both user companies and vendors are being offered:

    Regular Member - $25,000 (1st year)
    Research Member - $75,000 (1st year)
    Senior Research Member - $125,000 (1st year)

At this time, industry support is behind the Corporation for Open Systems. NBS has withdrawn their proposal.

The Industrial Technology Institute (ITI), a non-profit organization formed with seed money from the State of Michigan, has been the first organization to commit to industry compliance testing. They have also been actively developing testing tools. ITI performed much of the vendor testing for the Autofact Show held in Detroit in November 1985. Several vendors have begun to use the services of ITI to test their products. ITI will require industry funding to continue their functions. Indications are that COS, once organized, will provide some funding for ITI.

While a basic foundation has been laid for dealing with compliance testing, a great deal of work still remains to be done in order to accomplish the objectives of this function. Greater industry support is required.

SUMMARY

Over the past two years, industry has shown a high interest and willingness to support the MAP/TOP objectives and strategy. This support has been driven mainly by the need to improve productivity within our factories.

Successful implementation of the MAP/TOP strategy requires a close, cooperative working relationship within and among the user community, the standards community, and the vendors.

22

Pennwalt Attains Productivity via Information Systems

Robert Rubin

We have heard and will hear much about the undeniable
need for productivity improvements in American business
if we are to survive as an industrial power. The
definition of quality, however, is not so easily come
upon as is the certainty that we need to improve it.
Quality means different things to different people.
The fact is that quality is a subjective attribute.

The quality of the service that an information
systems function provides, in the eyes of the consumer
of that service, is dependent on how the service is
used. To a business unit that is in a commodity business selling fungible goods, low cost and reliable
machine service may be much more important than responsiveness to requests for systems analysis and low error
rates, a key factor in the rapidly moving and highly
regulated pharmaceutical industry.

The salient aspect is that while all attributes of
quality in information systems work are important,
their weightings are not the same for everyone and may
change for a particular client, often without anyone,
including the client, realizing it. In other words,
the rules of the game are not the same for everyone and
can change without the rulebook being reissued.

Now, having stated that quality is subjective, let
us keep in mind that Lord Kelvin once said, "When you
can measure what you are speaking about, and express it
in numbers, you know something about it: but when you
cannot...your knowledge is...meager and unsatisfactory."

MEASURING INFO SYSTEMS QUALITY

There are metrics, ways of measuring quality, for
information systems. People who tell you that working
with computers is an art and not a science, that firm
estimates on projects or guarantees of availability of

service cannot be made, are not doing their jobs.
Metrics can be defined. What are some of them?

They group into availability of operational
systems (those that you depend on to run your business)
and development of new systems. Some of these specific
metrics are, from the analytical to the more
subjective:

- Mean time to repair

- Mean time to failure

- Response on a terminal

- Number of re-runs

- Number of errors

- Adherence to programming or operational schedules

- Adherence to cost estimates

There are other metrics that are not so apparent,
but which can be used internal to the information
systems function to assure quality. For example, as
systems get old they cost more to maintain. Just as
older cars require more work, so do systems. The
concept of "spoilage" which can be expressed as the
ratio of maintenance cost to replacement cost, is a
measure of how fast a system is wearing out.

All metrics, however, come down to the same basic
measurement in the eyes of the client: Are you perceived to have helped more than you cost? The amount
by which your value has exceeded your cost is your
productivity. Productivity is creating economic value.
It is the ratio of the revenue you produce, or the cost
we would bear for equivalent services, to the cost of
your service:

Productivity = Rev gained (alternative cost avoided)
                            Our costs

Information systems is a necessary part of all of
our businesses. It is a business within a business.
After all, it has products to sell, and customers who
are charged for services and who have the right to
purchase alternative products. Productivity within
information systems is essential.

Now, productivity does not mean more paperwork.
It does not mean the ability to generate forms, or
packs of paper for others to read (or, more frequently,
ignore). It is not the debilitating exercise followed
in too many offices of having individuals whose func-

tion in life is to check on whether others are following the procedures and filling out the forms generated by yet another staff group.

Productivity means the ability to increase revenue or lower our costs for providing needed services. I stress the word "needed" because the simple use of the word "why," over and over, can eliminate a great deal of perceived need. After all, the least costly way of doing a function, whether manually or via computer, is often to eliminate the "need" for it.

The computer can be a great tool for improving productivity. In the past, the computer has been used to reduce the cost of many clerical activities. And, without the computer, businesses could not function. Can you imagine a bank trying to process checks manually, or a large manufacturing company processing orders manually? Computers have been used in the back room for several decades.

COMPUTER TECHNOLOGY: TOOL FOR SIMPLIFYING ORGANIZATIONS

The computer is moving into the front office where it is even more visible. Information technology is changing, and changing significantly. Information technology is changing from a tool for managing the complexity of an organization to one that is simplifying the organization itself. This distinction is the key. It will make a more significant impact on American business than the introduction of the computer for clerical use ever did. It is changing how we do business, not just computerizing it.

Let's look at some of the ways that this change in the use of information technology is being applied to productivity. In the office, information technology as a tool for simplifying the organization will be applied to both the secretarial/clerical functions and the management chain. Technologies such as voice-mail, electronic mail and the personal computer are, and will be, used to eliminate the non-decision makers from the chain of command. The management chain can be shortened by giving executives the capability to gather data and analyze information themselves, providing them the opportunity to eliminate middle management positions that are not decision makers, the data gatherers, the staff positions.

An entire cadre of people has developed in corporations whose function in life is to gather information, distill it, and dole it out, carefully and painfully, to those above and below them in the organization. Their jobs will no longer have meaning in an

environment where information is easily transmitted and
aggregated.

The sales force will see information technology
making an impact in two directions radiating from the
salesperson.   The first, from the salesperson to the
customer; the second, from the salesperson to the home
office.   Salesmen and saleswomen can use computers to
help them sell their products to customers.   Such tools
have been used by major companies to sell insurance or,
in the case of Pennwalt, industrial products.   In some
companies, customers are given terminals that allow
them to enter orders directly.

Information technology will also be used by the
sales force to communicate with the home office.   Voice
mail and portable computer terminals for call reporting and order checking offer the opportunity to
increase productivity significantly, not just for the
salesperson, but the organizational entity.   After all,
given that reporting and order processing can take
place directly, what are the implications on the needs
for the physical district office?

It wasn't so long ago that the jet airplane
offered the opportunity to change significantly the
autonomy of the regional office.   When face-to-face
meetings could take place with a trip of six hours
coast to coast, rather than several days, the face of
business changed.   The use of information technology to
simplify the organization will make no less a change.
The overall result will be the reduction of the overhead costs of selling and the opportunity for the
salesperson to spend more time selling with customers
who are more likely to buy.

A third area of impact is in the factory.   The
computer is, and will continue to be, used to speed up
the product cycle as well as to lower labor, material,
and inventory costs.   Much has been written about the
use of the computer in the factory.   We are all familiar by this point with Just-In-Time and the concepts
of computer integrated manufacturing.   The most significant impact, however, will probably come through the
speeding up of the product cycle.   Our factories will
be able to be more responsive to our customers' perceived needs by using the data gathering ability of the
computer and coupling it with the CAD/CAM capability to
design and manufacture products more quickly.

PRODUCTIVITY GAINS: INTERNAL EFFICIENCIES AND CUSTOMER
SERVICE

The opportunities that exist for productivity
increases cluster on those that are internal to a

company and those that are in the eyes of the customer.

Internally, we have the ability to flatten organizational hierarchy, to eliminate middle management layers and clerical support requirements. The computer will allow this if we control it and do not let it become the end-all instead of the tool. There are those, for example, who question whether General Motors might not have committed itself too rapidly and totally to computerization of their plants instead of first concentrating on the work groups and management overhead of the plants. The most effective GM plant is the Fremont, California, joint venture with Toyota. It is not particularly computerized and did not enjoy a particularly high reputation when it was a GM plant without Toyota.

We have the ability via the use of information technology to achieve success in our relationships with the customer of our businesses. We can lock the customer into us by making the cost of changing too high. After all, if a bank were to offer every type of financial record keeping you might want, you would be unlikely to leave them.

I deal with Fidelity Funds. I have several accounts with them. They offer the ability to transfer funds quickly, to check balances at all hours of the night via a computerized telephone service. They provide an extremely comprehensive report to me each month detailing each account and aggregating them. It would take a great deal for me to leave them, and indeed, they have a leg up with me whenever I think about adding another financial account.

Systems that provide additional services to the customer warrant higher prices. This is especially true whenever a commodity can be turned into a branded product. Take a look at Frank Perdue who turned the chicken into a Perdue chicken. A bank is no longer interchangeable with any other bank when it offers services such as a strong presence with automated teller machines when others do not. Take a look at what CitiBank did in this area. The list is growing longer. Rosenbluth has grown rapidly along the East Coast as a corporate travel agency primarily because of their ability to provide reporting such as the lowest cost fare and trip summary information.

Information technology can give you the opportunity to change the way your business is viewed by your public. American Airlines changed the airline business drastically with an on-line reservation system and the mileage bonus club. No major airline today can operate without these conveniences. Federal Express's use of computerization is featured in their advertisements.

IMPLEMENTING INFO TECHNOLOGY TOOLS

How do you get started in getting your company to use these innovative techniques, to improve the value of your service, to enhance its quality? How do you build an organization that strives for productivity through quality?

Quality is defined by the customer; and for the information systems manager, your customer is the line management of your company. There are some key points:

1. Define your constituency and what is important to them. Remember that what is important to a vice president of your corporation may not be important to a division manager, and vice-versa. In addition, different business units have different cultures as well as different needs.

2. Develop your organizational focus. In Pennwalt, for example, we decided that we wanted our systems function to be perceived as providing high value to users; to be a user-driven service organization. The organization focus of the systems function must be identifiable with the direction of the company itself.

3. Build bottom up and top down. Make the technical vision understandable. Explain the result, not the technology. Instill an absolutely unrelenting insistence on quality. Do not compromise, no matter what else you do. And, finally: communicate, communicate, and then communicate some more. Just when you are sick of hearing your own voice, you probably are starting to reach the folks you care about reaching.

If you are able to work with the management of your company, if you are able to understand the culture of your corporation, you will then have the opportunity to develop the technical and management tools to provide quality. The technical tools are relatively straightforward. The tools improve the ability of the systems designer to prepare, rapidly and accurately, the systems that the user wants. Since many clients are not really sure what it is that they need until they see it, the proper tools provide the ability to modify and maintain easily. They are such items as:

- Fourth generation programming languages

- On-line, sub-second response time access for programmers

- Structured design of systems

- Prototyping of designs

     Even more important are the managerial tools that keep you aware of the changing needs of your clients and in control of the quality of the information products that you produce. These are:

- Review boards for projects

- Separation of maintenance and operation of systems

- Peer level review of projects

- Service level definitions

- Personnel performance evaluations

     They are important. As the ad says: Don't leave home without them.

ISSUES AFFECTING SUCCESS OF INFO TECHNOLOGY

     The above items are important in providing productivity through quality in information systems. But there are major issues beyond technology and organizational structure that will determine the ultimate success of the application of information technology in your company.

1.   Success occurs because of the change in how the business operates, not in the automation thereof. In other words, no matter how important I think, or you think, information technology is in a business, it is still a means to an end, not the end itself.

2.   The limiting factor is how fast people can accept change. Alvin Toffler, in <u>Future Shock</u>, pointed out that people can accept change only so quickly. Hundreds of years before, Niccolo Machiavelli explained how difficult it was for people to accept change when he said, "There is nothing more difficult to take in hand, more perilous to conduct, or more uncertain in its success, than to take the lead in the introduction of a new order of things."

3.   Technology is changing too rapidly for people to assimilate the impact of the technology. Did our grandfathers realize the impact of the automobile on our culture? Did we envision, five years ago, the impact of the personal computer?

4. Advanced use of computer technology calls for a
    radical change in traditional work practices.

5. To be successful requires potentially painful
    organizational change.

    What is the risk of commitment to information
technology to achieve productivity? If we do nothing,
conflicting technologies will be installed and costs
will increase dramatically. In other words, people in
your company will still install the word processing and
the computer networks and the personal computers. Your
costs will increase and you will wonder what happened
to you. Without a technology base, you will be unable
to compete in hiring competent people. People who grew
up with computers will expect them as almost a birth-
right when working. Most important of all, competition
that uses technology wisely achieves the critical
advantage.
    Finally, I would like to leave you with this
thought. Productivity in systems is the creation of
economic value for the client. By making computer
technology a management tool, as opposed to a clerical
tool, we provide the capability to flatten the organizational hierarchy and to make the organization more
responsive to the customer.

23

Exploding the Information Management Myth

Duncan B. Sutherland

"ROUND, AND ROUND, AND ROUND SHE GOES..."

In recent years corporate America has been acting a lot like a greyhound chasing a mechanical rabbit around a racetrack. In this case, however, the racetrack is the office, and the mechanical rabbit is improved office productivity.

Pundits have estimated that the efficiency of the office, as it exists today in most large organizations, may be as low as 30 percent. Whether or not one is inclined to believe the "numbers" on the drooping performance of U.S. office workers, even a cursory glance around a typical white-collar workplace suggests that there's plenty of room for improvement! Unfortunately, the problem is much more serious, and much more complex, than simply finding ways to jack-up the output of the nation's office workers, although this is the tack companies have most often taken in recent years. If this were all there was to it, significant improvement could probably be achieved by nothing more radical than simply reintroducing some management discipline into the white-collar workplace! More to the point, corporate America's massive investment in new office technology of all kinds over the past couple of decades would have worked--which it hasn't.

I don't take issue with the fact that the modern office is not contributing as much as it should to the competitiveness of American companies and, indeed, to the competitiveness of the nation as a whole. It clearly is not. However, it is becoming more and more obvious, at least to me, that the "problem" of flat office productivity growth lies more in the nature of the office itself--in how we organize and marshall people, tools, and the environment in which they come together to accomplish the intellectual work necessary to meet some commonly agreed upon mission--than it does in the performance, or lack thereof, of the white-collar workforce. In other words, the problem lies in the very way in which America, as a nation, offices.

American enterprises, along with their industrialized counterparts around the globe, have entered a new era in which a company's (and ultimately a nation's) success depends more and more on its ability to leverage human intellectual potential, rather than physical effort. At the same time, companies (whether they realize it or not) find themselves saddled with outmoded and largely ineffective forms of organization and management which, although they worked amazingly well on the shop floor, have too often turned out to be counter-productive when applied in the office!

Whether the true "efficiency" of today's office is 30 percent or 130 percent begs the question. If management steps back far enough so that the trees no longer obscure the forest, it's clearly the office itself that's the problem. To remain competitive, American management must find new and more effective ways of organizing people and technology to accomplish intellectual work, or suffer the ignominy of yet another defeat--this time in the "front office" rather than in the factory--at the hands of some enterprising nation that does find an answer. Unfortunately, a major stumbling block which stands in the way of achieving these new forms of officing--new approaches to integrating people, computer-based tools, and the places in which they work, and which focus on extending individual human creativity and endurance--is the nation's own past success in achieving massive productivity gains on the farm and in the factory.

WE ARE WHAT WE KNOW

Why has it proved so difficult for U.S. companies to come to grips with the office productivity "problem"? The answer lies, at least in part, in understanding how management views the "office" and "office work," and how management's perceptions have misshaped U.S. strategies for investment in office technology of all kinds, including so-called "information management."

In The Day the Universe Changed (Little, Brown and Company, 1985), James Burke points out that people are what they know. For example, prior to Galileo's invention of the telescope, people knew that Aristotle had been correct: The stars, including the sun, were attached to crystal spheres that revolved around the earth. This knowledge constituted the "truth" and, as a result, formed the basis for both construing--and, perhaps as important, verifying--people's day-to-day experience. There is another, more colloquial way of saying the same thing: If the only tool you have is a hammer, every problem begins to look like a nail!

Let's jump back for a moment to the mid-1970s and put ourselves in management's shoes. What was the "truth" about slumping office productivity? Well, management knew that the office is a system of sorts and that a system is only as efficient as its least efficient subsystem. This is a lesson U.S. companies had learned the hard way in manufacturing. Thus, it stood to reason that the way to increase the productivity of the office was to increase the "productivity" of individual office workers. What was the easiest and least risky way to do this? Obviously, by adopting the same strategy that had worked so well in the factory: leveraging (and, wherever possible, replacing) individual human effort with machines. True, office workers do "knowledge work" rather than physical labor. But they still produce something, even if it's called "information" rather than the hard goods people usually associate with manufacturing. Nevertheless, the same principle ought to apply.

This "knowledge" about the underlying cause of the nation's growing office productivity conundrum, too little mechanization, was confirmed and verified by decades of experience on the farm and in the factory. For example, as recently as 1880 almost 50 percent of the U.S. labor force was still directly engaged in agricultural production. In no small part as a result of ongoing mechanization, the percentage has dropped precipitously until today less than 3 percent of American workers are farmers. Ongoing mechanization-- and more recently automation--has had a similar effect in the factory by continuously driving down the direct labor component in the nation's manufactured goods. Thus, when the business press (largely at the instigation of the technology vendors themselves) severely chastised management in the late 1970s because the average industrial worker in the United States was "backed by" $15,000 or more of capital investment while the comparable figure for the average American office worker was $2,000 or less, their admonitions struck a resonant chord.

The result was nothing short of infomania: a buying spree for new "office productivity" tools of all kinds, beginning with the word processing "revolution" of the mid-1970s, and continuing largely unabated until the so-called "computer slump" of the mid-1980s. The numbers really are astounding. According to the Computer and Business Equipment Manufacturer's Association (CBEMA), corporate America's total expenditure for office equipment of all kinds in 1986, the latest figures available, was a whopping $358 billion. While this figure does include support, software, maintenance, and so forth, it is still a rather impressive commitment of economic resources. All the more so, of course, because notwithstanding decades of massive investment in new "information technology," America still has a serious office productivity problem!

What went wrong? Why did the impact of all this new technology fall so far short of management's expectations? In hindsight, the answer is almost embarrassingly simple: The universe has changed, but with all due respect to "Jethro Tull," management is still "living in the past."

TOWARD A COPERNICAN VIEW OF THE OFFICE

With the invention of the telescope, people quickly discovered that not only were there a lot more stars and planets in the sky than anyone had imagined, constrained as they had been by the limited power of the human eye, but that other planets had satellites revolving around them. This was, in Burke's words, the day the universe changed. The Aristotelian view of the universe persisted for a long time, of course, largely because it threatened the established dogma of the Catholic church. Nevertheless, it was eventually replaced by the Copernican view in which the earth, and the other planets of our solar system, revolve around the sun.

When it comes to addressing the "problem" of office productivity, corporate America has been gazing through the wrong end of the telescope. By immediately zooming-in on the "problem" of finding ways to increase the productivity of individual office workers as components of a larger system, the office management missed the most important point: individual office workers are the system. The office is not an "information factory." In fact, it is not a factory of any kind. More important, "information" itself is not something tangible that can be changed.

Information is an abstract concept that describes the intellectual product of human information processing. Information remains locked deep in the human mind. The things people normally think of as "information"--the memos, letters, telephone messages, etc., which are usually associated with the office--are simply by-products or artifacts of human information processing. They are symbol buckets, convenient ways to store and move around conventional tokens--words, charts, images, and so forth--for the purpose of communicating abstract ideas.

As human beings, our ability to create information is dependent on knowledge--captured experience that allows us to interpret data or facts about the world in the hope of using this increased understanding to reshape the world to our ultimate advantage. As a result, it's really knowledge, not information, that's critical when it comes to gaining a sustainable competitive edge in the world of business. Francis Bacon clearly knew this when he pointed out in <u>Meditationes Sacrae</u> (1597) that knowledge is, indeed, power. Unfortunately, it's something management has seemed to have lost sight of in recent years.

The key point to keep in mind is that the purpose of the office is not to "produce" information, but to accumulate knowledge. Where does knowledge come from? Knowledge is acquired when companies hire people. Knowledge is also acquired as a result of human information processing--through people's day-to-day experience. If the office must be cast in a mechanistic metaphor, it would be that of a knowledge machine, not an information factory.

There is no question that so-called "information technology" can help in the formation and dissemination of experience within a company. It's not technology that's the problem. It's how management has applied technology that's been the problem. The key to achieving a sustainable competitive advantage lies not only in the creation of a common base of experience, but in the continual creation and application of new organizational knowledge. Unfortunately, current technology-based approaches to "information management," focusing as they do on manipulating the artifacts of human information processing, offer companies precious little help in this regard. In fact, they have often ended up costing companies more than if they had done nothing at all!

PROBLEM SEEKING VERSUS PROBLEM SOLVING

The first step toward successfully solving any problem is to make sure that we have defined it correctly in the first place. In many cases, problem seeking turns out to be more important than problem solving, because a search for the correct problem definition often leads to a total reformulation of the "problem" management thought it was trying to solve in the first place. Sometimes, the original "problem" even disappears altogether. This is a particularly important point to keep in mind because human beings are incredibly good at coming up with innovative solutions to non-problems! As we have seen, this appears to have been the case in management's mis-directed search (no pun intended) for a solution to the "problem" of flat office productivity growth.

There is nothing inherently wrong in management using new technology to address current office produc-tivity issues so long as they don't stop there. The key point here is that management must also view tech-nology as an enabler--as a tool to help define totally new approaches to organizing people, computer-based tools, and the environments in which they work together--which can help companies accomplish three simultaneous objectives:

1. Maximize the intellectual contribution of individual workers.

2. Minimize the number of people a company employs.

3. Eliminate unnecessary (and therefore non-productive) capital investments in technology.

These should be the goals of any company in any era. However, competing in the officing age--an era in which organizational success will increasingly depend upon working smarter, not harder--will force organizations to reevaluate, and greatly simplify, their current approaches to planning and managing the integration of human and technical resources.

What this means in practice is that companies, once they have a clearly defined mission and understand their knowledge requirements, must begin from the "bottom" up. They must first focus on the individual knowledge worker and the role that technology can and should play to enhance the process of individual human information processing. Again, the individual knowledge worker is the system. This will require a much closer marriage of behavioral scientists and tech-nologists than presently exists in most companies today.

Next, individual office workers must be woven into "neural networks" or lattice organizations. Unlike today's centralized, hierarchal corporate structures, these new forms of organization may not even have "offices" as we think of them today. For example, people may work primarily at home, traveling, when necessary, to local (or perhaps regional) switching centers--communication nodes--that serve as concentra-tors for both individual and interpersonal knowledge work, as well as a resource for shared technology. This may eventually lead to the reemergence of the "family enterprise" that existed in many societies prior to the Industrial Revolution.

Finally, both visible "overlays" and invisible "underlays" of technology infrastructure must be added to support the communication and information processing requirements of individual knowledge workers, as well as to provide the tools necessary to effectively manage the overall system. As I am sure you have already concluded, this is exactly the inverse of how most companies go about addressing office productivity today where the emphasis is on technology rather than on people, and on artifacts rather than on the process of human information processing.

"...AND WHERE SHE STOPS, NOBODY KNOWS!"

The evolution of radically new approaches to officing raises many complex questions. For example, how does a company build and maintain a tight "corporate culture" in a lattice or networked organization? I have no doubt that questions like this one can be answered satisfactorily. In fact, once management is freed from trying to "fix" today's ailing approaches to officing, they should have much more time to spend on really critical management responsibilities like leadership, vision, and fostering ongoing innovation. Most important, the emphasis will be back where it belongs in an increasingly intellectually competitive age: on extending the endurance and creativity of the individual knowledge worker.

The mind naturally rebels at the very thought of change, and this is certainly no less true when that change involves thinking about a future "office" that may be radically different from today's "reality." However, there is no question that the office must change. The "modern" office is an anachronism, an artifact of an era that knew neither modern transportation nor modern telecommunications. In recent years, technological advances have virtually eliminated the traditional constraints imposed by time and space on how enterprises organize themselves. This is already causing many companies to challenge much of the conventional wisdom about "offices" and about the nature of "office work." It is now possible, quite literally, to "office" wherever one happens to be--at home, in an airplane, in a car, on the beach, or even in traditional, purpose-built "office" buildings.

For an increasingly large percentage of workers, the immediate impact of the emerging "virtual office" has been a blurring of the traditional separation between home and work--a separation that did not exist for most people prior to the Industrial Revolution. Homes are beginning to look and function like an extension of the office. Offices are beginning to look and function like an extension of the home. But this is only the tip of the iceberg! It is entirely conceivable that, in the not too distant future, there may be no need for offices as we know them today--and perhaps no need for traditional office buildings, as well!

To repeat a point I made earlier, and in conclusion, American management must find new ways of organizing people and technology to do intellectual work that produces a sustainable, long-term intellectual advantage--and quickly--or run the unacceptable risk of suffering yet another defeat at the hands of some enterprising nation that does find "the magic bean." Make no mistake, other nations are looking.

Take Japan, for example. To ensure its continued role as an international economic leader, Japan has embarked on a major and extremely farsighted initiative to upgrade the performance of the nation's offices. Called the "New Office Concept," the program is an ambitious effort by Japan's Ministry of International Trade and Industry (MITI), Ministry of Construction (MOC), and leading Japanese companies to position the Japanese white-collar workplace--and Japanese white-collar workers--to capitalize on the unique opportunities associated with a twenty-first-century global economy.

As part of this initiative, Japan has been carefully studying the office planning, design, and management practices of other highly industrialized nations, including those of the United States. Even to skeptical Japanese eyes, the offices of most large American companies do look very innovative, and perhaps even very efficient, with their ergonomically designed systems furniture, high-touch interior architecture, and plethora of information technology. But, as Japanese businesspeople well know, looks can be very deceiving. The real test is whether or not a company's offices make a direct and substantial contribution to the intellectual performance and well-being of its most important asset: its people.

Japanese management finds itself once again in the unique position of being able to "leap frog" other industrialized nations, including the United States, in defining the form and function of the office of the future. Certainly, the obstacles facing Japanese management are formidable. Japan is going through wrenching socioeconomic change at the same time it is attempting to confront the white-collar productivity issue head-on, and only time will tell if they will ultimately be successful. Of course, many executives in Detroit and Silicon Valley didn't believe that Japan's automotive and high-tech industries could become as competitive as they have in such a short time, either.   How does that old saying go? "Forewarned is forearmed?"

24

The Impact of the Electronic Office at Mead

Ronald P. Carzoli

As a human resources professional, I won't pretend to speak to you on the finer points of electronic office wizardry, advise you on state of the art equipment, or predict the next revolution in office automation. What I will share with you are several observations on the profound impact the electronic office is having on the way people do their jobs and what that means to us as human resources managers.

I've observed the growth and development of the electronic age first at Ford, where I served in a variety of human resources positions in this country and abroad, and more recently at Mead. Mead Corporation is a world leader in forest products and in the rapidly growing high-tech field of electronic publishing. During that time I've seen the development of office automation progress through three different phases.

OFFICE AUTOMATION: PHASE I

In Phase One, which began in the 1970s, the improvements made through office automation were those of <u>efficiency</u>. The emphasis was on making people more efficient at the activities that they were already doing--typing, bookkeeping, and copying. Miniature dictating equipment, electric typewriters, eventually some with limited memory, and finally word processing helped our clerical people become more productive by working faster and eliminating unnecessary steps. For example, at Ford, we used to have three people who typed letters for applicant rejection. With the advent of memory typewriters, we were able to prepare four basic letters: a polite "No Way", a "Maybe", a "We are investigating the possibilities", and a "Come On In!" As a result, one person could do it all.

OFFICE AUTOMATION: PHASE II

In Phase Two, which probably began in the early 1980s, the emphasis shifted to <u>effectiveness</u>. We began to take the next step, to find new and different ways to do things using the advancing technology. We began to take advantage of the technology. We developed <u>decision</u> support systems. We used the new technology to improve production planning, finding the most effective way to load machines and reduce the expense of set up and down time, for example.

In making capital investment decisions, we learned that we could use financial modeling technology to evaluate more alternatives and compare their rates of return. We could dare to ask and get answers to questions of "What if...?" Sometimes we didn't like the answers.

Electronic mail programs within companies, like the IBM PROFS system we use at Mead, let us send notes, information, drafts of presentations and documents instantly to people at their desktop terminals between floors, across town or across the country. It's a far more efficient and effective way to share detailed information than the telephone. It's far quicker and easier than the telecopier or fax systems, and distribution can be done by the manager alone with a simple keystroke. I can provide management development alternatives over this system to managers in different time zones. We can "print" our conversations for the record.

OFFICE AUTOMATION: PHASE III

Today we are seeing the beginning of Phase Three. That's the point at which the advancing technology in office automation begins to <u>actually change the way we conduct business</u>. Let's look at some examples of this.

The speed and efficiency of electronic mail means you are more likely to do things you would not have even considered before. Our managers are more likely to share information with those in the field, ask for input and use it. For example, I can share the draft of a document with division managers around the country and get their feedback, not in days or weeks as with the mail or telephone, but in hours or even minutes.

Computers have helped us improve customer service. We can handle customer inquiries, check inventory, place orders, trace shipments, issue billings all with a few simple keystrokes. In some cases we are placing our computers in the customers' hands. American Airlines' Saber reservation system was developed by the airline as a better way to make ticket reservations and

print boarding passes.   Then American began leasing the
system to travel agents so they could issue the tickets
and the boarding passes themselves.   It's a great convenience to travelers, and of course, in the hands of
travel agents the system gave American Airlines a real
competitive edge.

American Hospital Supply connected its computer
system directly to hospitals, allowing the hospitals to
do on-line ordering themselves.   By doing so AHS eliminated much of the buyer's activity and allowed a once-
complicated order to be placed by lower-skilled, and
lower-paid, workers.   Now, the night before an open
heart surgery procedure, for example, a clerk can order
up a "kit" of all the materials needed for the
operation--from scalpels to gowns.   The order is filled
from the hospital's inventory, and everything else--
inventory records, reordering, billing, etc., is generated by the computer with a few simple keystrokes from
a hospital unit clerk.

At Mead, our rapidly growing electronic publishing
business, Mead Data Central, literally changed the way
legal research was done, and in just a few short years.
We have developed a unique system that lets lawyers
conduct legal research at their desks on a personal
computer.   They can search through the full-text of
case law from the fifty states and the federal government, thousands and thousands of cases, looking for
specific points of law and find exactly what they need
in a matter of seconds.   Before Mead developed its
unique search and retrieval software and built its huge
database, the largest privately owned database in the
world, researching through libraries of law books could
take days or weeks.   Our LEXIS service is used by
virtually every major law firm in the United States.
Our NEXIS service gives that same instant access to
news and business information.

All of these examples point out how technology
used in office situations has evolved.   These are not
just refinements of efficiency and effectiveness.   They
are major leaps forward, so significant that they are
changing the way we do business.

LESSONS FROM OFFICE AUTOMATION

What have we learned in the first phases of office
automation?   First, we have learned some tactical lessons.   We have learned that you begin by fitting the
technology to the organization, you don't try to change
the organization to fit the technology.   For example,
we put the technology of word processing on secretaries' desks.   We didn't reorganize the office into word
processing centers.   We have learned the importance of

making the technology as "user friendly" as possible,
so people feel comfortable with the technology.

The new "Fifth-Generation" computer languages,
like the "Intellect" program we've now begun to use at
Mead, will let people get the answers they need from a
database by typing in inquiries in plain English.
Programs like "Intellect" develop a personal lexicon or
dictionary for each person and convert your question
automatically into the proper computer instruction. No
fancy computer commands to memorize, just ask the question in your own words--that's the ultimate in "user
friendly."

Second, we found one of the best ways to improve
office productivity was to prevent interruptions. Productivity is lost in the starting and stopping--waiting
for the secretary to retype a letter, waiting for
someone to return your phone call. New office technology allows managers to bring more tasks to closure by
themselves. At one time a manager would conceive of a
memo, write it out or dictate it, have a secretary
prepare a draft, change it, approve it, have the final
copy typed, sign it and send it. Time was spent in
each step and lost in between each step. Today, a
manager often creates the memo at his or her desktop
terminal, and with a keystroke sends it to a number of
people around the company and even prints out a copy if
needed.

As advancing technology begins to change the way
we do business, it also has had the effect of changing,
or at least blurring, many of the traditional jobs
within the office. Secretaries are becoming true administrative assistants. Instead of simply typing
department budgets, they're using spreadsheet software
such as Lotus 1,2,3, entering data, and preparing the
reports, charts, and graphs themselves. Others are
using the time created for them by office automation,
and the opportunity created by ever-shrinking staffs,
to take on additional responsibilities. As a result
many offices are now doing the same or more work with
fewer people than just three or four years ago.

We're also seeing the rise of the "functional
specialist," the people in particular disciplines such
as accounting, compensation and benefits, who also have
a keen knowledge of computers and software. These
people are helping us find computer solutions to the
information needs of their disciplines. At Mead we
have people filling these roles in such areas as
payroll, retirement, and management recruitment.

We are seeing, and will continue to see, a gradual
flattening of our organizations, and a corresponding
broadening of the span of control. The head person
will become much closer to his or her employees, to the
line people or the lowest level professional. We'll

see more and more contribution by individual managers as the nonproductive distractions of their daily routine are gradually eliminated.

FUTURE IMPROVEMENTS OF AUTOMATION

Advancing technology will continue to challenge our basic ways of doing business. In the 1990s, companies will increasingly turn to the strategic use of technology to gain a competitive advantage over domestic and international suppliers.

Michael Porter, in his book <u>Competitive Advantage</u>, explains that technology can reduce a firm's cost structure, differentiate it from its competitors, raise or lower economies of scale, create advantages in timing, and make interrelationships possible where they never were before. Perhaps this last point about creating new interrelationships is the most compelling one for those of us in human resources. These interrelationships, between business units within our organization and also with our customers and suppliers, will be fundamental to the new, flatter, more horizontal organization.

These new, leaner, flatter organizations, Porter says, will require human resource policies that support interrelationships. These policies include: firmwide hiring and training, promotion from within, rotation of personnel among business units, forums and meetings for managers that cross business units, and education about the concept and importance of interrelationships, as well as new incentive and conflict resolution practices.

To help accomplish this, Mead's Human Resource MIS personnel have developed a series of computer "screens" containing information on all individuals in the corporation. Managers can access the database easily and instantly get the following information about their employees:

<u>General Information</u>: SS, DOB, DOH, FLSA status, address.

<u>Education</u>: degrees, year, major.

<u>Current Job Info</u>: Title, SG, Range, Compa ratio, monthly and annual salary.

<u>Salary History</u>

<u>Management Personnel Review</u>: Performance Rating, Potential positions

<u>Employment History</u>

Incentive History

Stock Option History

Total Remuneration: last two years

        The evolution of the electronic office is
unfolding before us. Two things are vital: that we
manage the transition to the new equipment with the
objective of using the technology to become increasingly competitive; and, most of all, that we understand
the impact that this new technology will have on the
effective employment of our human resources.

25

Human Interaction with Computer Technology at Control Data

Rodney D. Becker

Today, we have certainly heard a great deal about the possibilities associated with the introduction of decision support systems (DSS) design. However, right at this particular moment, I am not so much concerned with what we implement but how we implement. The fact of the matter is that there will continue to be an incredible pressure to improve productivity, and technology will continue to be seen as the solution to the problem. Both the pressure and solutions are legitimate.

JOB DESIGN/MOTIVATING EMPLOYEES/EVOLVING TECHNOLOGIES

We have learned that it will be the synergistic interaction of people, work environments, and technology in the context of an organization's objectives that determine the success of office automation efforts. In my opinion, however, the place where people, technology, and work environments come together is in the design of a job. Our challenge is to so structure the design of jobs that the new technologies will actually motivate employees to use more of their skills and talents, and even to use abilities that perhaps even they did not believe they had. To me, this is the kind of motivation that will result in the most dramatic productivity improvements, with the accompanying payoff for the organization.

The good news is that many of the decision support systems we are hearing about today offer us just exactly these kinds of opportunities. The bad news is that unless we take control of the implementation plans, we may never see these benefits.

Now we talk and read a great deal about the psychology of motivation, but, in my opinion, we must refocus our efforts on the motivational aspects of the job itself--that is, the very nature of the tasks

themselves and the organizational context into which
that job fits. Employees, even more so in the future,
will insist that the job be described, measured, and
compensated accurately.

The nature of a job, that which we do at work, is
very dependent on the kind of technology we have at our
disposal. As new, more powerful technologies are
introduced into the office, everyone's job--from the
clerical worker's to the manager's--will change. So
while the technology evolves, the fundamental nature of
the job must also change. We must understand the job
content "before and after" the introduction of decision
support systems in the office environment.

I see significant management attention being
focused on building networks to integrate the various
technologies and particular emphasis on the dollar
payback to the company. And that certainly must be
done. But, I fear, not enough attention is being given
to the effect these evolving technologies have on the
nature of work, and the impact they have on job design
and accompanying employee motivation and reward. And
unless we begin to address these issues seriously, we
will not realize the potential that these supporting
systems and technologies offer.

Now, before you accuse me of preaching, let me
observe the obvious--that this is not at all an easy
task. We face the same problem that our parents (or
perhaps for some of you, your grandparents) had with
the automobile. The car was there first. Only later
came the improved roads to take advantage of the car's
technological superiority. In today's office, we have
a new type of "car," but not enough attention is being
paid to the "roads"; that is, the specifics of the
jobs, their design, and the organizational context in
which they are performed.

MISTAKE OF MOVING OFFICE WORK TO COMPUTER CENTERS

Since the general theme of our panel is solving
problems with DSS design, I certainly must first define
the problem before we begin discussion of the solution.

Much of today's "wisdom" in how we design jobs--
although it is now being challenged by many authors--is
based on the work of Frederick Taylor. You know, he
was the division of labor and the stopwatch guy.
Unless we change some of our assumptions, the office of
the future will look very much like the factory of the
past!

The tendency of the past has been to apply
technology to the office by moving the work out of the

regular office and into a technology center.   We have
identified potential computer applications, programmed
them, and then physically moved the processing out of
the office and into the computer center.   And we have
repeated   that   with   word   processing   centers,
reprographics centers, message centers, and heaven
knows what else.

As a result, the capabilities of the technology
are available to a specialist in a technology center,
but not to the general office work force and certainly
not to the managers and professionals.   Now this was
not because we were stupid--or callous.   In the early
days of the technology, computers were so expensive
that we had to keep them running twenty-four hours a
day or we would be losing money.

So while that may make for a very high utilization
of the machinery, it often does not do much for worker
satisfaction.   As on the assembly line, so in the
office, division of labor may result in boring, unin-
teresting jobs.   And if   you isolate the technical
specialist from the mainstream of the business, you are
also extremely unlikely to come up with the sort of
productive use of the technology that we expect.

I have found both within my own company and in
discussions with colleagues in similar large organizations that these forces are very real.   Good decision
support systems properly implemented can stand in
opposition to these forces.

Today microelectronics, personal computers,
network technology, and decision support systems are
available cheaply enough that we now can create jobs
that are interesting, challenging, rewarding, and of a
different worth than those jobs were before.   In fact,
it is the phenomenon that personal computers provide
these fulfillments that I believe is fundamentally
responsible for the incredible growth of personal computers.   And   if we make the job interesting and challenging, we will have more success in reaching the
goals that management expects of all this new technology.   While much is being written about the electronic
revolution or how integrated software will solve all
our problems, I am more concerned about the things that
I do not read much about: How are we going to change
our approach to the job?   What is the job?   What are
its tasks?

This is a matter of concern because, as I talk
with my colleagues, I find that most data processing
and office technology planners are not trained to
aggressively participate in job redesign as offices are
restructured about the decentralization of technology.

JOB REDESIGN FOCUS: EMPLOYEE SATISFACTION

But we, the management, need to take
responsibility for job redesign, facility design, and
some aspects of organizational design as we plan
systems for the future, or else we cannot realize the
productivity, the bottomline contribution, that we
believe is achievable.

Why is this significant? A recent survey of
office technology applications indicated that non-
managers have a higher degree of recognition for the
importance of job and job relationships (worth, values,
tasks) while managers had the tendency to believe the
benefits would come from the technology alone.

And this is not limited to office technologies
only. In my previous job at Control Data I also had
responsibility for a traditional data processing group
and the office technologies department. While there, I
tried to instill the idea that the system designers'
work is not done until they have considered how the
user's job can be restructured in order to make it
fundamentally more interesting, more challenging, and
more rewarding. Job design is not a new strategy. It
has been around since the advent of the Industrial
Revolution and organizational theory. We have seen
various approaches to job design, from the division of
labor principles of Frederick Taylor to the job enrichment and job enlargement strategies of the 1960s and
1970s.

Yet all of these tend to focus on only one aspect,
either productivity or human relations alone. In this
era of the development of more "user friendly" hardware
and software, we can achieve a better fit between both
the high productivity and quality of work life dimensions by concentrating on how the job content and
context is structured. The job itself must enable the
employee to experience more personal achievement,
accomplishment--and success--which fortunately new DSSD
software makes possible.

Job design and measurement is also the vehicle to
solve some other emerging issues, specifically, the
health and safety concerns, i.e., the stress and the
need for frequent rest breaks associated with sitting
in front of a terminal for long periods. We have the
opportunity and the vehicle--in design of the job--to
ensure that employees are not sitting in front of a
terminal for long stretches. We can reduce the need
for breaks in the job by designing breaks in the form
of task variety.

Another issue that must also be addressed when
implementing decision support systems is the change in
the inherent value of the job. Often an employee using

these new systems is adding a significant value because of his or her skill with the data and the system. Yet, in many operations, this value is unrecognized. Employee awareness of this new value must be recognized. Therefore, we at CDC have developed another decision support system that measures changes in job content and value with far more accuracy than ever before.

Well, the question arises, "How do we go about doing this?" After all, if it was easy we would have done it long ago! Some experts have pointed out that it's really the style and value system of a company that usually determines whether or not a given proposal is approved. Ideas such as providing more motivation through good job design must find a basis in the culture of the company if they are to gain acceptance. We must therefore learn to understand and articulate the basic value system of our organization and translate these values into the very design of a job that will be affected by technology.

Our job, then, is to hold the jobs that we have designed up to the mirror of these values and principles. Conversely, we can also leverage these principles to introduce decision support systems more effectively in the context of a given company. Within Control Data we are starting to do just that. The good news is that tools and methodologies are now available to help us with the evaluation before it's too late. Let me cite our own experience.

EXPERIENCE WITH JOB DESIGN/NEW TECHNOLOGY

Approximately two years ago, we held ourselves up to this mirror and found that one of our pilot implementations of the "Office of the Future" was beginning to look exactly like the factory of the past. We had relegated all word processing to a typical word processing center. All service work had been moved to a different service center; the overflow of the telephones was being handled by a message center; and we had taken all the secretaries away from their managers. We then had all the secretaries report to a separate group made up of all the service functions I just mentioned.

Basically the result was that we took people who were trained as secretaries, and reduced the scope of their jobs to filling out time cards, getting coffee, or making xerox copies. Over time we had reduced headcount, optimized specialized jobs, optimized CPU usage and, of course, almost forgotten the human element. Separating these employees from integrating the technology with their jobs also limited their creativity and frustrated their interest in learning more about new DSSD.

After substantial manager and employee complaint about the system, we did an intensive, very special survey to determine how our employees really felt about the new technologies and the effect it was having on them. The interesting thing is that, in general, most employees welcomed the new technology and did not see it as a threat to job security. But the job designs and organizational structure for the clericals had left the large majority very dissatisfied with their stature, with their career perceptions, with the nature of their work, and with the use that was being made of their skills. Conversely, the managers believed that clerical work was being passed on to them. They, too, had no real ownership in the concept.

INTERDISCIPLINARY TASK FORCE ON JOB DESIGN/PRODUCTIVITY

These conclusions led us to rethinking how the office should be structured to take advantage of the productivity potential of the new technologies while at the same time challenging and motivating employees through the use of the new equipment and processes. We realized also that our model of the "Office of the Future" was not being adopted elsewhere within our corporation, but that virtually every new facility being planned was structuring offices in a different way. Fortunately, we recognized all of this before it became a free-for-all. With senior management backing we created an interdisciplinary task force including specialists from telecommunications, office technology, facilities, and very importantly, personnel research.

The result was agreement among these disciplines that we would not achieve the improvement in productivity unless we achieved the synergy of all these critical functions. This group also made a set of recommendations on the various behavioral, environmental, and organizational factors in the Control Data office. The critical factor in this whole change process was that data, hard data that management respected, data that came from the special survey I mentioned above, was available! What we learned from this and other similar experiences throughout Control Data has now been packaged into new products of CDC Business Advisors.

I cannot help but stop here and put in a plug for our new products. They consist of a change management survey and an office automation profile. The first one helps operations plan for the introduction of office automation, and the second helps you follow-up to determine if you met your original objectives, as far as employee acceptance is concerned. It allows you to then make adjustment in your implementation program quickly to deal with any problems you might be experiencing. Fundamental to the process, of course, is an understanding of the job. We recently have also introduced a sophisticated task-based job evaluation system far different from traditional systems.

Returning to the results of our interdisciplinary task force, we are now in the process of implementing their recommendations in a number of pilot projects. Ergonomics has become a key word, not just in furniture design, but in all aspects of the Human Interaction with Technology. In particular, we are looking harder at office jobs and how they should evolve, and we are working harder to stay "close" to the people that actually have a business to run in those facilities where we are introducing new technology. We have developed new tools to define the work and tasks performed, and to develop a pay structure that reflects the redesigned jobs. It is already clear that we are consciously moving away from the concept of division of labor. And rather than adapting people to the machine, we are--to the largest extent possible--looking at adapting our system to the people.

SUMMARY

In summary then, while we must concern ourselves with the integration of the technology into solving real business problems, we cannot allow ourselves to focus exclusively on the technology itself or on the optimization of its use without considering the human factor. The lessons of recent experience in traditional work places with employee participation, quality circles, involvement teams, and task integration apply to most jobs in the office.

We cannot oversubscribe to industrial engineering principles that make jobs as simple and routine as possible. Many of us have tried it and scrapped it. While we achieved some short-term productivity, the high turnover, low morale, and under-utilization of employees proved too costly.

Rather, job design must result in work for both the professional and clerical that is a personally rewarding experience. If it is, there may be little, other than to support the system, that management need do to foster high motivation and satisfaction. On the other hand, if work is not personally rewarding to a fault, then there may be little that management can do to create really high productivity.

26

Unisys Sees Competitive Edge in Fourth Generation Language

Dagnija D. Lacis

Arnold Toynbee once described the rise and fall of nations in terms of challenge and response. "A young aggressive nation," he said, "when confronted with a challenge, finds a successful response. The country then grows and prospers. But, as time passes, the nature of the challenge changes. And if a nation continues to make the same once-successful response to the new challenge, it eventually suffers a decline." The same analogy applies to business.

Today there is a revolution in information systems management. The challenge has changed, and to be successful, businesses must realign to respond to this challenge.

In business, opportunities rapidly appear and disappear. New products or services are introduced and mature much more quickly than in the past. The opportunity for success can come and go virtually overnight. Today more than ever before, information is the key to success. Timely information can enable astute management to make the right decisions. Lack of proper information can result in poor decisions, but more often result in indecision and lost opportunities. However, getting this timely information is easier said than done.

I will discuss some of the latest software tools, often called fourth generation languages, which can dramatically speed the development of complex application systems. I would like to stress that even though these tools by themselves result in increased programmer productivity, really significant productivity increases will only be realized when we change the way we design and build these systems.

APPLICATION BACKLOGS

Today, most companies rely on computer-based data systems for accurate and current information.

Obtaining new information usually requires new computer programs. Since computer application programmers in many organizations spend over 80 percent of their time maintaining application programs, it should come as no surprise to us that the backlog of new problems is growing at an ever increasing rate. And even when the programmers find the time to attack new problems, analyze management's requirements, develop the programs, write the documentation, incorporate last-minute changes, and eliminate program bugs, they find that a different problem now has management's attention. Charles Lecht, in his classic book, "Waves of Change," described the phenomenon as "stagnation," or, more simply, when it takes longer to develop the solution than the expected life cycle of the problem, the solution is doomed to obsolescence even before work begins. Many data processing departments today are faced with application backlogs measured in years.

Much of this application backlog exists because the complexity of information systems has grown almost beyond our control. Today, many MIS managers spend a great deal of time pigeonholing new application systems into the organization's database, verifying a relationship here or adding a new database item there. How many times have you heard someone say "If I could just develop the whole thing from scratch, it would be so much easier"? But of course, it is just too risky to throw away something that works, especially in a complex system that no one completely understands. Modern MIS systems include not only the business applications themselves, but also complex programs to support large communications networks, shared data base, access security, audit, backup, and recovery.

FOURTH GENERATION LANGUAGES

Today, so called "Fourth Generation Languages" can assist in reducing this backlog by making it easier to develop complex systems. Now what exactly is a fourth generation language? There does not seem to be a universally accepted definition. Vaughn Merlyn of Merlyn Corporation said, "A fourth generation language is whatever a vendor wants it to be." James Martin defined them as programming languages that provide a ten to one productivity improvement over Cobol or Fortran. That means a programmer should get the job done in one-tenth the time it would normally have taken using Cobol. Martin also states that a fourth generation language must be appropriate for use by both end users and professional programmers.

For our discussion, fourth generation tools are software packages designed to facilitate development of

application software systems and can be mastered by
non-technical staff. They speed development of interactive, transaction-oriented, database systems and
include application development tools, report generators, query languages, "What-if" modelling languages,
and business graphics.

Application development tools can create single
programs or complete systems. Report writers and query
languages allow rapid development of reports and
browsing through system databases; "What if" modelling
languages facilitate analysis and simulation of corporate data; and business graphics and make information
more easily understood. These tools make it possible
for organizations to build sophisticated information
systems.

EVOLUTION OF DATA PROCESSING

We can better appreciate some of the present
trends and changes in data processing if we consider
its brief but rapid evolution since the early 1950s.
Data processing automation started in the accounting
and administrative areas, since at this time only high-
volume, repetitive applications were cost effective to
automate. As hardware became less expensive, other
business operations and services were computerized.
Data processing staff who designed and developed these
systems usually were skilled bit manipulators.   The
business applications they developed were very simple.
Middle managers did not view themselves as an integral
part of the application development process. Although
middle managers were responsible for the integrity of
the information they managed, they had little or no
control over information from other departments on
which many of their transactions were based. The end
users, who had not been involved in the software development process, had to coordinate these information
transfers.

Traditionally, the data processing staff followed
this approach to develop these systems:

1. They defined and analyzed the problem and
    determined the required information.

2. They specified the functions and features of the
    proposed system.

3. They designed the physical system taking into
    consideration technical issues like database struc-
    tures, machine implementation and which software
    building blocks to use.

4. Then they implemented the physical system using a
    programming language like Cobol.

    It is important to recognize there are problems
with the traditional software development approach.
    First, since a system is partitioned into many
small modules, checking and debugging inconsistencies
between these modules is labor intensive and error
prone. Often, changes made to one part of the system
introduce errors in other parts of the system.
    In addition, the specification of these many
modules is a time-consuming job. Analysts often
include too many technical details, which makes it
difficult for users and managers to understand. The
users, not understanding the system design, can't
determine if all their business requirements will be
met. Usually, it seems that no one in the organization
really understands the total system requirement or has
a good grasp of the subtle interactions between the
different components of the system.

IMPROVING THE TRADITIONAL SOFTWARE DEVELOPMENT APPROACH

    We need to improve this traditional approach.
Using fourth generation tools with old programming
methods is not enough; significant productivity gains
will only come when we also change the way we design
and develop systems. At Burroughs, we now train our
systems analysts in a new process for designing business systems. We teach analysts how to define requirements on a logical level, by working directly with the
end users. The requirements document clearly describes
the business problem and the proposed solution from the
end user point of view. It is not filled with implementation details and technical issues and will be easily
understood by users and managers. Our analysts also
learn to take into account the current system,
including data, report, and functional flow diagrams
since this helps establish the direction for the
proposed system.
    We can also improve the traditional approach by
providing end users with working models of the system
called prototypes. This is now feasible using fourth
generation tools. Prototyping can insure that design
requirements are correct before additional resources
are committed to the project.
    Bernard Boar, in his book, _Application
Prototyping: A Requirements Definition Strategy for the
80's_, notes: "Users must be actively involved to
ensure the functionality and acceptability of the
system. In conventional prespecification, users may or
may not find reading the documents and attending the

walk-throughs exciting.   They may sign off after a careful analysis of the proposal or simply sign off to get rid of it.   Users can't wait to see a prototype. Their eyes light up and ideas swirl as they experience the imperfect model.   The users actively participate because they have a comfortable medium to participate through."

Prototyping allows end users to try working models or application systems before final delivery.   They help make necessary changes and influence the shape of the final system.   The working model gives everyone a better feel for how the proposed system will work in the end user environment.   It helps determine if the initial design is correct.

You may ask, if prototyping is so easy, won't users request new systems even more frequently and with every feature and function imaginable?   And, since it will be easy to add more features later on, won't preliminary designs get sloppier?   Of course, proto- typing is not a cure for bad system design habits or lack of organizational discipline.   But when properly used it can be a very powerful tool.

TECHNICAL FEATURES/BENEFITS OF FOURTH GENERATION TOOLS

Now that we have discussed how we can improve the traditional software development approach by using fourth generation tools, let's look at some technical features of these tools:

1.   An interactive design and development environment prompts the end user or programmer to enter system specifications and immediately reports errors and inconsistencies.

2.   A corporate data dictionary controls standardiza- tion of common data items.

3.   Capabilities for development of on-line documenta- tion facilitate initial end user acceptance by helping them learn how to use the system. In the longer term this will help reduce input errors and improve productivity.

4.   Video screen and report "painting" allows rapid development of attractively formatted reports, customized to any organization's unique require- ments.

5.   Video "browsing" allows most system reports to be accessible on line and easily perused by the end user.   This feature can replace the need for

printed reports and eliminate delays in obtaining critical information.

6. Synchronized recovery allows the complex application system (all the programs, terminals, computers, and data bases) to restart automatically without loss of information in the event of hardware transmission, electrical, or other failures.

7. Inter-application communications enables different systems to share data or transactions.

By now, we should have a better appreciation of the benefits of using fourth generation tools:

1. Reduced application development and maintenance costs.   Experts estimate that even a very basic application generator can improve programmer productivity five-fold and integrated systems generators can increase productivity ten times. In an actual study of six companies using fourth generation development tools, Dr. Eberhard Rudolph of the University of Auckland, New Zealand, found that these tools improved programmer/analyst productivity up to twenty times.   Using these tools can help reduce application backlogs, and enhance management's ability to react to new business opportunities.

2. Better communications between user and DP departments.   The mystique of the computer will be reduced as users recognize its value as a business tool and accept their responsibility for information management.   The DP staff will better understand the business needs and become more responsive to end users.   Because end users will help create the applications, training time will be greatly reduced.

3. Information systems can be adapted more easily to the dynamics of the business environment. As new information needs are determined, these can be added without introducing errors into existing systems.   Applications can be readily customized to individual departments and organizations.

4. The corporate investment in application software will be   protected.   As the organization grows, the corporate information systems resource can be regenerated on future hardware configurations.

The bottom line is that fourth generation tools can dramatically improve organizational effectiveness. As better information systems become available, managers can spend more time managing the business and less time trying to collect and analyze the data needed to make decisions.

EXAMPLES OF FOURTH GENERATION LANGUAGES

Some examples of popular fourth generation languages include:

Language	Developer
ADR/IDEAL	Applied Data Research
Mantis Application Development System	Cincom Systems, Inc.
Application Development System/On Line	Cullinet Software, Inc.
NOMAD2	D&B Computing Service Co.
NATURAL	Software AG
RAMIS II	Mathematica Products, Inc.
FOCUS	Information Builders, Inc.
SQL/DS	IBM
MAPPER	Sperry Corp.
User Language	Computer Corp. of America
Information/Expert	Management Sciences Associates
Use-It	Higher Order Systems
PROGENI	Progeni Systems
PROXY	ESI
LINC II	Unisys

I couldn't resist saying a few words about our system, LINC, Logic and Information Compiler. LINC is available on all mainframe computers. Our new version, LINC II, allows data, relationships, screen layouts, and report formats to be defined interactively using menus, prompts, and on-line "Help."

LINC II checks these specifications for consistency and reports any errors. The programmer can correct these interactively.

LINC II generates the "environmental" software components to support the communications network and terminals, the shared database, security, audit, backup and recovery as well as the on-line application systems, report and query programs, end user prompts and on-line documentation. LINC also generates data validation and consistency checks to ensure that the defined rules and relationships are met.

In fact, the LINC II system is itself written by LINC. This allows Burroughs to easily enhance and maintain LINC on current and future Burroughs hardware architectures.

Like most large companies, Burroughs has significant information needs. My company employs about 60,000 people in 100 countries. We have twenty-four wholly owned subsidiaries overseas. One-half of our employees are involved in technical operations. We are a five-billion-dollar company with over 40,000 computers installed in customer sites. We now use our LINC II fourth generation product to develop our internal information systems as well as application systems that we sell to our customers.

GUIDELINES FOR CHOOSING FOURTH GENERATION TOOLS

I would like to close with some management guidelines on how to choose fourth generation tools. First, we need to distinguish between our operational areas. Some only generate parts of an application system and require a sophisticated systems staff to integrate the individual pieces into a working system. Many fourth generation products on the market today are query language tools geared toward informational needs. They often are used in conjunction with current operational databases, but usually do not address the problem of expanding these databases. Many programmers will favor these types of products, since they still can maintain control over current operational systems and will be able to continue to utilize traditional technical skills.

But the foundation of a good information system is the underlying operational system. Decision support tools and management information systems are only as good as the underlying operational databases. Organizational priorities should establish a balance between investing in new operational systems versus additional management information. It is the responsibility of top management to help establish these priorities. Senior managers are starting to understand that corporate objectives determine corporate information, not the other way around.

As you know, my interest is education. Vendors like Burroughs are involved in partnerships with educational institutions. We look to education to train the next generation of systems designers to properly use technology. As Sidney Harris said, "With our technology changing so fast, what we used to call a practical education has totally changed: It's no longer learning how to do something, but learning to prepare for a future that does not yet exist."

Advances in applying information technology will be accelerated as long as we continue to train our students in the latest technologies. We at Burroughs are committed to providing our higher education institutions access to the latest technology such as LINC II to prepare for this future.

Information, corporate databases, and management information systems are some of our company's most precious assets. Management of these assets will be essential to maintain our competitive posture and to ensure our future success.

27

Strategic Use of Information Technology

Richard L. Chappell

I will be discussing change. I will discuss how you can take advantage of change to create a sustainable competitive advantage, particularly through the use of information and information technology.

Managing has always been a difficult process, but over the past few years it has become increasingly difficult as change in our economy and each of our businesses has accelerated. Abe Lincoln once said there are three kinds of people: those that make things happen; those that watch things happen; and those that say what happened.

I would like to give you some insight into the future from Arthur Andersen & Co.'s perspective. Arthur Andersen & Co. is best known as a major international accounting firm. However, the firm has been helping clients make productive use of information technology since the information age began back in 1952--thirty-five years ago. The first computer application for business was installed that year in Louisville, Kentucky, at the GE Appliance Park--and AA & Co. was the consultant. We have continued to maintain a leadership position ever since. Our Management Information Consulting Division has grown during the last thirty-five years and we are now the world's largest consulting organization with over 9,000 professionals.

At AA & Co., we learn a lot--year in, year out-- in our work with clients who are pioneers and who are trying innovative and creative ideas in their quest to understand and take advantage of the information resource. I'd like to share with you what we see the pioneers doing with information and information technology. My remarks will cover four points:

1. What is the linkage between Information Technology and strategy?

2. What are some emerging information technology trends?

3. What are some of the recommendations that we make to our clients in helping them prepare for this emerging information technology?

4. To share some specific examples where companies have used this emerging Information Technology for strategic purposes.

LINKAGE OF INFO TECHNOLOGY/STRATEGY

In my opinion, the key link between information technology and strategic systems is--sustainable competitive advantage. Information technology presents one of management's most significant challenges--it can be the key to gaining sustainable competitive advantage in your marketplace if used innovatively; but, for companies and managers who don't stay abreast of these trends, there is a risk of falling behind the market.

During the past few years, the competitive balance has been altered within many industries--auto making, airlines, telecommunications. One list shows how hard it is to preserve the competitive status quo in an industry: the Fortune 100. Sixty-one of the top U.S. companies in 1958 were no longer included in 1983. Of the thirty-nine that remained, twenty had dropped in the rankings and only nineteen of the top one hundred had improved their competitive positions over the twenty-five-year period. Statistics don't really tell us everything, but these do suggest that complacency and the failure to adapt to change may lead to an organization's decline or even demise.

The impact of information technology is one of the primary causes driving the change we have experienced in recent years. The advances to the silicon chip have given us the capability to alter and enhance product capabilities and to improve our overall productivity dramatically. In the process, information technology has eroded the competitive strength of organizations that chose to believe that the marketplace still demanded the proverbial buggy whips.

Information technology, in fact, has turned what had been a gradual industrial evolution into a virtual revolution. Over the next decade, technology and the use of computers are expected to account for almost half of the total productivity gains realized by U.S. industry as shown below:

Sources	% Productivity Gain in U.S.
Technological Innovation	25
Computer Power	15
Intangible Capital	16
Scale Economies	16
Education	12
Better Resource Allocation	12

Information technology is being used as a strategic weapon by more and more organizations. The airline industry, for example, is totally dependent on information technology for its reservation systems. Together, the American Airlines Saber system and the United Airlines Apollo system have captured more than 80 percent of the reservation market by using their systems to gain competitive advantage.

In the distribution industry, a prime example of information technology being used to gain competitive advantage is American Hospital Supply, now part of Baxter Travenol. About ten years ago they began to offer a system to gain competitive advantage by making their inventory information available on-line to major hospitals through terminals. Hospital personnel appreciated the convenience of ordering with this system, developed a habit of using it, and began ordering more from AHS than other suppliers. Hospital personnel felt comfortable with it and, as such, did not want to learn another order processing system. As a result, American Hospital Supply gained control over the primary channel of distribution, creating a barrier to competition, giving them a significant competitive advantage over their major competitors.

We are seeing a trend develop. In the 1960s and 1970s, we were automating basic applications. They tended to be unintegrated, stand-alone, fundamental business applications. Good examples would be payroll and accounting. In the 1970s we got into integrated transaction and control systems. In the last five years, we have seen more emphasis on strategic systems, systems that differentiate you from your competitors.

We're in the era now where we are using information technology as a vehicle for gaining sustainable competitive advantage. That requires a fundamental shift in management attitude toward managing the information technology resource.

The only way you are going to survive in the 1990s is to use information technology to its fullest extent, because if you don't, your competitor will and your competitor can put you out of business. This is especially true for industries with high information intensity such as financial services companies. In short, how you use information technology is going to have tremendous impact on the success of your business.

Not all organizations are affected equally by information technology. The matrix in Figure 27.1 measures something we call INFORMATION INTENSITY. It reflects both the degree to which information is incorporated in the product or service itself and the degree to which information is used in the process of creating the product or service.

For example, the financial services industries are highly information intensive. There, product is an exchange of value between two or more parties and value today means, above all, the exchange of information about money. Their operating processes are entirely information driven; the heart of a bank or brokerage firm is no longer its vault but its computer and operation center.

At the other extreme, a tire manufacturing company is relatively low in information intensity. There is essentially no information content in the product

Figure 27.1.   Information Intensity

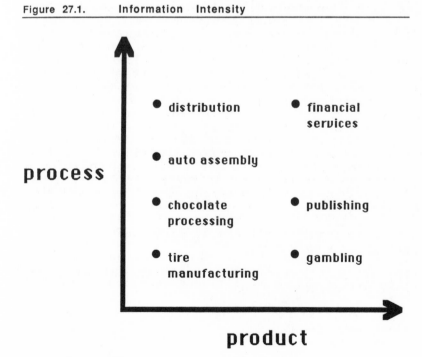

itself. Likewise, relatively little information or information technology is required to create the product.

The matrix is not static. Organizations are increasing their information content as part of their overall strategy. I think a good example of this would be Federal Express. I'm sure many of you have seen their recent ads which extoll the virtues of their being able to tell you within X minutes where your package is. Essentially, they are adding information as an integral part of their service, increasing the value of their service and differentiating their service from the competition.

It is clear that many companies are attempting to exploit changing information technology for their own advantage and that your present success is no guarantee of future success. Will Rogers once observed, "Even if you are on the right track, you'll get run over if you sit there."

More and more, we are being confronted with competitors who are aggressively investing in information technology. Not only is competition intensifying, but proliferation of new information technology opportunities greet us. This leads us to address the issue of how an organization can use information technology to gain competitive advantage.

In attempting to answer these questions for ourselves and our clients, Arthur Andersen has been working with Professor Michael Porter of the Harvard Business School. Porter's first book, Competitive Strategy, is one of the most widely read publications on strategic planning and the analysis of your existing or potential competitors. His recently published book, Competitive Advantage, takes the next step in identifying the sources for building and sustaining the long-term competitive advantage. We have worked with Porter to develop a framework that can help you assess your competitive posture and determine the ways in which you can use information and information technology in enhancing your overall competitive position.

We believe that information technology can play a significant role in the strategic quest to develop that sustainable competitive advantage enabling you to do three things:

1. Alter the industry structure, shift the competitive balance in your favor,

2. Improve your relative position in the existing lines of businesses, and

3. Create entirely new business opportunities and new industries or markets.

COMPETITIVE ADVANTAGE: ALTERING INDUSTRY STRUCTURE

Let's first evaluate how the use of information
technology can help you alter the structure of an
industry. Information is altering the rules of
competition among the Five Forces described by Porter:

1. The power of buyers

2. The power of suppliers

3. The threat of new entrants

4. The threat of substitute products

5. The rivalry among existing competitors

Information can be pervasive enough to alter each
of the five forces and, therefore, even the structure
and attractiveness of the industry. Technology is
unfreezing the structure of many industries. Information technology has altered bargaining relationships
between suppliers and buyers. Operating information
designed to cross company lines is becoming common. In
some cases, the boundaries of industries themselves
have changed.

My firm installed a system for the new Nissan
plant in Smyrna, Tennesse, that automatically integrates the supplies into the production schedule. It
does this so well that the plant carries inventories
well below those of the typical manufacturer; it also
reduces the cost of indirect labor associated with the
assembly line. For example, the seats for the truck
assembly line are finished and loaded in the supplier's
plant in color sequence so they can be unloaded at the
same time and in the same sequence as the trucks with
which they are associated. The cost of inventorying
these seats is virtually eliminated at both companies.

Charles Schwab offers customers on-line access to
information, as well as on-line capability to place
orders. They were attacking a different competitive
problem than American Hospital Supply--one unique to
the brokerage business. Customers in this business
have been notoriously fickle--moving from firm to firm
as their salesmen change firms. Schwab's investment
information on-line differentiates their firm from
others in the brokerage business and at the same time
isolates the customer from Schwab's most threatening
possible future competitor--their own salesmen.

Xerox supplies its master manufacturing schedules
on-line to its major suppliers, notifying them of
exactly what Xerox plans to produce and when, making
the suppliers responsible for helping Xerox maintain

efficient inventory control. This information reduces
Xerox's cost structure and allows its suppliers to do a
much more effective job of production scheduling,
reducing their inventory levels.

COMPETITIVE ADVANTAGE: IMPROVING EXISTING POSITION

What about more traditional means of achieving
competitive advantage? To enhance your existing position, you effectively have two choices--you can either
become a low cost provider in your industry, or you can
become a differentiated provider, responding to the
buyer needs uniquely in a way that commands premium.
Being "all things to all people" is a recipe for strategic mediocrity and below average performance--it
probably means a company has no competitive advantage
at all.
These are two distinctly different strategic
approaches and, as a rule, it is exceptionally difficult to have both low cost and differentiation. The
reasons for this are straightforward. Differentiation
says you are adding value in a way that is important to
your buyer. That buyer value may be satisfied through
better materials, quality of service, or through the
perceived value which is often reinforced through the
costly advertising and marketing programs. This value
added approach is generally inconsistent with being the
cost leader in an industry. Technology may be one of
the few opportunities you have to both differentiate
new products and reduce your cost structure at the same
time.
Low Cost--Information technology can reduce a
company's costs in many ways. Historically, its impact
on cost was confined primarily to repetitive information processing. These limits no longer exist. Even
assembly processes with extensive physical processing
now have a large information processing component.
Canon, for example, built a low-cost copier assembly
process around an automated parts-selection and
materials-handling system using automated information.
Differentiation--The contribution of information
technology to differentiation strategies is equally
dramatic. Information can be used to customize products and services. For instance, Digital Equipment
has developed an artificial intelligence system, XCCN,
to improve accuracy in customizing computer configurations. This competitive innovation was designed to
differentiate Digital from its competitors by enhancing
its image of quality.
These examples of the use of information to gain
competitive advantage in existing lines of business
represent the most pervasive of the measures used by

businesses to reap the harvest of the information revolution.

We have found that the best tool for finding opportunities to reduce cost or enhance differentiation is the value chain--developed by Professor Porter. By analyzing linkages in the value chain, companies can discover new ways to use information to reduce their costs or enhance their differentiation.

COMPETITIVE ADVANTAGE: CREATING NEW OPPORTUNITIES

Another way in which you may find a new source of business is a by-product of an existing one. For example, food retailers use bar coding for inventory control purposes. By capturing this information, they now have a very valuable source of market research information which they have, in turn, been able to sell to market researchers around the country.

There are many examples where information technology can create a number of new business opportunities and this may, in fact, be one of the most exciting sources of competitive advantage which could have the ultimate impact of altering industry structure as well.

EMERGING TECHNOLOGY

With that background, let me now shift to emerging trends in information technology. In general, the cost of hardware has come down and the outlook is a continuing downward trend. At the core of these cost decreases is the microminiaturization of computer circuitry. It has accounted for decreases of 20 percent per year across all types of computers with much more dramatic drops in the cost of small computers as compared to large ones.

The unit cost of data processing is much lower on smaller rather than larger computers. The reason? Fast circuits are disproportionately more expensive than slower ones--just as a supersonic jet is dramatically more expensive per passenger mile than one of the slower models.

The inversions in the economics scale are causing the move away from centralized computing and spreading small and medium-sized computers throughout the organization. The result is a multi-tiered network. While these networks have all the appearance of enhanced economy and responsiveness to end-user computing, these benefits don't come easily. Making the best use of a multi-tiered network takes careful design, well educated and attentive management, and thoughtful hardware and software selection.

Looking at the hardware marketplace, the electronic work stations will likely be in place for every white-collar worker by 1990. This machine will cost less than five percent of the manager's annual salary and will be as powerful as a machine costing twenty times that manager's salary back in 1970. That is a 400 to 1 improvement in cost effectiveness. The potential for the PC is positive as we move into the control phase of its use. Lack of planning that was an obstacle in the past will be replaced by sound planning and implementation techniques by companies who use the PC as a powerful, competitive tool. PCs will also see increasing application as they replace dumb terminals in high-volume transaction processing systems.

Multi-tiered network and electronic work stations are two key trends that management is going to have to come to grips with.

RECOMMENDATIONS FOR USING NEW INFO TECHNOLOGY

How can managers turn information technology into a strategic tool? No matter what your business or industry, information technology can be an important strategic tool for senior managers. Arthur Andersen has developed some specific recommendations for companies who want to make the fullest use of information to manage change and gain competitive advantage.

At the core of the recommendations is the new member of senior management, the chief information officer, or CIO. In most companies, it will take a catalyst to bring about effective use of information technology. The average business has no such catalyst--no one who is responsible for information. With a CIO in place, companies can become effective and more competitive.

What will the CIO contribute to the organization? This is the senior executive who can:

- Harness information to build business strategy that will make the organization more competitive.

- Put an information plan in place to ensure that information technology contributes to the success of the organization.

- Enhance productivity of systems, the systems' users, and the information produced by the systems.

- Make optimal use of the end-user computing and all the efficiencies that it offers.

- Manage the implementation of telecommunications systems to enhance the use and management of information.

- Use state-of-the-art tools and tactics to move the organization toward office automation.

    The second recommendation for companies using information to make them more competitive is the integration of systems architecture. With a CIO in place, companies can choose wisely from the options that are available for integration. Some of these options include:

- Develop a common data strategy by determining the information needed to carry out the business as a whole.

- Develop hardware and software strategies that will allow for a common technical architecture.

- This infrastructure may feature a three-level hierarchy, with a:

    1. Mainframe computer at the center managing corporate data and providing broad information distribution services;

    2. Departmental computers at the next level managing local data; and

    3. Intelligent work stations like today's PCs providing the human interface.

- Develop a network strategy to connect these workstations by a LAN (Local Area Network). Systems will be designed with components operating on all levels of the networks.

    Implementation of the steps we recommend depends on establishing Information Services with new responsibilities that include:

- Managing corporate databases;

- Acquisition of new technology;

- Improvement of systems development productivity; and

- Facilitating end-user computing.

The last two responsibilities for Information Services--enhanced systems development productivity, and end-user computing--may be among the most important. Keys to this enhanced productivity would include:

- Prototyping--a technique for demonstrating how the system will work before it is fully implemented. When users have input, costly redevelopment can be avoided.

- Application software--a technique for such common business functions as payroll, accounts receivable, or order processing and production scheduling.

- Fourth generation languages--new tools that enable end-users and programmers to develop programs in a fraction of the time required when using a lower-level language such as COBOL.

- Integrated development environments--systems development package that provides all the tools the systems designer needs--in a single system.

At Arthur Andersen we have a designer's workbench that is microcomputer-based and we have multiple versions of a programmer's workbench to fit the particular technical systems environment.

End-user computing--the other key responsibility of Information Services must also be used to harness information technology. To accomplish this, companies must:

- Identify a recommended personal computer (PC) and the appropriate software.

- Provide the required data in a shared database using the information center concept.

- Implement effective micro-to-mainframe links to enable data to be downloaded to PCs for manipulation.

- Provide mainframe-based products for applications that exceed the capacity of PCs.

- Provide education and training for end-users.

Will this thing work for you? Of course there are no guarantees, but track records show that companies who have implemented these tactics have built the information infrastructure that is required to meet the demands of today's market.

Peter Drucker says it takes forty years for a new technology to be accepted and incorporated into daily business activities. Information Technology is now entering that same time frame. We are at a fundamental transition point--at the beginning of an information technology revolution.

Information Technology can be used for competitive advantage--as a strategic weapon. Information technology can:

- Raise barriers to new entrants or substitutes.

- Reduce costs.

- Increase buyer switching costs.

- Increase differentiation, product quality, service, and higher prices.

- Enhance distribution channels.

Managing strategically is critical in times of change. I think we all believe the rate of change is likely to accelerate in the years ahead, placing even greater pressure on us as competitors in our own indus-tries today. Our challenge is to recognize and embrace change, particularly technology-based change, taking advantage of these new opportunities to enhance the value we provide to the marketplace, gaining for ourselves a SUSTAINABLE COMPETITIVE ADVANTAGE.

28

Managing Creative Programs at Unisys

Michael A. Brewer

Ask ten people to tell you how they manage change, and you'll no doubt get ten different answers. Some liken it to dancing as fast as they can when the record keeps changing. Others take it on simply because it's there.

However we define it, change is as much a part of being in business as receiving a paycheck. (Although today, some cynics would say we can count on change with more regularity!) Handled properly, and welcomed rather than dreaded, change is the necessary refresher to a tired way of looking at things. No matter how gruesome it may appear at the outset, change will always bring benefit, because we grow and learn from it. It never fails to challenge and to extract greater creativity from us.

As I see it, change presents the ultimate challenges; it dares you to manage it. If we look at the great successes in business, we'll see that they were brought about by people who dared to face change and master it and most important--turn it to their advantage. Since my base is in Detroit, examples from the automotive industry come to mind. Henry Ford mastered change when he created and implemented the production line and the five dollar a day wage. More recently, Lee Iacocca overcame many obstacles and employed great change to rescue Chrysler from bankruptcy. Within Unisys, my company, innovative engineers have mastered change and pioneered breakthroughs in the computer industry. Bob Barton defied all odds when he successfully pushed for a machine that was the first to use the concepts of multiprocessing, multiprogramming and virtual memory on the B 5,000 computer. That was twenty-five years ago and even today, many systems strive for the same standards of flexibility and user performance.

STYLES FOR MANAGING CHANGE

Like many of you, my job is beset by change; from
employees, company, the customer, industry, technology
design and manufacturing. All must be juggled successfully. The trick is how: Do I manage change with
an iron fist, or with kid gloves?

The knee jerk reaction is to take total control.
The power of control conveys a comforting superiority--
and is the easiest way out. If I was to describe my
style of managing change, it would be as a little of
both--a firm hand in a kid glove. That style works for
me because it lets me acknowledge the sensitivities and
needs of those who make it happen, without being perceived as "too soft."

I manage a staff team responsible for product
programs for intelligent workstations, personal computers and office information systems. Our products are
marketed in nearly one hundred countries, and support
twenty-one different foreign languages. Our task is to
conceive the product and deliver it to market--on time
and at the right price, until obsolescence. To make
life even more interesting, all this takes place in an
extremely dynamic marketplace where companies are born
and die within a year, and even the giants take their
share of bruising.

Managing people with a firm hand in a kid glove
takes a certain amount of detachment. Let me give you
some examples. During the course of a product program,
we go through several steps that must occur to get the
product out the door. There's the research mode, the
creation mode, development mode and implementation.

MANAGING CHANGE DURING RESEARCH/CREATION STAGES

During the research and creation modes, we define
our market and its requirements. We usually target one
of three market potentials: A solution to a problem
using existing technology, improve on an existing solution, or create an entirely new solution using new
technologies. As tempting as it might be, our technical creativity cannot get the best of us to the point
where we have developed a batch of solutions out
looking for problems. The innovative manager has to be
firm and objective enough to overlook the elegance and
creativity of the design proposal if it does not solve
a problem. The manager must also know when to move the
program from phase to phase. Many a good idea has died
wallowing in research mode and many have died emerging
unready for development.

A direct outgrowth of knowing our market
requirements is knowing which and how technologies

properly apply to the product.   We must approach
technology objectively, bearing in mind my earlier
comments about solutions searching for problems.   Our
eye must be creative, yes--and also businesslike.   Product managers are often smitten by a preoccupation with
engineering elegance and the glamour of technology.
It's only natural.   However, the manager must view the
technology area as an investment of the company's
money, people and other resources.   The job is to
realize that investment.

If there's anywhere change is more prevalent and
needs to be better managed, it's in technology.   The
available technology for workstations, PCs and office
systems changes every day.   Processing speeds move up
at an alarming rate.   In the great data storage race,
we've gone from 12 megabytes supporting 2,000 terminals
twenty years ago to 300 megabytes supporting a single
workstation today.   Continuous downsizing of components
has changed the physical profile and ergonomics of the
industry.

How do we manage technology on a product program?
First, we need to determine which technologies are
required.   Does the user ultimately want a very fast
processor? Ability to integrate text and graphics? or
PC networking?   Notice I didn't say, "What would Mike
Brewer like technology to be?"   The key person is the
user of the product and the user is looking for key
results.

Second, we need to know if new technologies are
present and mature enough, or if there is a shorter
term solution using a current technology.   Third, we
need to know what's new and worth pursuit, and fourth,
how to get there.

MANAGING CHANGE DURING DEVELOPMENT STAGE

Only then can development begin.   At Unisys,
development of a product happens in one of two ways--by
our engineers, or contracted outside to an original
equipment manufacturer, or OEM.   We happen to have a
longstanding relationship with a specific OEM in the
workstation product area, and this presents its own set
of challenges, which I'll describe in a moment.

The development scenario in the workstation
product area is a balancing act.   The rapid pace of
technological change has shortened our product life
cycles considerably, in comparison to its larger mainframe cousins. A mainframe computer, for instance, has
enough horsepower and is a great enough investment to
merit a product life cycle of twenty-four to thirty-six
months, although even that is narrowing today.

In the workstation marketplace, new generations of microprocessors are arriving on an annual basis, with Intel and Motorola leapfrogging each other. This is important because it is the microprocessor that provides the speed and power that many of our workstation and PC users crave. Certain other features, like storage, change even faster. We're moving from a 20 to 40 megabyte disk today to 80 and 170 at the end of the year—double and redouble capacity. To put things in perspective, the 10 megabyte disk born in 1983 is technologically and market dead. Memory capacity quadruples and its price drops to one-fourth every four years. Screen size, resolution and type have all undergone immense change.

The result of these technology changes is a shorter product life cycle. I'd estimate that a workstation's product life today is twelve to twenty-four months, at best. To sustain an existing product it must be performance upgraded annually and there must be a strong technological upgrade every two years.

It is within this scenario that we must balance internal and external product development. For example: engineering is working on a product that will integrate voice and data over a network, and this product is scheduled to ship in a year. However, a vendor has it now, the price is attractive, and the hardware and software are compatible with your products.

We then must make some choices...choices that involve managing a fast major change in a very short time. Do we forgo internal development to catch the elusive market window before our competitors? Or should we wait it out? How strong is the demand today? The cost and risks must be weighed against the significance of the early offering. An even more sensitive issue we continue to grapple with involves our internal resources and skills. Should we continue to split development between our engineers and an OEM? By aligning solely with the OEM, we may risk losing our reputation for innovative engineering and worse yet the engineers who built it. We must continue to balance our decisions based on the economics of time, cost and resources, but ensuring preservation of critical skill sets in strategic product areas.

As you have probably concluded, workstation and PC product development rarely takes a straightforward, synchronized path. More often than not, it is asynchronous, a high-powered effort driven by market demand. About six years ago, we incorporated major changes in human factors, or ergonomics, into our workstations and terminals within six months because our European customers considered this important. Within a year, our terminals had been certified by Professor

Grosjean of the Technical University of Zurich as ergonomically "perfect." That high-powered effort paid off. We gained a six- to nine-month lead over the competition, which was then forced to play catch-up. Six to nine months may not sound very significant, for us it was half a lifetime!

MANAGING CHANGE DURING IMPLEMENTATION STAGE

The development cycle of a product is followed by implementation. I consider implementation as the following: the path from development to manufacturing and from manufacturing to the marketplace. Additionally, implementation includes establishing the necessary product assurance and support functions.

Once again the economics of internal versus external resources come in to play. Our manufacturing plants face two formidable challenges in the workstation/PC area. First, they must keep product cost down in the face of strong competition from sources such as Korea, Singapore, and Japan. Second, they must maintain highly competitive quality and performance. Cost/performance is a prime measure of the product.

Within my group, our job is to see that the product is sourced at the best possible cost, with reasonable margins, without sacrificing quality.

Whether we're dealing with internal or OEM, our role also includes managing delivery schedules based on orders from marketing, supply sourcing and inventory. Responsibility for asset management embraces both finished goods and work-in-process because the workstation/PC market is so dynamic, we're constantly on the watch to see that we avoid excess inventory and provide plans and programs to control it. This can take many forms; pricing actions or special package programs are just two commonly used tools. When we're dealing with an OEM, we contractually commit for their product availability and this source must be managed as carefully as our own direct sources.

When the product moves from manufacturing to market, we are responsible for developing all of the various marketing and promotion programs for the sales force. These include providing input and often writing the raw copy for advertising, brochures, product announcements, technical documents, training, and even the public press announcement, if that option is taken.

The sales force is our toughest audience to please. They see a lot of workstation competition in the marketplace, so we have to make sure that our products have distinguishing features and are attractive for them to sell. Their customer presentation

material must be first class.   The technical
information we provide must be accurate and timely.

This leads me to another important facet of a
successful product implementation--our support functions. We are responsible for ensuring the plans and
programs are in place for the product in terms of field
engineering service, customer training, and documentation.   Any problems that arise in these areas will
invariably come back to us, since we are the focal
point for everything about a particular product.   In
Unisys the program manager owns the entire product
program.

BUSINESS PLAN: TOOL FOR MANAGING CHANGE

How do we keep track of this inexhaustible supply
of issues, problems and just plain details? And how do
we manage changes that affect various facets of a
product program?

Before anything is begun, we start with an agreed-
to, thorough business plan for our products.
Certainly, it changes often, but we have built into the
plan mechanisms to manage change.

What's in our business plan, and how do we go
about fashioning it?

First of all, some thought on what the plan is.
It is:

1. Written down for all to see.   There are no hidden
    agendas that cause the program staff to try and
    guess what it is they're responsible for.

2. Fashioned by all of us, and shared among us. That
    way, we all have a greater stake in its success.

3. Presentable at any time to senior management, if
    they should request it.

4. Continually updated.

5. A means of handling change. How? We do a lot of
    contingency planning..."what ifs"...when the plan
    is written.   What if this market doesn't materi-
    alize?   What if the customer doesn't like this
    software feature?   We must be prepared--in
    writing--and know what our Plan B is for as many
    instances as possible. It is a valuable exercise
    that we have found avoids "brush fire" or crisis
    management a good deal of the time. It also builds
    a lot of resilience into the plan. If someone gets
    sick, leaves the team, or gets transferred, we
    know where to pick the program up and keep it

going. A good contingency program constantly tests
the logic of the main plan.

What else does our business plan contain? We
include dynamic financials for our program. Here we
borrow lessons from the financial community. We examine the product from all relevant financial measurements and apply sensitivity tests based on experience
with this type of program. For instance, what will
happen to the financials if we are three months late to
market? Time lost at the front of a program is gone
forever. And, as often happens, what if our competitors slash prices 20 percent? How does that affect the
financials?

We look at what the product will cost in terms of
design and manufacturing and its competitive position,
and then determine the pricing. We work out the revenue targets and margins based on the corporation's
goals. Will our program stand up and secure funding
justification?

The program business plan must include cost
results projections on promotion plans, training and
field support arrangements to round out the complete
launch.

MANAGING PEOPLE REQUIRES OPEN SYSTEM

As you've heard, this entire process deals with
change as a matter of course and not as an unwelcome
visitor. The evolutionary change of the product program can have dramatic effects on the program staff.
This is a creative group of people from many different
disciplines. Some are engineers by training. Others
are from sales. Still others are administrators. Each
brings his or her bias to the table when a product
program is formed. Engineers may want the product to
be technically perfect. Those with a sales background
may want bells and whistles. Administrators may want
the cost contained and a tight financial picture. All
these views must be harmonized, and the energies chan-
neled positively and constructively. Listen carefully
for those who emphasize the ideal program plan, for
they may not be prepared for change.

Sometimes, the differences are more pronounced
because of cultural backgrounds. During time spent in
Europe it was, for me, a good experience bringing
together engineers from France, Belgium and Scotland on
a product program for financial terminals. In addition
to the language and cultural differences among them,
there were differences in how they were taught in
engineering school, not to mention a little national
pride in approach. After a good deal of discussion

across the borders, we eventually came to agree on the program direction. It was an altogether challenging experience, the result was wonderful, a tight team performance and product result second to none.

Creatively managing people with diverse backgrounds requires that you maintain an "open system." Let people know what you're doing at all times. Get everyone interested in the program to establish a sense of ownership. Try and show the team that there is something in the program for them, that they share in its success. Establishing this environment takes nurturing and encouragement. There must be no such thing as a crazy idea or dumb question. All have a say in the development and direction of product programs. In fact, many of the program managers grow attached to their product programs in such a way that the product becomes their "baby" to be nurtured every step of the way. It is important to be sure that this commitment does not color their judgement.

CONCLUDING SUCCESS STORY

My conclusion is a success story that includes many of the elements we discussed. We first introduced an intelligent workstation known as the B 25 back in 1984. It was the first time the company went outside to an OEM, and we had our share of struggles and concerns on how to manage what was a major change. Our marketing attitudes and methodology were confused and lacked a clear sense of direction.

Early in 1985, we drew marketing, manufacturing, engineering and the supplier together and began to encourage team ownership in the product program. It was agreed that the internal engineers would do the value added for our lines of business. Manufacturing now had a stake in the program by negotiating manufacturing rights with the supplier. Marketing strategized a position for one product family built on ascending levels of microprocessor technology. Everybody had a role to play, a goal to achieve, a vision.

It 1986, it showed. Our unit growth was up 50 percent, and our margin improvement strong. The quality and performance standards were a testimony to excellence.

We were successful because we got all the players to own a piece of the eventual success--they literally had a stake in it. We kept the lines of communication open, which helped us effectively manage the many changes that came our way.

That's the key, really, to <u>welcome</u> change and manage it properly, with a firm hand in a kid glove.

29

An IBM View of Information Processing in the 1990s

Michael Connors

Most speakers want their audience to remember at least one thing. Sometimes all the audience can remember is, "Thank goodness that's over." I hope that's not the case here. The one thing I would like you to remember is that the rate of change that has taken place in the information processing industry for the last ten to twenty years is a rate of change that I believe will accelerate in the next ten to twenty years and persist to the millennium.

DRIVER OF CHANGE: SOCIAL-ECONOMIC STRUCTURE SHIFTS

I'd like to talk to you about why I think that's going to be true and maybe give you some perspective on what the rest of this decade will be like and what the decade of the 1990s might be like. I'll introduce three subjects that are drivers of change. One of these drivers is transitions that are going to take place in our social-economic structure, the second driver is changes in technology, and the third driver is the integration of the media of communication. These three thoughts form a basis for discussing a number of opportunities that will be created for the information processing industry and for the customers who use information processing equipment as we go out through the remainder of the decade of the 1980s and turn over into the decade of the 1990s. Let's start with the driver of change: transitions in our social and economic environment. It's certainly no secret that over the last century there has been a change in the basic structure of our economy. The change has been from an agrarian to an industrial to a service economy. Presently, the service economy is changing into an information economy with an increasing proportion of the workforce being knowledge workers.

DRIVER OF CHANGE: NEW TECHNOLOGY

The second driver of change is technology. Technology is the big bang from which almost all industry events have been derived. The rate of techno-logical improvement, expressed in terms of price per-formance improvement, has run along in the range of 20 to 25 percent per year and all indications are that it will continue to be in the range of 20 to 25 percent per year until the turn of the century.

You can put some quantification behind this estimate for large processors.   Large processors are composed of logic chips. Those logic chips are mounted on ceramic packages.   The result is a processor that has some level of capacity.   Industry jargon for capa-city is mips.   Mips is an abbreviation for a million instructions per second.   All of this is a bit irrele-vant to the customer.   What the customer pays is a price per unit of capacity, and one way of looking at that is the price per mips--price per millions of instructions per second.   And what the industry has experienced is that in exchange for taking ten years worth of progress in time, you get approximately a thousand-fold improvement in circuit switching time. In other words, in 1970, logic circuits switched in the range of a microsecond, which is a millionth of a second.   Ten years later, they switched in nanoseconds, which is a billionth of a second.   By 1990, switching times will be easily within the range of a trillionth of a second and for those of you that are interested in absolutely spellbinding audiences at Saturday night parties you can tell them with certainty that, by the year 2000, switching times can be in the range of femtoseconds (which is one quadrillionth of a second). In other words, ten years yields a thousand-fold improvement in circuit switching times.

There are very few other industries that press the rate of change at a thousand-fold every ten years. When you can switch circuits faster and faster and faster, you can put more circuits on a single piece of silicon.   You can put more circuit chips on a single package and the result is that you can build single system images of larger and larger capacity.   A very large processor in 1970 would be a one mips processor. A large single processor in the year 2000 will be a 100 mips processor.

But that's really not the end of the story.   The pay-off to customers is that, in 1980, one mip cost $500,000.   By 1990, I'm suggesting that one mip may cost $100,000 and, by the way, this is 1987 and you can buy a mip today for about $150,000 to $160,000 so we are on the curve, which means by the year 2000 one mip could cost $10,000. This is what it means to share with the customer environment a price performance improvement of between 20 and 25 percent a year, year after year after year, decade after decade after decade. All this I believe to be technically quite easy to do.

For those of you who do not wish to discuss femtoseconds on Saturday night, the short version of all of this is that while computer logic is improving at 25 percent per year, computer memory costs have been improving at a much higher rate, maybe 40 percent per year. Communications costs have been improving at 11 or 12 percent. These are all pretty good compound growth rates of improvement because these are repeated year after year after year. At the same time, labor costs have been going up. Clearly, technology is much more productive than labor.

So far all we've discussed is the large processor environment. Let's talk for a few moments about the small processor environment, and when we talk about small processors, we should think of a personal computer or whatever your version of a desk top processor really is. A couple of years ago, personal computers were comparatively new, and I think it is fair to say that they were complex and difficult to use, at least I found them that way. By the 1990s, we think that personal computers will be increasingly easier to use and therefore more ingrained in our day to day life. Not only our day to day business and educational life, but more ingrained in our day to day personal life.

We can put some quantification behind the idea of personal computing or end user computing technology now and in the 1990s. Today, a personal computer is perhaps four-tenths of a mips in processing power. A large real memory would be 640 kilobytes. A large random access memory would be perhaps 20 million bytes of information. By the 1990s, the computer sitting on your desk--I hope my desk for sure--will be a ten mips processor (that's a twenty-five-fold improvement in performance), it will have 16 million bytes of real memory, and it will be a common thing to have 100 megabytes of random access storage. One of the major physical differences would be, that instead of taking up the majority of your desk, it would be comparatively small and perhaps even built-in. Another difference would be that while today we depend upon keyboard for input, by the 1990s and now I am really describing the late 1990s, it would have non-coded information as input. It would be normal to have image information as input and by the late 1990s it would be reasonably common to have voice as input to the computer.

DRIVER OF CHANGE: MEDIA INTEGRATION

     Let's talk about the third driver, media
integration. There have been traditionally four separate kinds of media. There's been data, text, image,
and voice. The concept of media integration is that
traditionally computers have processed data, typewriters have dealt with text, copiers have dealt with
image, and telephones have dealt with voice, but
increasingly as we go out through time, one product
will deal with multiple media. We see examples of
media integration. We have intelligent copiers. In
other words, a device that is traditionally an image
processor having computer-based logic so it can be a
data processor and a text processor as well. We have
examples with cash registers that are traditionally a
kind of local data processing device, but very often
nowadays you see cash registers with magnetic readers
so that you can take your Visa Card or your American
Express and drag it through the device. The cash
register will use the voice media (the public switched
telephone network) to dial a computerized data base,
engage in data processing to check whether or not your
particular card is valid and flash back to the cash
register a light indicating whether or not credit
should be extended, whether or not credit limit is
being exceeded or whether this is a lost card--a
combination of data processing and voice processing.
     Last but not least, media will be integrated by
translating text into voice and that's possible right
now. As a matter of fact, an experimental computer
that I have on my desk can dial up an external
database--an example would be the Dow Jones data base.
I've programmed into it a profile of my personal interests.   It will search the Dow Jones database for
today's news stories concerning my personal or business
interests and then read them to me while I'm doing
something else, perhaps reading the non-electronic mail
or writing a memo, so that if something comes along
that I'm particularly interested in, I can either
listen to it or I can turn around and watch it being
played back on the computer screen itself. So text-to-
voice is here today. The complicated thing to do is
voice-to-text. In other words, take your voice, translate it into digital information and have that digital
information made a permanent part of the magnetic media
which could be transmitted to somebody else, stored,
used or retrieved. Nonetheless, voice-to-text will be
a reality in the 1990s and that will be the ultimate
integration of media.
     We are beginning to see today the emergence of the
first aspects of media integration and that can be
thought of as a compound electronic document. In other

words, a document upon which you can get text, data, graphics and image. Further, you can integrate that with voice, not completely today, but through voice annotation kind of techniques.

OPPORTUNITIES IN OFFICE SYSTEMS

Let's talk about several categories of business opportunities that are generated by the three drivers that we have been talking about so far. I am going to talk briefly about three kinds of opportunities. One of these is office systems. The second subject is mechanical design automation. The third subject is artificial intelligence or knowledge-based systems.

The first subject is office systems. The office systems opportunity is really driven by the fact that there are almost 50 million office workers in the United States today and by the end of the decade there will be almost 60 million office workers, spread between the categories of secretarial personnel, clerical personnel, managers, professionals, and sales people. So there are a lot of people that we would think of as office workers.

By the end of the decade, a telephone on the desk is going to cost about 80 cents a day. A non-intelligent terminal will cost about $1.50 a day. A programmable or intelligent terminal will cost between $4.00 to $6.00 a day. A typical clerical salary, however, will be about $85 a day and a typical white-collar salary will be in the range of $200 a day. The point is it doesn't take large percentage productivity improvements in clerical productivity and it takes even less improvement in white-collar productivity in order to fully pay for a non-intelligent terminal on virtually everybody's desk.

People who look at the industry take these kinds of statistics and suggest that out of the approximately 50 million white-collar workers in 1989, more than half of them will have some form of an electronic workstation on their desk. When you think about the economics that we're talking about, that is probably not an unreasonable forecast. The real pay-off from office systems is whether or not the particular activity that you have pays for itself when you install an office system, and my point of view is there is so much opportunity to displace inefficient office activities with electronic media that this process will take place fairly rapidly over a period of time.

In my previous job, I was the chief financial executive for one of IBM's largest manufacturing and development sites. The site ran an operation of approximately 50,000 manufacturing part numbers. We

had about 650,000 invoices per year. On a good day the
inventory was $200 million. On a bad day it was more.
The thruput to the general ledger was about $20 billion
a year. Obviously, a very large amount of computing
was required to keep track of all of this. At the same
time all of that computing was keeping track of this in
a rather centralized fashion. What was really the case
in the office (which is the subject we're discussing--
office automation) was that the the cost accounting
department did absolutely everything in a manual
fashion because they did not have access to the central
computerized files. What I found out, what I wasn't
particularly pleased with, was that this site used
8,000 key punched cards weekly for doing the payroll,
and they used 90,000 key punched cards in order to keep
track of physical inventory. Clearly, there was a
large problem associated with sorting cards, having
carts filled with cards in the aisle and so on. It was
difficult to get to the coffee machine because at the
end of the month when it was cost accounting time, all
of these carts arrived in the aisles of the cost
accounting department. The carts were approximately
four feet high and three feet wide. These aisles were
five feet wide so, if you were a little on the hefty
side, you would have to postpone your coffee break
until cost accounting closed the books for the month.

When edits and audits were performed, great piles
of paper were consumed. On one occasion the cost
accounting manager took me into a large room and it was
absolutely stacked from the floor to the ceiling with
paper. There were seventy people in the cost accounting department. I invited the cost accounting manager
to do something about that; he was a very creative
young man and applied what today might be a rather
simple concept. What he did was take all the centralized files that contained all this data that I was
watching physically wander around the hall, stripped it
off the central files, created small files using end
user techniques, and made them available to the cost
accountants on their desk with the following results.
About ten months after the beginning of the project,
the project had broken even and by the end of the year,
the cost accounting manager felt quite confident of two
things. First of all, the quality of his work had
improved substantially, and, second, he was able to
take a 10 percent reduction in headcount for the subsequent year. The next question, back to social economic changes: what happened to the seven people left
over? Three of them went into another function short
of headcount, and four of them were reinvested in
subsequent office automation, productivity applications, and the process was repeated in the subsequent
year.

The message that I'm trying to deliver about office automation is that very, very large improvements in quality and productivity can be embarked upon successfully in very brief periods of time. That's the promise against which professional market researchers forecast that by the end of the decade, approximately one desk in every two will have an electronic workstation on it.

OPPORTUNITIES IN MECHANICAL DESIGN AUTOMATION

Let's switch gears for a moment. This time talking about mechanical design automation. Many of us are familiar with mechanical design automation in a two dimensional context. In other words, many manufacturing organizations use electronic techniques for doing mechanical drawing, mechanical drafting and renderings. One of the better known tools for doing that is CAD/CAM which was developed perhaps over a decade ago by Lockheed Corporation. What CAD/CAM does is basically automate the processes that a draftsperson would use in drawing the front view, the top view, and the side view of a mechanical part.

Many firms take that design, done electronically in CAD/CAM, and release that electronically to the manufacturing floor. Release is a manufacturing term that means give them the plan and kiss them on the cheek and tell them good luck--"It looked okay when it left here."

The new world of mechanical design automation is going to step up one dimension. As you might guess, the new world is going to involve three-dimensional modeling. Clearly, you can do everything in three-dimensional modeling that you could have done in two-dimensional modeling, so you can do detailed drawings, assemblies, manuals, dimensioning, etc.

But there are large numbers of things that you can begin to do in three dimensions that you can never touch in two dimensions. For example, you can do a variety of static analyses, and one of the most important is tolerance analysis--in much simpler terms, "Does it fit?" Design engineers are notorious for designing the part that they have been asked to design without regard for making sure that it fits in the assembly that it's supposed to fit in. In other words, some other designer in Department A is designing a collection of components that fill up that space or make it difficult to assemble from a manufacturing point of view. When you design in two-dimensional technology, nobody ever knows that. Of course, the manufacturing floor knows it eventually because they take this thing and make it fit. With three-

dimensional modelers, the processing of assembly can be
simulated as well.

Why is that important? In the old days, the time
between designing the power supply and finding out that
the darn thing doesn't fit can be six to nine months,
and at that point, you cycle again and what have you
lost? You've lost six to nine months of development
and/or manufacturing cycle time. If you can simulate
the fact that it does or doesn't fit now and make
changes right now, you have automatically shrunk the
development cycle time of that particular product by a
very, very large factor. And cycle time is one of the
key ingredients of the competitiveness of any
organization.

Further, you can do dynamic analysis with three-
dimensional mechanical design automation. The real
pay-off of these kinds of techniques is in the manufacturing environment. With a three-dimensional modeler,
a product can be taken from the development laboratory,
and turned over to manufacturing engineering in electronic form. Manufacturing engineering has to define
the process by which this part will be manufactured.
They have to write the direct numeric control machine
instructions for controlling the drill press head to
remove metal, create a mold and subsequently state the
instructions to the robotic arms for how to take the
bolt from this box and drop it into that hole. With a
three-dimensional modeler and with computer programs
that simulate either numerically controlled machine
tools or robotic machines, you can actually get a
picture on a computer screen that is the three-
dimensional part that you have just constructed. You
can take a subroutine that calls up a robotic arm and
take a light pen and cause that robotic arm to drop the
bolt in the hole. At that point you can push a button
saying "Please record what just happened to that robotic arm and translate those instructions into robotic
arm control instructions and when I get ready to manufacture it, ship those over to the robotic arm." Those
kinds of processes are not 1990 processes, those kinds
of things are beginning to be seen in the late 1980s.
Again these are extraordinarily simple things to do on
a computer and very time-consuming things to do by
hand.

Recently I hosted a reporter from a national
financial publication who is interested in how IBM was
using information systems in manufacturing and development. One of the subjects that I wanted him to see was
mechanical design automation. I can tell you that he
basically napped through most of the theoretical presentation. But then we took him over and gave him a
demonstration. The individual who was demonstrating
was an ex-tool and die manufacturer. He did not happen

to have a college education, although he was obviously
quite an intelligent, and very capable individual.
After about ten minutes of demonstration, he invited
the reporter to sit down and do it himself.   The
reporter designed a rather simple part and then was
asked if he would like to design the mold for casting
that part.   And he did that by pushing a few buttons,
and then the problem began because that was the end of
the demonstration and the reporter was so intrigued by
all of this he didn't want to leave.   So we had to take
him by both elbows and march him out to the door.   The
message is: "It is not technically complicated to use
these kinds of devices, but it does produce enormous
savings in people time and it creates productivity."

OPPORTUNITIES IN ARTIFICIAL INTELLIGENCE

     The last subject is a new opportunity, artificial
intelligence.   Artificial intelligence is a slippery
term to most people.   Some researchers define artificial intelligence as performance of a task by machines
which would require intelligence if performed by
humans.   My eyes tend to glaze over on statements like
that because that includes robotics, it includes computer vision, it includes natural language processing, it
includes expert or knowledge based systems, and all
that's good but it also includes all of the computing
processes that you and I are familiar with as well.   So
it doesn't help us to understand anything by talking in
that fashion.
     I prefer to discuss artificial intelligence or
knowledge based systems by drawing a distinction
between traditional data processing and the problems
that can be solved by it, and artificial intelligence
and the problems that can be solved by that.   Traditional data processing requires fixed and bounded
problems.   The kinds of problems we would think about
as general ledger, payroll, and so on.   The data is
well known.   The mechanism for translating the data
into output is well known.   And the answer is fixed,
precise and definite.   There is only one trial balance
to the accounts, so to speak.
     The distinction between traditional problem
solving and artificial intelligence is that there might
not be a fixed, precise answer, that the organization
of the analysis has to be done using rules of thumb or
inference.   A reasonable distinction between problems
that can be solved by these two processes can be made.
On the one hand, the traditional problem could be
exemplified by engineers who want to calculate how a
bridge ought to be designed.   There is a way of
starting at one end of that problem and calculating all

the way out to the other end. Contrast that with the
problem that a doctor might face in diagnosing a blood
disorder. What a doctor does is look at a series of
symptoms and construct relationships: "About 80 percent of the time, this collection of symptoms relates
to this disease..."

Artificial intelligence has a potential for giving
larger and larger portions of our population a certain
degree of technological literacy over time. We have
probably gone past stage one, which is experimental.
There are a lot of things going on in artificial intelligence and knowledge based systems which are certainly
in stage two. An example would be artificial intelligence or knowledge based system applications that act
like a senior loan officer in granting a loan. A
senior loan officer goes through some quantitative
material but will also go through some qualitative
material--all of which can be embodied in a set of
rules that can, through a computer, be played back to a
less senior loan officer thereby creating efficiency or
convenience.

There are some applications already in stage
three. One of the best-known applications and one of
the most comprehensive is a program called Internist
that was developed by the University of Pittsburgh. I
understand it covers 85 percent of internal medicine.
It covers 500 diseases. It covers 100,000 symptomatic
associations. The usual difficulty with these kinds of
systems is "Gee, that's great but a practiced internist
can do it better." So the real test is how does this
kind of system stack up in terms of diagnostic reliability against a practiced internist. Tests have shown
that it rivals the capability of a trained internist in
solving a particular collection of refresher or test
pattern cases published by the New England Journal of
Medicine Clinical Pathology Conferences. In other
words, in the ways that internists continue to update
themselves and test themselves on their diagnostic
capability, the artificial intelligence based system
does as well as the internists themselves. So it is
increasingly possible, and increasingly likely that we
can get knowledge based systems to perform on these
unstructured contextual problems as well as people with
intelligence.

SUMMARY

Our society is undergoing social-economic change
toward an information society. Technology has produced
change at a very rapid rate, and this rate will continue. It's possible to increasingly integrate the media
used for communication. These drivers create new uses

and, therefore, new users and that all implies that there will be a great deal of further opportunity for the information processing industry.

Part IV

PRODUCTIVITY, QUALITY, AND MANUFACTURING TECHNOLOGY

30

Overview

Y. K. Shetty and Vernon M. Buehler

Chapters in Part IV concern productivity, quality, and manufacturing technology.

Theodore A. Lowe's essay, "Excellence in Manufacturing at GM," explores the experiences of General Motors Truck and Bus Group in improving quality. GM in this division uses the concepts from three major quality experts: W. Edwards Deming, Joseph M. Juran, and Philip B. Crosby. The company tries to understand and integrate the three approaches in order to accelerate its quality improvement journey.

Major commitment to quality improvement was initiated in 1982-1983 by providing Deming Process training to management. Since then a series of actions has been taken including Juran training, establishment of Quality Council, Crosby training, union involvement, and continuous employee awareness and training initiatives. After outlining the process involved in implementing the quality improvement initiatives, Lowe reviews his company's experiences in applying the approaches of experts.

Lowe feels that Deming's philosophy challenged the company's past practices and helped to introduce the statistical thinking in managing quality. The company learned from Deming and Juran that 85 percent of the quality problems are management controllable. Therefore, it is management's job to initiate action on the system in correcting the common causes of variation and improving the process capability of operation.

Whereas Deming's fourteen points are more of a road map, Juran provided for all parts of the company a more specific systems approach to quality control, improvement, and planning. Juran provided an assortment of problem-solving tools. In addition, Juran recommended four major thrusts in the quality improvement process:

1. Establish annual quality improvement goals.

2. Establish hands-on leadership toward quality improvement.

3. Establish an executive steering committee to lead the quality improvement process.

4. Establish a vigorous education and training process.

To further accelerate the company's quality journey, it needed to achieve an awareness and involvement of the entire work force, from top to bottom. Philip Crosby's program provided the mechanism for transforming the quality of the whole company. Crosby's fourteen steps gave the company guidance for building a quality improvement attitude throughout the organization. Lowe feels that the ideas of Deming, Juran, and Crosby are complementary. For example, the Deming process is reinforced by the Crosby fourteen-step process by assisting management in transforming the organization. It also needs Juran's breakthrough process as a framework for applying statistics. The effectiveness of Juran's problem-solving approach is enhanced by the application of statistical tools that the Deming process promotes. By integrating the three approaches, the company has made tremendous advances in its quality improvement efforts.

"Xerox Gains from Productivity Innovations," by Ed Finein, discusses the experiences of his company in improving productivity to meet the challenges of the worldwide competition. To confront the competitive challenge, Xerox in 1980 started to assess its corporate strengths and weaknesses. The company found that: it took too long to develop new products; its products cost too much; and it did not fully satisfy customers' requirements. The company recognized that in order to overcome these problems and to eliminate their competitors' advantages, it had to engage in an internal revolution in product design, production, and distribution. Specifically, company renewal efforts were based on reducing product development time by one-half, reducing product development manpower by one-half, and improving unit manufacturing cost by 50 percent--an aggressive productivity improvement target. In order to achieve these targets a number of actions were taken such as the use of customer need-based product planning, co-locate design engineers and manufacturing, and encourage continuous supplier involvement. As a result of these actions, Xerox's product development schedule has become shorter and engineering productivity has more than tripled. Along with productivity, the company has taken measures to improve quality, mainly in the area of process control

and automation.   Finein feels that Xerox has made the
most progress in the area of quality.
    The next chapter, "Ford's New Business Ethics:
Quality Is Job #1," by Lewis C. Veraldi, discusses the
efforts of Ford Motor Company in achieving world-class
quality and improving productivity.   Back in 1980,
Ford's image for quality was not very good and that
caused the company to examine its existing design process, which was sequential in nature: design-
engineering-manufacturing-marketing.   Under this system
each group of specialists operated in isolation from
one another.   Recognizing the problems involved in the
traditional process, Ford replaced it by a simultaneous
"team" approach in their new cars.   This approach is
designed to promote continuous interaction between
design, engineering, manufacturing, and marketing along
with top management, legal, purchasing, and service
organizations.   Overall, coordination or direction is
by a car product development group, or program manager
whose primary function is getting all of the team to
work together.
    This approach was a major factor in the success of
Taurus and Sable.   It has helped Ford to improve its
product quality enormously--over 60 percent since 1980.
The future is going to see this team approach in all
new Ford Motor products.   For six years in a row, Ford
has been first in quality among all domestic auto
manufacturers.
    "Met Life's Quest for Quality," by John J. Falzon,
describes his company's efforts in improving service
quality.   Met introduced a process that requires every
organizational unit to identify the major service being
rendered and the customers to whom they were being
directed.   Each area then developed a quality network
composed of Quality Improvement Teams for each identified service.   These teams established performance
measures and, through a structured problem-solving
technique, identified opportunities for improvement.
    Using the available research, the company is
developing survey instruments to measure the customers'
view of service.   Also, the actions aimed at improving
service quality are closely tied to the company's strategic planning.   Using these--survey technology and
strategic planning--all organizational units are being
requested to provide as part of their annual planning
objectives, "an assessment of the gap that exists
between customer expectation and perceived level of
service," and to provide an explanation of how they
expect to go about reducing these gaps.   The company
hopes to gain significant improvement through its continuing efforts aimed at systematic measurement of
customer needs and providing an enhanced level of
service to meet those needs.

"Computer Integrated Manufacturing: Allen-Bradley's Experience," by J. Tracy O'Rourke, advocates that a whole new philosophy of business must be adopted to gain the full benefits of manufacturing technology. Electrification gave birth to modern manufacturing. Then came the computer control, making automation possible. Computer Integrated Manufacturing (CIM) is the next logical step in this evolution--a step that requires no great leap of faith. It is an advanced business philosophy that unifies a company's administration, engineering, and manufacturing. To manufacturers inundated by the tides of global competition, CIM is an attractive strategy for survival, growth, and prosperity.

The world-class company of the future will integrate all three areas--the office, engineering, and manufacturing--into a closed loop system for the entire production cycle. This system can provide quantum improvements in a number of areas. CIM can improve human productivity, capital resource productivity, and quality. It offers manufacturers a rapid response to the marketplace. These benefits will make a company the low-cost producer. Considering the benefits, a decision to invest in CIM should be an easy one. Even though there are a number of barriers to CIM, management must rethink and find new ways of doing business.

Some guidelines derived from Allen-Bradley's global experience with CIM include: don't start the CIM trip without a roadmap; remember, what is tailor-made, fits best; set your sights high--make zero defects your goal; make your system right, then automate; only automate things that add value; plan from the top and implement from the bottom; and automate the system step-by-step and don't implement without trained people. O'Rourke's message is simple: The global economy is here. Yesterday's ways of doing business are passe. The time to start CIM is now. It's the way to survival, growth, and prosperity.

Alexander MacLachlan's essay, "Emerging Technologies, New Materials and Optoelectronics at DuPont," addresses several aspects of emerging technologies and advanced materials as viewed by the DuPont Company. Some of these new technologies include: plastic automotive body parts; barrier plastics in food packaging; optical storage technology; and lightweight transmission of voice and data signals. These technologies have tremendous potential for improving productivity, quality, and competitive position in many industries. They are not only creating new products and whole new industries but also are revitalizing existing industries by improving product quality and manufacturing processes. They have the potential for transforming markets and reshaping the companies that serve those markets.

To gain the full advantages of these technologies, first we must have a continuing supply of well-educated researchers and scientists. Second, we must have clear understanding and appreciation of the value and contribution that technology can play in our lives. Third, there must be a commitment on the part of the society to continue to compete in a high tech race. And, finally, we need government policies and economic practices that will enhance the capability of American business to compete and prosper in the high tech race.

Wayne R. Pero's chapter, "Dow Chemical's Quality and Productivity Improvement," describes the four key elements of his company's Quality Performance Improvement Process. This process is simple and starts with the customer. The second element of the process is having a good working relationship with the suppliers. The third element is focused on the process and the recognition that continuous improvement comes from working on the process, and that the process is under management's control. The focus here is on identifying, through the use of statistical tools, which processes are in control and which are not. Then everyone in the organization works on reducing the variability to bring the processes into control. The fourth element of the company program is the need to measure. These measurements, to be effective, should measure how well the customer requirements are being met, how well the suppliers are meeting company requirements, and how well the key processes are running.

Our challenge, Pero says, was to develop training programs that would help the managers and employees implement the necessary tasks to accomplish our improvement goals. Dow used a "Toward Excellence" program that rests on the fact that most successful companies have an obsession for quality products and services, and the customer is where it all starts and ends.

The next challenge was to develop a training program to make quality and productivity a way of life at Dow. This program, "Quality Performance Workshops," emphasizes dissatisfaction, vision, and application to create the change in attitude that is necessary for quality and productivity improvement. In simple terms, this training program works as follows: (1) foster dissatisfaction with the current situation, (2) create a vision of a better way, and (3) encourage application of the concepts, tools, and techniques that will result in change for the better. The business teams that have gone through these training programs report significant cost reductions and quality

improvements. One team credits savings of $17 million
in the first year. In another, the company found ways
to save $2 million in the first year.

Jim Shadler's chapter, "Pillsbury's Productivity
Through Motivation and Participation," analyzes
Pillsbury's Profit Improvement Program. The objective
of the program was to formalize ongoing cost reduction
efforts in realizing greater savings, improved quality,
and productivity.

The key elements of the program include: First,
naming coordinators who are responsible for seeing
through the program. Second, creating a cost
accounting system, which is the important means for
tracking the program and its effectiveness. Third,
assigning a productivity MBO where everybody in the
system has a productivity objective. Fourth, measuring
improvement. The finance area is charged with the
responsibility of measurement and reporting results of
the program.

The source of the productivity improvement ideas
came from the people in the field. Free communication
along with continuous training is also essential to the
success of the program. The company had excellent
results in using this program for improving productivity and quality. The tenth anniversary of the
program is in 1987 and it will be an all-time record
year for improvement. There is a culture change in the
company to reinforce the commitment to quality and
productivity throughout the organization.

John R. Black's essay, "Boeing's Quality Strategy:
A Continuing Evolution," provides an overview of the
quality improvement process within the Boeing
Companies. The formal efforts to enhance quality
started in 1980. The company goal is to deliver products that meet its commitment of excellence--to make
Boeing the recognized standard for the quality after-
sales support, and for technical and economic performance of those products. As a part of the program,
Boeing made a comprehensive attempt to change company
culture. Along with it, the company established a
quality improvement program. It focused on developing
a comprehensive and cohesive strategy for implementing
employee involvement.

Quality circle process is a part of the employee
involvement program and is responsible for systematically locating and solving problems. One-half of the
projects provided cost savings, while the other half
resulted in improvements that are highly significant,
but are not easily convertible to dollar figures.
According to Black, Boeing expects to achieve
significant results in the future. They expect to have
(1) all employees trained in the use of statistical
tools, (2) a pilot project on self-regulating work

teams, and (3) layers of management reduced. The company is also planning to use suppliers that have their processes under statistical control.

The process of continuous improvement, to which Boeing is committed, is not one that can suddenly be realized in a company. Every company must make it their own. Top management must be committed, employees must be involved, and the process must be locked in for the long term. Only then, will the program achieve the most results.

The next chapter by Bill Bradshaw, "Measuring White-Collar Productivity," discusses the myths associated with white-collar productivity and provides some guidelines for improving measurement methods.

Bradshaw believes that there are a number of myths about white-collar productivity. These are: creativity and judgement cannot be measured; service jobs do not produce tangible outputs and hence are difficult to measure; the available measures fail to consider things beyond control; and measures aimed at measuring intangibles are corruptible. After dispelling these and other myths, Bradshaw suggests guidelines for improving measurement of white-collar productivity. These include: (1) clarify organizational objectives; (2) examine internal outputs, inputs, and activities; (3) identify key departmental customers; and (4) understand customer expectations. Finally, he discusses a technique called "Productivity Objectives Matrix," as an effective tool for measuring white-collar productivity.

In "Managing Business: Today and Tomorrow," Richard A. Jacobs examines the findings of his company, A. T. Kearney, on how to improve productivity. A number of conclusions can be drawn from the study that identifies the attributes of successful companies. To cite a few, these companies operate with fewer staff personnel per $1 million of revenues; have flatter organization structure and increasing spans of control; reflect the entrepreneurial spirit; are making cultural changes in areas such as product quality, risk taking, and rewarding performance; have leaders who are more concerned about the future; and understand their industry and what it takes to win in that industry. These and other findings showed how some companies were able to achieve superior productivity and consistently outpace their competitors.

31

Excellence in Manufacturing at GM

Theodore A. Lowe

I would like to start my presentation with an old Buddhist parable.

> It seems that three monks were having a long and fruitless argument over some matter. . .and they decided to present their views to the wise old master. The first monk spoke long and well, supporting his case with illustrations from the life of Buddha. The master listened and said, "Why you are quite right." The second monk presented his arguments, and they were truly eloquent, and when he finished, the master told him, "You, too, are right." Now, a third monk had been following all this, and he became very troubled, "These two men were in complete disagreement, surely they both can't be right." The master turned to him and said, "And you too are right."

This presentation will make the point that we must be like the wise old master and use the concepts of all of the gurus in developing a quality improvement process.

BACKGROUND FOR QUALITY JOURNEY

On our journey to world-class quality, we will need to follow the directions of three leading quality experts: Dr. W. Edwards Deming, Dr. Joseph M. Juran, and Philip B. Crosby.

While others have debated what's lacking in the direction from either Dr. Deming, Dr. Juran, or Phil Crosby, we looked for the right things from each. It was our position that they were 90 percent common in philosophy and that they complemented each other. Therefore, we tried to understand and integrate the three approaches. By using the strengths from Deming,

Juran, Crosby, and others, we believe that we have been
able to accelerate our quality improvement journey.

Before I highlight some of the most valuable
directions and the lessons learned from the masters
along our journey, let me provide you with the
background for the quality journey at General Motors
Truck and Bus Group. This, headquartered in Pontiac,
Michigan, is responsible for the planning, designing,
engineering, manufacturing, assembly, and marketing of
General Motors trucks and buses worldwide. As a result
of consolidating all of GM's truck and bus operations
in mid-1982, the truck and bus group has become one of
the largest and most important parts of GM's worldwide
business.

In 1982-1983, commitment and involvement in
quality efforts were begun by providing Deming Process
training to management, SPC Training to production
people, and we established a groupwide SPC council. In
February 1984, Juran training was initiated. In March
1984, an executive quality council was established, and
in October, the quality improvement process was
launched at an off-site with over one hundred of our
senior managers. In 1985, we established our quality
improvement structure--which included forming customer
satisfaction improvement teams--and we also initiated
Crosby training.

In early 1986, we took strong action to involve
the union in our process--a major step forward. In
July of last year, we believe we came to a real turning
point. With union involvement, we jointly redefined
the quality improvement process and developed a quality
road map. The process consists of eight elements, each
with a tenet, a vision of where we want to be regarding
our quality culture, and initiatives to be taken to
attain our objectives.

To continue the quality improvement process
momentum, the executive quality council has designated
1987 as the year of employee involvement, application,
and achievement. We have spent the last four years in
building awareness and commitment and providing training. Now we need to apply the lessons learned from
the masters if we are to achieve world-class quality,
but what are the lessons that we learned from Deming,
Juran, and Crosby? Next, I'll review our experiences
in applying the approach of each expert, look at how
and why we integrated the three approaches, and
conclude by summarizing the lessons learned.

APPLYING THE DEMING APPROACH

We were well into the Deming Process when the old
GM Truck and Coach division was merged into the Truck

and Bus Group in September 1982.   In fact, our first Deming overview for our executive staff was titled "Deming's Ten Points." A few months later, when Deming came out with his fourteen points, we were hard-pressed to explain how we had missed four points.   One of our first initiatives in Truck and Bus was to provide Deming Process overviews to all of our plant managers and staff heads and their staffs and to initiate SPC training for the work force.   Later we added concepts from Myron Tribus and others to reinforce the training.

Dr. Deming triggered our quality renaissance.   His new philosophy challenged our past practices. While many of us were not very comfortable in confessing our sins in public, Dr. Deming's denunciations gave us the license and courage to question our traditional practices in the open.   In addition to sparking the renaissance in thought, Dr. Deming gave us a way to get at "truth"--through statistical thinking.

We learned about variation and the difference between common and special causes.   We learned to use control charts to separate the special causes of variation from the system problems.

We learned from Deming, and later Juran, that 85 percent of the problems are management controllable. Therefore, it is management's job to initiate action on the system to reduce or correct the common causes of variation and improve the process capability of an operation.

To understand whether our processes were stable, capable, and on target, we learned to use statistical tools, starting with seven soft tools of Ishikawa and progressing to the Taguchi and other design of experiments techniques.   We're still learning how to apply these tools.

To spread our experiences, and efforts, we established an SPC council. We learned from each other's experiences in training and applying the Deming Process.

Dr. Deming emphasizes the need for ongoing improvement.   His fifth point tells us to constantly and forever improve the system of production and service.   This concept is illustrated by the Deming circle, which symbolizes the problem analysis process. The plan, do, check, analyze, and act circle also serves as a model for the quality improvement cycle of a company--planning and designing a new product, making and selling it, checking customer satisfaction, and acting to further improve customer satisfaction.

Dr. Deming's principles helped us establish a learning environment, which in turn gave a boost to the effectiveness of the training from the other experts that followed.

However, not everybody was ready or willing to learn. Many managers turned away because of the criticism of their past practices and others were put off that Dr. Deming focused more on what needed to be done than on how to do it. Deming himself scoffs at the type of executives that ask for the recipes--there is no instant pudding.

Dr. Deming warned us not to jump into SPC training for the masses until we had removed the roadblocks to quality. We did not listen, however, and we learned the hard way that plants that trained selectively with a purpose were more successful than those that trained everybody top to bottom and were done with it.

The use of SPC was more difficult to apply in our assembly plants and staffs. The journey in these areas was certainly tougher than in traditional manufacturing plants as there were no documented experiences or directions to follow. Often we faced the response, "How can I use it, we're different," from people in functions that used little variable data.

The implementation of the Deming Process and SPC in general was more successful at the lower levels. The workers and quality engineers could make use of the SPC tools. So, even though Dr. Deming directs his fourteen points at management, the Deming Process at Truck and Bus became a grass roots movement, centered around SPC training and application.

But we feel the grass roots movement was successful for us. If we revisited the fourteen points today, we would be amazed as to how far we have addressed his roadblocks. Dr. Deming triggered the reaction by helping to establish a participative and learning environment and by providing a means for getting at the truth, through the use of statistics.

APPLYING THE JURAN APPROACH

Whereas Dr. Deming's fourteen points are more of a final destination than an actual road map, Dr. Juran provides a more specific systems approach to quality control, improvement, and planning for all parts of the organization.

We learned from Dr. Juran about this more scientific approach to quality management. We learned the value of a project-by-project orientation and the application of the "Breakthrough Sequence." Dr. Juran provided an assortment of problem-solving tools in addition to SPC. With his quality spiral and 'fitness for use' definition of quality, Dr. Juran gave us a strong orientation to meeting the customer's expectations.

When Dr. Juran addressed our top eighty executives for the first time in November 1983, he recommended four major thrusts in our quality improvement process:

1. Establish annual quality improvement goals.

2. Establish hands-on leadership toward quality improvement.

3. Establish an executive steering committee to lead the quality improvement process.

4. Establish a vigorous education and training process.

We acted aggressively on his recommendations, starting with the creation of our executive quality council, and the implementation of Juran training.

Dr. Juran's education process is structured around the Juran trilogy of quality planning, quality improvement, and quality control. Quality planning focuses on creating a process that will from the start have a very low cost of poor quality. Quality improvement strives to lower the cost of poor quality in existing processes, and quality control's intention is to hold the gains and keep the process in control.

"Juran on Quality Improvement" is a training process that is based on a project-by-project approach, with project being defined as a problem scheduled for solution. It addresses the <u>vital few</u>. Many of our plants continue to use the "Juran on Quality Improvement" video tapes to facilitate floor-level project teams.

"Juran on Quality Planning," on the other hand, facilitates projects that are processes scheduled for improvement. Because systems are responsible for 80 percent of our problems, "Juran on Quality Planning" helps address what Dr. Juran calls the "useful many" and what we at GM call the "significant many." We participated in the field test of this training program and plan to use it to help staff project teams improve the quality of their processes.

Dr. Juran stressed the need for establishing the problem-solving machinery required to achieve improvement. The third step in his Breakthrough Sequence is to organize for a managerial breakthrough in knowledge by creating problem-solving steering arms and diagnostic arms. The steering arm guides the overall problem-solving effort by establishing the direction, priorities, and resources to accomplish the task. The diagnostic arm is the work group with the investigative skills and mobility to follow the trail wherever it leads until the root cause is identified.

Dr. Juran divides the problem-solving effort into two journeys--a journey from symptom to cause and a journey from cause to remedy. He states that the most difficult journey is from symptom to cause because it is not clear where the responsibility lies.

To steer our quality improvement efforts, we established customer satisfaction improvement teams in 1985, for each one of our product lines. The teams, which include representatives from all key disciplines, focus on increasing customer satisfaction. They have assessed our current quality position, established objectives, and developed plans and projects to accelerate our rate of improvement. The customer satisfaction improvement teams operate on a project-by-project approach and have effectively addressed many of our vital few problems.

Dr. Juran urged our management to establish an annual quality improvement program, setting objectives and seeing that specific projects are chosen, year after year, with clear responsibility for action. His approach conformed readily with our five-year business planning process. Dr. Juran advised us that our quality objectives must be set according to the marketplace. They should not be limited by elements outside our immediate control, or by what we think can be realistically achieved with our current resources.

He told us that after our upper managers set the broad improvement goals, it is up to the middle managers to establish the teams, resource requirements, measurements, and project to meet the goals. To support our commitment to becoming a customer-driven organization, our executive quality council has established the broad overall objective of reducing the discrepancies in our processes and problems experienced by our customers by at least 25 percent in the 1987 calendar year. In this regard, we are asking all plants and staffs as well as our customer satisfaction improvement teams to align their goals, plans, and efforts with our group quality objective and cascade the objective and strategy downward to cover all functions and levels in the organization.

APPLYING THE CROSBY APPROACH

Dr. Juran's direction accelerated our knowledge and also got more members of our organization involved in our quality process. His training proved to be most valuable to the managers and quality professionals responsible for implementing and managing the quality improvement process. With his total comprehension of quality management, Dr. Juran and his material served as our teacher. To further accelerate our quality journey, we knew we needed to achieve an awareness and involvement of our entire work force, from top to bottom. The principal strength of Philip Crosby's program is the attention it gives to transforming the quality culture of the organization.

By stressing individual conformance to requirements, Crosby involves everyone in the organization in his process. His fourteen steps, a 'how-to' for management, provide an easy-to-understand, structured approach to launching the quality improvement process and starting the journey to world-class quality. Although there were pockets of the organization that were well on their way, the Crosby Process helped us to get almost everybody in the organization started on the journey. The fourteen-step process provided a very explicit and structured approach to implementing a quality improvement process--a common road map for everyone to follow. It also gave guidance for building a quality improvement attitude throughout the organization, and for establishing a uniform quality vocabulary.

In following step two of Crosby's fourteen-step process, we implemented a group quality improvement team, as well as plant and staff quality improvement teams, to run the quality improvement process. The quality improvement teams manage the "soft" or nonproduct areas covered in his fourteen steps such as awareness and communication, quality education, cost of quality, and recognition. The fourteen-step process provided us with the most defined, simple-to-follow road map for our quality journey.

Like Deming and Juran, Crosby starts his process with management commitment. But we were more successful achieving management commitment with the Crosby Process. Perhaps it is because these four absolutes of quality required by Crosby to achieve management commitment also provide a new philosophy and a breakthrough in attitude that Deming and Juran require in their approaches. An example of the commitment from our management is our executive quality council visit process to our plants and staffs. These visits, which occur twice each month, are solely for quality and for discussing progress and any obstacles to the quality improvement process. The theme for these visits is "How Can I Help?" Crosby also asks for a management commitment to quality training and awareness for all levels and functions. It does not go into the depth of Juran's training but it provides broader coverage. Whereas Dr. Juran's project-by-project approach attacks the vital few problems, Crosby's error-cause-removal step and emphasis on conformance to requirements have helped us address the category of problems that Dr. Juran calls the "useful many."

We have used the Crosby training to help everyone focus on the quality of their business process and to understand their internal customer/supplier relationship. We all have individual products and customers. How do our personal customers perceive the quality of our individual products? And what are their requirements for us?

We also have internal suppliers who provide us with products and services that we use to complete our tasks. What are our requirements for our suppliers? How do we provide feedback to them on the quality of the work they provide for us? This understanding of the customer/supplier relationship helped to further crystallize the concept of quality in our staff functions.

INTEGRATING THREE APPROACHES: BENEFITS/PROBLEMS

In adding the Crosby Process to the previous focus on Juran and Deming, it was important for us to overcome the perception by some parts of the organization that the Crosby Process was only a means of cheerleading workers to achieve zero defects. In that regard we were fortunate to have the depth of knowledge gained from Deming and Juran so that we were prepared to get the train accelerating quickly once Crosby helped us finally get almost everybody on board. The quality management practices and problem-solving techniques that we learned from Juran and the statistical techniques and management principles that we learned from Deming increased the effectiveness of Crosby's training in developing a new quality culture and implementing the quality improvement process.

In conducting quality education programs on the three approaches, we had to overcome differences in terminology used by the experts. Often we had to serve as interpreters, translating Juran's language into Deming's or Crosby's terminology or vice versa. For example, Crosby, Juran, and Deming all define quality differently.

To Crosby, quality is conformance to requirements; whereas Juran defines quality as fitness for use. Although Deming does not give an explicit definition of quality, he describes quality as a predictable degree of uniformity and dependability, at low cost and suited to the market. Dr. Juran relates his definition to Crosby's by stating that the quality mission of a company is fitness for use whereas the quality mission of departments or individuals is conformance to specifications. In their quality definitions, Crosby's emphasis is on doing things right while Juran is stressing the need to do the right things.

Finally, General Motors has developed a definition of quality that encompasses all three ideas: "Quality is conformance to specifications and requirements that meet customer expectations."

Understanding and following the directions from each of the leading quality experts is necessary, but not sufficient, by itself, for us to reach our destination. Crosby, Deming, and Juran are interdependent. Companies using one of their processes need to borrow concepts and techniques from the others to make their own processes more successful. A Deming Process, for instance, needs a Crosby fourteen-step process to assist management in transforming the organization. It also needs Juran's breakthrough process as a framework for applying statistics. Juran states that the breakthrough sequence must start with a breakthrough in attitude. Crosby helps achieve this breakthrough with his four absolutes. The effectiveness of Juran's problem-solving approach is also enhanced by the application of the statistical tools that Deming promotes. A company using the Crosby Process needs these tools and Juran's techniques. It also needs the teamwork that Deming and Juran emphasize to address the system problems that keep the individual worker from reaching zero defects.

In establishing our quality improvement process we wanted to fit Crosby, Deming, and Juran into our process and not try to fit our process into one of their programs. In incorporating the best features of all three into our process, we tried to avoid the perception by the organization that we were jumping from one "prophet" to the next. To avoid the conflict that comes when an organization tries to choose the proper "champion," we tried to show the benefits of integrating the concepts of Deming, Juran, Crosby, and others into our quality improvement process. We used the metaphor of three preachers, one religion.

Dr. Deming has been called the "fire and brimstone" preacher. He lays down the fourteen commandments for management. He tells management that they are responsible for 85 percent of the sins and that they must repent or their businesses will go to hell. Dr. Deming also provides the congregation with an SPC "prayer book." Dr. Juran is the theologist who has extensively researched the scriptures of quality management. He provides the quality "bible." Philip Crosby might be viewed as the evangelist of the three...exciting, positive, generating enthusiasm. His message is simple: the four absolutes. He preaches that no level of sinning is permissible but he provides management with a way to get to heaven.

Management and the work force make up the
congregation at General Motors Truck and Bus. We are
learning to sit together and be more than Sunday
Christians. As we practice our quality religion, our
process is becoming homogeneous. The different concepts and techniques that we have picked up from
Deming, Juran, and Crosby are losing their identity
with the preacher.

GM's TENETS OF QUALITY IMPROVEMENTS

Therefore, the lessons that we have learned from
the masters are best summarized by reviewing the tenets
that we have established as the guiding principles of
our quality improvement process. The definition of a
tenet, as we're using it, is: "a principle, belief, or
doctrine generally held to be true; one held in common
by members of an organization."

Our first tenet is that quality improvement
requires management and union commitment and leadership
at all levels. Leadership means comprehending the
quality improvement process, developing a shared
vision, communicating clear direction, gaining environment, developing a sense of trust among the people, and
establishing a "can do" attitude.

The second tenet is that a quality improvement
structure and strategy are necessary for providing a
systematic approach to continuous quality improvement.
The structure requires the networking, teamwork, and
cooperation of all groups within Truck and Bus, as well
as throughout General Motors, our suppliers, and
dealers. The strategy must be well defined, encompass
all the elements in the quality improvement process,
and provide a systematic and uniform approach that is
clearly understood and followed throughout Truck and
Bus.

Our third tenet is that awareness and open and
free communications are necessary to create a climate
of continuous improvement. All employees must have a
shared understanding of our quality issues, challenges,
goals, commitment, and accomplishments if we are to
establish a new quality culture. This groupwide awareness can only happen in an environment of open, free,
and honest communications--in all directions.

The fourth tenet in our quality culture is
contingent upon an environment where all employees are
learning to apply the quality concepts and techniques.
A climate of continuous quality improvement process.

Commitment and leadership, structure and strategy,
awareness and communication, and education are all
necessary prerequisites for the attainment of our fifth
tenet: we (everyone) will continuously improve the

quality of all of our products, services, and business processes. Continuous improvement requires the following actions:

- Measurement, analysis, and continuous improvement of our business processes.

- The use of cost of quality as a management tool to help gauge the effectiveness of our quality improvement process.

- The use of statistical methods to identify, understand, and continually improve process capability.

- A corrective action process that includes an error-cause-removal system.

- A focus on preventative actions and on planning and providing capable processes.

The sixth tenet is that employees will be recognized and rewarded on the basis of their contributions to a team approach as well as their contributions as individuals for their continuous quality improvement.

The first six tenets all lead up to and support the seventh tenet: a group quality culture will be achieved when each and every employee at Truck and Bus is constantly trying to improve the quality of his/her processes. A work environment must be established where each and every employee understands and can contribute to the quality improvement process.

32

Xerox Gains from Productivity Innovations
Ed Finein

I'm delighted to be here today to share with you our experiences in improving Xerox's productivity to meet the challenges of the 1980s. Xerox today is a company that is confronted with significant worldwide competitive pressures in all aspects of its business. I hope our experience in meeting those challenges will provide some insight into what it takes to be a world-class company.

PRODUCTIVITY BACKGROUND

First, let me discuss the issue of productivity and its impact on Xerox. It's no secret that in the United States, growth of real wages has declined and productivity, as measured by output per worker, lags many industrialized countries. Current headlines reflect the frustration felt by Congress, industry, and the American consumer with our ability to compete on a worldwide basis.

Much of what has happened to industry is the result of arrogance and complacency. Americans, in the past, felt they could produce anything, their products were the best, and they could sell whatever they made, at home and abroad. At the same time, foreign competition was felt to be inferior in quality, performance, and value.

By the 1970s, we faced high levels of inflation, oil shocks, and significant foreign competition capable of meeting the needs of the marketplace. By the end of 1986, we faced a situation where real wages were equivalent to their 1969 level. Industry has fought back by closing plants, cutting wages, and exporting jobs. While companies may have achieved short-term improvements, these actions may slash individual living standards in the long term. The real culprit has been declining productivity.

Since 1973, output per worker in the United States, as defined by gross domestic product per employee, has been growing less than 1 percent per year compared to a rate of 2 percent in the 1960s. That statistic places the United States as twelfth among the leading industrialized nations of the world for the period of 1981-1985. The lag in productivity is clearly a significant factor in America's declining competitiveness.

Today we face a world of open markets, rapid technology transfer, and shortening product life cycles. To compete requires more flexibility than shuffling financial assets or moving jobs, production, or technology overseas. We live in a world economy where technology, information, and money can move across borders with the speed of light.

The company that can play--and win--in this complex environment must be fleet of foot and efficient to the extreme. It is my feeling that any company that is not already engaged in an internal revolution of seeking productivity in all aspects of its business-- product design, production, and distribution--could well find itself assigned to slow but certain decay.

Having said this, let me say that many--if not most--American companies are woefully behind the curve in recognizing, let alone implementing, what it takes to continually improve their productivity to a point where they can successfully compete on a worldwide basis for the long term. We need to capture the innovative spirit that used to be reflected in our products and focus it on new ways to improve the management of productivity throughout the business.

XEROX IN PERSPECTIVE

Xerox, like many other companies, suffered from arrogance in the 1970s. Xerox created the plain-paper copying industry in 1959. That was the year we introduced the Xerox 914 copier. Prior to that, you'll recall, the most popular office copying techniques were relatively messy--either wet processes or carbon papers.

The 914 changed all that. It transformed the Haloid Company of Rochester, New York, into the Xerox Corporation. For the next fifteen years, we were, in a way, victims of our own early triumphs. The company's--and the world's--first plain-paper copier, the 914, was one of the most successful, if not the most successful, new product ever introduced in corporate history.

We had such a stranglehold on the copier market throughout the 1960s and early 1970s that we hardly paid attention when IBM and   Eastman Kodak   began marketing high-speed copiers, the most lucrative part of the market.   Nor did we worry when the Japanese began to offer small inexpensive copiers in the mid-1970s, an area we ignored until recently.

The Xerox of the late 1970s was a bureaucratic company in which one function battled another, and operating people constantly bickered with the corporate staff.   Disputes over issues as relatively minor as the color scheme of machines had to be resolved by our CEO. The result was painfully slow product development, high manufacturing costs, copiers that were hard to service, and unhappy customers.

In the mid-1970s, as I've said, the Japanese camera makers entered the low end of the market.   They used aggressive pricing to gain a foothold there, facilitated by lower cost, and proceeded to own a sizable market share.   Their strategy was similar to the one they used so successfully in automobiles, cameras, home appliances, calculators and watches.

The Japanese strategy worked well.   Through the late 1970s, we saw our market share erode at an alarming rate.   But the problem was masked for awhile because the industry was still growing.   Xerox continued to place more units, and the company's revenues and profits continued to rise, peaking in 1981.   In spite of this, the problem had become readily apparent by 1980. We had to rapidly improve our productivity to deliver products to the customer if we were to effectively compete with our competitors in the global market.

It has been said that "self-reflection is the school of wisdom." If so, Xerox got quite an education during those few years.

The courage and willingness to be introspective is a prerequisite for any organization that aspires to carry the banner "world-class." The nineteenth-century philosopher, James Russell Lowell, said: "No man can produce great things who is not thoroughly sincere in dealing with himself." I believe that same holds true for companies.

We began in earnest in 1980 by assessing our corporate strengths and weaknesses.   Many of the problems we identified relate to points I mentioned before. We found, for instance, that it took Xerox too long to develop new products.   Our products cost too much, and that did not fully satisfy our customers' requirements.   In fact, we were horrified to find that the Japanese were selling their small machines for what it cost us to make ours.

After an in-depth study, we recognized the need to overhaul the very way we managed our business.   We found that to play--and win--in the new competitive

market, we had to be efficient to the extreme. We had
to engage in an internal revolution in product design,
production, and distribution to eliminate competitive
advantages.

COMPETITIVE BENCHMARKING

     Let me first talk about competitors. Here we
began to understand and evaluate our problems through a
process called competitive benchmarking. Benchmarking
in Xerox is a tool to identify industry cost and
performance standards and to set goals. It also
provides us with insights into how these cost and
performance standards can be achieved or exceeded and
to develop internal action plans. Most importantly,
benchmarking is an ongoing learning experience, both
for the people involved and for the company as a whole.
It defines the productivity improvements we need to
compete effectively.
     Forms of benchmarking have been used in industry
for years. Early in the twentieth century, for
instance, Walter Chrysler used to tear apart every new
Oldsmobile that came off his competitor's assembly
line, to determine precisely what went into the car,
how much it cost, and how it was made. Armed with this
information, Chrysler had a better sense of what he was
competing against.
     At Xerox, our competitive benchmarking looks both
inside and outside the reprographics industry. In the
beginning, we visited and studied several leading-edge
American and Japanese companies. We visited Japan and
studied several of our competitors there--Canon and
Ricoh--toured factories and visited R & D facilities,
and reverse engineered our competitors products. Our
studies revealed the need for at least some change in
virtually all phases of our business, from product
design to sales and service.

CREATING PRODUCT DELIVERY TEAMS

     One of our problems was the way we were organized.
We had a matrix management structure in our Product
Development Organization. That meant that any one
development effort had to flow through separate
functions--product planning, design engineering, manufacturing engineering, and service engineering--each,
essentially, operating in a vacuum. There was then no
individual clearly responsible for the end product.
The reason this structure was created in the first
place was to prevent errors. But it had the unintended
effect of almost preventing product delivery.

With so many different organizations involved, there were costly overheads and inevitable slowdowns. There constantly was need to review and gain concurrence across various groups, necessitating program management and time-consuming committees to address cross-disciplinary issues. Virtually all issues were cross-disciplinary and fell victim to this process. The cycle was so long, in fact, that product development programs sometimes became obsolete in midstream because of changing market needs. To address these problems, we were forced to improve the productivity of our product delivery organization. The first thing we did was to disassemble the matrix organization I described a moment ago and we created Product Delivery Teams under one person, called the chief engineer. The chief engineer is totally accountable for a product development project in total, including quality, cost, and schedule. The chief engineer manages all the design teams, the model shop, and pilot plant. He has complete authority to modify the development schedule and to make go and no-go decisions along the way.

I am happy to say that nowadays the bureaucracy--and much of the corporate staff--is gone. In their place are entrepreneurial "product-development teams" and "problem-solving teams."

PRODUCTIVITY TARGETS

Using the competitive benchmarking techniques, we also set very aggressive targets for ourselves in all areas of product development from product planning to product development cost, schedules, unit manufacturing cost, product quality, marketing, service, and customer acceptance levels.

Specifically, our renewal based upon industry benchmarks called for nothing less than reducing product development time by half, reducing product development manpower by half, improving manufacturing quality by 90 percent, and reducing unit manufacturing cost by 50 percent; clearly aggressive productivity targets!

REDUCING PRODUCT DEVELOPMENT COSTS

In addition to these objectives an equally critical challenge was to insure that we fully satisfied our customers. Accomplishing that was contingent upon understanding a customer's existing and latent requirements.

The process begins at the product planning stage. We identify market trends by conducting focus groups and customer surveys. In essence, we have our customers "build" their ideal machines. They make their own trade-off decisions. For instance, sacrificing a little speed for an additional paper-handling feature. In the end, we're able to pair appropriate technologies with customers' requirements.

In product development, many of our competitors were using only one-fourth the labor we required for the same level of engineering output. Matching this was essential to our goal of accelerating our rate of new product introductions, and of reducing our product development costs.

To pare down those costs, we took several important steps. First we took a page out of the Japanese book and began promoting the concept of the engineering generalist, or multi-functional engineer. Generalists are capable of handling a greater variety of tasks. Engineering specialists, on the other hand, constantly pass pieces of projects from one to another throughout a development cycle, causing delays and adding cost to the process.

Another action we took was to co-locate design engineers and, where possible, manufacturing. Putting the entire design team responsible for a product under one roof facilitated communication among the team members, saving time and enhancing the creative process.

A seemingly simple and obvious action was to stabilize design goals. In the product development process, changes frequently force retooling, which is expensive and time-consuming. Whereas Xerox routinely used to change about 50 percent of the parts over the course of a design cycle, that figure is now down to 20 percent--still not good enough, but it is better.

We have had a reputation as a technological innovator, so it should come as no surprise that, in reassessing our product development process, we decided to incorporate the latest technology. Xerox now uses state-of-the-art computer-aided design systems, which reduce the amount of time it takes to bring a design concept to prototype stage.

Another idea we've borrowed from the Japanese is continuous supplier involvement. Xerox encourages its vendors to share their expertise, not just sell us their product. In this way, they become more than suppliers to Xerox. They become strategic partners in the design process. Many of the suggestions they make directly improve parts quality or availability, impacting the product's manufacturing cost and deliverability.

And finally, Xerox--an office systems vendor--is practicing what it preaches in the area of office automation. Many of our people have used the ethernet local-area network for several years now, and they wouldn't give up their electronic mail for anything.

As a result of all this, our product development schedule is 20 to 50 percent shorter than it used to be.   Engineering productivity has more than tripled since we began our efforts nearly five years ago. Specifically, on our recently introduced 1065 copier, we have been able to cut the development time by at least 20 percent, our development resources by one-third, and in the process have reduced our unit manufacturing cost by half! And all of this has not come at the expense of product performance either. Our field testing shows that the 1065 is meeting all critical customer expectations in terms of reliability, copy quality, features, and productivity.

CUTTING MANUFACTURING COSTS

In parallel with our progress in reducing our product development costs, we also moved to reduce manufacturing costs. As I mentioned earlier, our goal was to reduce Xerox manufacturing cost by at least half to match or improve on competitive levels.

Our early supplier involvement program has paid dividends in this area, as well. Also contributing to lower manufacturing costs have been partnerships with the union; reduced manufacturing staff and overhead; multi-national sourcing of materials; and automated materials handling techniques.

While reducing manufacturing costs, Xerox did not sacrifice quality. While before, budgets and schedules were two key criteria for deciding bonuses and promotions, now product quality and customer satisfaction are just as important.   In fact, we took a number of measures to improve product quality, mainly in the area of process control and automation. We have perhaps made the most progress in the area of quality.

LEADERSHIP THROUGH QUALITY

All of what I've been talking about relates in some way to products--how we develop them and how we build them to compete in the world marketplace. But product is just one part of a much larger question, which is, how do we recast the COMPANY--the entire organization--to succeed in the global area?

This larger question deals with quality, not only in manufacturing, but also in how we think, how we

conduct our business, and how we relate to each other
and to our customers. We have addressed this through a
program called "Leadership Through Quality," which our
chairman, David Kearns, personally began and fostered
throughout the company.

This program has encouraged every Xerox employee
to identify and satisfy customer requirements for his
output whether the customer of his effort is another
Xerox employee or a marketplace customer. One of the
things this process has emphasized is a concerted
employee involvement program. We are attempting to
bring every Xerox worker at every level of the
organization--from Senior Management to the hourly paid
manufacturing people--into the problem-solving process.

The program has been very successful. In one
case, a team in our high-volume business unit worked
for nine months to reduce the cost and delivery time of
a copier part. What had been a $70.00 part with a
thirteen-week schedule was whittled down to $30.00 and
a ten-week hardware delivery.

What makes the accomplishment so gratifying is
that the team members represented a variety of
disciplines, and worked well together to solve a
problem that no individual could have tackled alone.
It took people from design, engineering, the model
shop, and administration--with their collective
knowledge and ideas--to make it happen.

In another case, a team in our low-volume business
unit took an infant technology and evolved it into a
production design, meeting all the customer requirements for product functionality AND driving down cost
from $12.00 to $5.00 per unit. Again, the sense of
ownership of a problem by a group of employees
encouraged them to surmount it.

There are more than 600 problem-solving teams like
these within the Xerox Reprographic Business Group
today. Some 70 percent of our people are involved in
them. Each person receives forty hours of classroom
training in problem-solving techniques and statistical
quality control, to enable him to help solve problems
in his immediate work environment. We have made the
investment in our "human capital." Our employees are
our most valuable asset.

Our "Leadership Through Quality" religion has also
emphasized giving the customer what he wants. We
sample each of our customers--at least once every
eighteen months--to survey their satisfaction with our
products and services. The output of this survey is
used to restructure our products, services, and, in
fact, corporate priorities.

Our efforts in meeting customer requirements have
paid dividends. Our surveys show that, in many market
segments, we are now the reprographic industry

benchmark.   If the new corporate approach sounds
similar to that of the Japanese, it is no coincidence.
We have gone to extremes to study our Japanese
competitors.   Our successful Japanese subsidiary,
Tokyo-based Fuji Xerox, helps us keep tabs on the
Japanese.   In fact, we visit Japan annually to
recalibrate our competitive benchmarks.

SUMMARY

This is a summary of what we have been doing for
the past five years--developing the strategies, and
putting in place the systems, to become a world-class
company.   To accomplish that required redefining the
way we do business and achieving productivity
improvements in all areas.   I think we have done what a
few other American industries have done.   We have
significantly narrowed the advantage that such
formidable Japanese competitors as Canon, Sharp, Richo,
and Minolta had enjoyed, thereby stemming their advance
in our market.

But for all our efforts, there is more to be done,
especially since the competition is hardly standing
still. To remain competitive in a worldwide market is a
never-ending process.   We may no longer be the company
we once were, but we are not yet the company we want to
be.   We are a company in transition.

33

Ford's New Business Ethics: Quality Is Job #1

Lewis C. Veraldi

Today, I would like to share with you what we at Ford are doing to achieve world-class quality and improve productivity. But, to understand where we are today and where we are going, it is necessary to understand where we have come from.

WEAKNESS OF SEQUENTIAL DESIGN PROCESS

Traditionally, Ford and other automobile manufacturers developed new vehicles in a "sequential" design process. What do I mean? Basically, throughout the design of a vehicle, each activity does its thing and then hands off to the next activity.

The designers do their thing--design--and then turn that over to the engineers. After the engineers do their job, manufacturing is told to go mass produce the product. And marketing is then told to go sell it. That's an oversimplification, but it dramatically points out the flaws in the traditional system. What you have is each group of specialists operating in isolation of one another.

Moreover, what someone designs and styles may be quite another matter to engineering, and by the time it reaches manufacturing there may be some practical problems inherent in the design that make manufacturing a nightmare. The people who actually build the vehicle haven't been consulted at all, and marketing may well discover two or three reasons why the consumer doesn't like the product and it is too late to make any changes!

As in football, sometimes the ball is fumbled during the hand-off.

The result of this traditional process was a lack of team work, poor quality, redundancy of effort, and extreme inefficiency.

TEAM APPROACH REPLACES TRADITIONAL SEQUENTIAL PROCESS

What caused us to change the way we do business? Back in 1980, Ford's image for quality was not very good. In addition, we were in the process of losing over $1 billion two years in a row. In this climate, we were beginning to plan a replacement for our mid-size and large cars. Of course, these replacements would become today's Taurus and Sable. The investment was estimated to be $3 billion and we knew the old ways of doing business would not work.

In their class, Taurus and Sable were to be designed to compete with anything in the world—foreign or domestic—in fit/finish, ride, handling, and vehicle ergonomics. In short, they were to be designed with world-class quality and provide customers with a compelling reason to once again shop and buy American.

As vehicle objectives were defined (and redefined), we knew upstream involvement and employee commitment were essential, and that the existing organization wouldn't support our needs. Therefore, the sequential organization was replaced by a simultaneous "Team" approach, TEAM TAURUS.

Team Taurus's organization is designed to promote continuous interaction between design, engineering, manufacturing and marketing along with top management, the legal, purchasing, and service organizations. Overall coordination or direction is by a car product development group or program manager whose primary function is getting all of the team to work together. All these groups work simultaneously to bring our new car to market. Instead of being last to be involved, for example, manufacturing is involved some fifty months prior to producing the first vehicles. The "downstream" people were factored into the program as early as five years ahead of introduction. Thus, all activities had an equity and an equal opportunity for simultaneous participation throughout the entire program.

To insure that the best ideas of the people who would be designing and building the car would be considered, our engineers developed a comprehensive "want" system. We visited the Atlanta Assembly Plant and spoke to the hourly and salaried personnel who eventually would build the car. We asked them to tell us how to design a car that was easy to build, and would avoid the design problems that had led to poor quality in the past. We visited the plant fifty months before Job #1 so that we could incorporate their suggestions in the basic design of the product. Their suggestions, along with those of other groups, resulted in more than 1,400 wants being identified. Over half of these were incorporated into the product.

To illustrate, the assembly workers told us that to achieve consistent door openings and tight door fits, a one-piece bodyside was necessary. So we reduced the number of components on the bodyside from twelve components to two. The doors are also one piece for improved quality and consistent build.

SUPPLIERS TREATED AS PARTNERS

Just as important is the early involvement of our suppliers. In today's automobile industry, the suppliers are very critical members of the team. They are our partners and their early involvement is essential to achieving our world-class quality objectives.

On Taurus and Sable, we initiated two supplier programs:

- Early sourcing

- System sourcing

Early sourcing identifies a component source early and brings them upstream in the design process. The supplier is responsible for the fit and functioning of his components. System sourcing is a process where components that must be coordinated both in color and fit are sourced to one supplier.

An example of early sourcing is the Taurus subframe, which supports the engine and transaxle. It was sourced to A. O. Smith three years in advance. This allowed the engineering and manufacturing team to make 137 design revisions to improve the variable cost, reduce weight, and achieve an automated assembly.

By bringing suppliers into the process early, we could take advantage of their expertise. Prince Corporation, who supplies some of the sun visors, suggested several features that were added to the Taurus. These features included a dual visor system, a new pull-down visor mirror, and a new dome light, which will not shine into the driver's eyes at night. Masland Carpet, the supplier of the station wagon load floor carpet, suggested a method to insure the carpet nap all ran in the same direction. The result is a uniform appearance for the load floor.

Early sourcing has another advantage. Prototype vehicles, which are built from components supplied by the production source, allow early resolution of fit and finish problems.

On a limited basis on Taurus, we initiated a process of system sourcing. The station wagon interior garnish moldings were all sourced to O'Sullivan. They

were responsible for coordinating the fit and color
match of all the interior garnish moldings.

Another aspect to system sourcing was the sourcing
of the die models and tools for the interior garnish
moldings to one supplier, Pro-mold. By having one tool
supplier, we could evaluate the fits and coordinate
revisions much more efficiently. The result--better
fits and better quality.

ATTAINING PERCEIVED QUALITY OBJECTIVES

Several new engineering processes were developed
during the Taurus program. By far, the most extensive
program to involve engineering was the Best-In-Class
Expanded Image Program. The intent of the program was
to focus more attention on details that affect the
perceived quality of the vehicle, including the
interaction   between   the   vehicle   and   the
driver/passenger.

To determine our objective, the team sought out
the best vehicles in the world and evaluated more than
400 characteristics on each of them to identify those
vehicles that were best in the world for particular
items. These items ranged from door closing efforts,
to the feel of the heater control, to the underhood
appearance. The cars identified included BMW,
Mercedes, Toyota Cressida, and Audi 5000.

Once completed, the task of the Taurus Team was to
implement design and/or processes that met or exceeded
those "best objectives." I am proud to report that as
we went into production, we had achieved Best-In-Class
status on 80 percent of these 400 items. And that was
accomplished with teamwork and paying attention to the
details.

Finally, we continually asked ourselves the
question "WHY BUY TAURUS?" In other words, why should
someone cross the street to shop in our store? We
asked customers, dealers, buff magazine writers,
service people, insurance companies, and professional
drivers like Jackie Stewart, early in the program, what
features they wanted. By the time we introduced Taurus
and Sable, this translated into a feature list thirty-
two pages long! These features included flush side
glass for reduced noise, polycarbonate bumpers for
rust-free life, and first-class seats for long-distance
riding comfort.

PROGRAM MANAGERS IMPLEMENT TEAM CONCEPT

Enough history! Where are we going today? As you
can see, teamwork was a major factor in the success of

Taurus and Sable.   Early and dedicated involvement by all members of the team was key.

The future is going to see the TEAM TAURUS approach taken to all new Ford Motor Company products. It has already been put in place and endorsed by the top management of the company who have signed a pledge of support for our new way of doing business.   The team concept has been designated as the "Program Management" organization.   We have identified "Program Managers" who are putting their teams together to take new cars from concept to customer the same way Taurus and Sable were brought to market.

The program managers assignment is to provide leadership for the planning, design, engineering, sales, manufacturing, quality assurance, and service for their products.   I said "their products" because it is "their product."   The objectives for each vehicle will be the basis for measuring the performance for all members of the team.

The program managers have the responsibility for the "WHAT" and "WHEN" decisions of a program and for working with the line activities who retain authority for "WHO DOES IT" and "HOW THE JOB IS DONE."   The program managers have the further responsibility of involving the line people as early or as far upstream as possible--in effect making them part of the team from the beginning, and involving them in the decision making.

This up front commitment by all activities is reflected in one of the tools we have developed for the program managers, a timing discipline chart.   The purpose of the chart is to get a commitment from all activities as to what tasks are required and when they must be performed.   Each member of the team signs the chart.   The chart allows each member of the team to see how his activity impacts other activities.   For example, one area of the chart shows the design events that must occur twenty-six months before production begins.   In addition, management can use the chart to assess how the program is proceeding.   If these key events are performed late, we know they will affect the quality and timing of the launch of the new product.

The program manager's scope encompasses all aspects of the program and all activities.   Each activity--design, manufacturing, sales, etc.--identifies one representative to work on the program manager's team.   Each representative is responsible for coordinating the activities of his or her component and ensuring that the objectives of the team are achieved.

Well what kind of person are we looking for as a program manager?   I have a long list of personality traits but let me mention a few.   He should be results-oriented, self-confident, product excellence-oriented, decisive, have a broad view of the organization, and, maybe most important, have a sense of humor.

IMPROVING AMERICA'S COMPETITIVENESS

Now, let me shift gears for a moment. We have made great strides at Ford by changing the organization and initiating new processes but people are the most important element. Our colleges and universities have a role to play in helping us achieve our world-class objectives. First, we need more generalists. Technical specialists should be exposed to all facets of the business. All disciplines need a better appreciation of the roles of others.

Second, we need courses designed to improve America's competitiveness. Students need a better understanding of how quality is achieved, such as through statistical process control or quality function deployment. Continuous study is required to identify ways to help the American worker be more productive and efficient.

If America is going to be competitive worldwide, we must improve our manufacturing productivity and increase our level of engineering technology. At Ford, the results of our new process are already being realized. The quality of Ford products has improved over 60 percent since 1980. For six years in a row, we have been first in quality among all domestic manufacturers. While we are pleased with our progress, we are not there yet. The Japanese quality is still better but the gap is closing quickly. There is much that remains to be done. We are developing new processes to reduce the length of time to develop new products, improve the flow of information, and increase the role of the supplier to include participation in product design.

In closing, let me note that the latest result of our team effort is our new Continental to be introduced in late 1987. The Continental is equipped with a unique combination of hi-tech features not found collectively on any other luxury car in the world. You'll hear more about the Continental in the coming months.

At Ford we have a single goal and that is to be manufacturers of cars and trucks that are responsive to our customers' wants and exhibit world-class quality in every segment in which we compete. With the Taurus and Sable we are starting to see it. And this is just the beginning. The sparkle in the oval is getting brighter year after year.

34

Met Life's Quest for Quality

John J. Falzon

After being given the assignment as the Metropolitan Corporate Quality officer, I did what I presume most of you would do, I visited the people who had a similar assignment, and almost universally, they gave me the same piece of advice--"Make sure you enlist the cooperation of your CEO, otherwise you are doomed to failure." In my case at Met this was not a problem. It was the CEO who enlisted my support. Our President and CEO, John Creedon, has been the driving force behind Met Life's quest for quality. He is absolutely convinced that, in a highly competitive environment, quality is the most distinguishing characteristic. It was he who coined the phrase, "Quality is more a journey than a destination," which I have come to recognize as a truism. It is certainly definitive of what has happened at Met Life so far.

Recognizing that improvement is always possible, the initial programmatic features of our quality improvement process have undergone a series of modifications. Fortunately, each improvement and every record of change has been consistent with the original design. Today I would like to tell you about Met Life's quest for quality.

FOUNDATION FOR QUALITY IMPROVEMENT PROGRAM

Once again, just like many other companies, Met Life during the initial phases of development formed a Quality Steering Committee. The committee was composed of the people who were in charge of our major staff departments and businesses. After a good deal of discussion and deliberation the committee developed a series of directives that became the foundation elements in the Quality Improvement Process (QIP).

From a strategic view point, the formation of a steering committee was very important. Their commitment provided example and leadership to middle

management. Their directives included the following
major points. They said that any program for quality
must have strong customer orientation; that all people
in the company must be involved. In fact, they emphasized the fact that each employee would identify with a
customer. They went on to say that the quality process
should not be seen as an academic exercise, but rather
should be viewed as something that provides a significant business advantage to the company. They emphasized the need to have the process implemented through
the line organization with a minimum of staff support,
and that it should be implemented in a unified way
throughout the entire company.

Using these directives as guidelines, a process
was introduced to the company that was designed to
alter both the way things were being managed and the
way the work was being done. The new process required
every organization to identify the major services being
rendered and the customers to whom they were being
directed. Each area then developed a quality network
composed of Quality Improvement Teams for each
identified service. These teams established
performance measures, and through a structured problem-
solving technique identified opportunities for
improvement. They were encouraged to establish
challenging goals and to open up dialogue with their
customers, where it was envisioned they would obtain
the most pressing motivation to improve.

DEFINING AND MEASURING SERVICE

These initial steps appeared to meet the criteria
established by our steering committee. However, there
were still some open ends that needed attention. The
first of these is that, although we encouraged customer
dialogue, we did not say how this dialogue was to be
conducted, nor did we say how to evaluate the results
of the dialogue in terms that would be instructive to
the establishment of new opportunities for improvement.
A second aspect of the design requiring attention was a
need to forge a link between the Quality Management
Process that was being introduced at the operational
levels of the company, and the strategic planning that
was essentially being conducted at higher levels.

Each of the issues have been addressed. Let me
describe the first of these, that is the issue of how
to define and measure the customers' view of service.
Fortunately, we came upon some original research
completed by a group of professors from Texas A&M. The
research was conducted under the auspices of the
Marketing Science Institute of Cambridge,
Massachusetts. The objective of the research was to

understand the nature of services and to measure
quality in the service sector. Needless to say we were
extremely interested in the findings. It was the first
thing I ever came across that specifically addressed
the issue of quality in a service type environment. I
think you will agree that most of the literature
addresses the manufacturing sector, and although many
of the principles can be applied to service companies
there are still many gaps that are left unfilled, and
many differences left to be explained.

Let me point to some of the more significant
findings in the Texas A&M research. The first research
finding noted that there are several distinctions that
set services apart from goods. First of all, services
are intangible; they cannot be measured, tested or
verified in advance of sales to assure quality.
Second, services have a high labor content-- they
reflect the behavior of the service personnel who are
intrinsic to service delivery. Therefore performance
of the service varies from provider to provider, customer to customer and from day to day. Third, the consumption of many services is inseparable from their
production--and the customer often participates in the
service delivery process. For example, in getting a
haircut, the customer describes the expected outcome,
which becomes critical input to the quality of service
performance. Finally, services are perishable. They
cannot be saved or inventoried. Once the opportunity
is missed there is no second chance to sell a service
to a customer. Obviously these four characteristics of
service pose problems to firms who desire to deliver
high quality services. They also lead to the conclusion that because of the people intensity, the process
of delivering a service is as important as the outcome,
and non-routine transactions deserve as much--if not
more attention--as do routine. This is an extremely
important point, and in some respects an eye opener for
managers who are looking to improve quality. Too
often, and I know it's true in our case, the measurements of performance for service transactions are built
around the routine actions and are concerned primarily
with outcomes. Somehow the non-routine transaction is
excluded--perhaps we view it as "the exception."
Similarly the process of delivery does not lend itself
to easy measurement, and therefore we tend to exclude
it from examination.

The second research finding concerned the basis
for measurement. The research was able to identify the
most significant factors influencing the overall
evaluation. These factors fall within five generic
classifications: Reliability, meaning the ability to
perform the promised service dependably and accurately;
Responsiveness, meaning the willingness to help

customers and provide prompt service; _Tangibles_, meaning the physical facilities, equipment and appearance of personnel; _Assurance_, meaning the knowledge and courtesy of employees and their ability to convey that trust and confidence; and _Empathy_, meaning the caring, individualized attention provided to customers.

The third research finding was based on the foregoing factors. The researchers developed a survey instrument that was able to quantify service performance. Moreover, the instrument was designed within a system that enables management to identify and isolate problem areas.

The fourth finding which I believe was one of the most significant of the research was an ability to apply a rather precise definition to the measurement of quality in the service sector. They said that the quality of services is measured by the difference between the expectation levels of the customer on the one hand, and the customer's perceived level of service delivery on the other. The gap between these two points represents the size of the service problem.

An interesting aspect of this study was that the researchers originally presumed that on average the better companies were those where perception of service delivery consistently exceeded customer expectations. In reality even the better companies rarely did. Instead the researchers found that the better companies merely had a smaller gap. It is interesting to speculate that providing excellent service is one of the ways we raise customer expectation levels: Once we exceed a customer's expectations, we've set a new and high level of expectation for the future. It also explains why the pursuit of quality is a never ending process, and makes a great deal of sense out of John Creedon's statement, "Quality is more a journey than a destination."

With the assistance of one of the Texas A&M professors, we structured a special seminar that was delivered to principle officers from all parts of the company, with the expectation that each of the various departments would proceed with the construction of appropriate survey instruments to measure customer satisfaction.

LINKING OPERATIONAL QUALITY MANAGEMENT TO STRATEGIC PLANNING

In addressing the second problem that is "how to make the connection between high level strategic planning and the quality improvement process which was instituted at the operational level," we again looked

to the academic community for help.   Dr. Russell
Ackoff, formerly of the University of Pennsylvania,
Wharton School of Business, has developed and written
about a very unique approach to planning that appears
to fit our needs.   It is an interactive planning
process that has many of the characteristics associated
with our Quality Improvement Process.   There is a great
deal of consistency and continuity between the two.
But the Ackoff Method adds a good deal more.   Most
important, it provides the means to bring together all
levels of planning starting with the very top layers of
management in the company.   Its introduction to the
company is not being viewed as replacement, but rather
as an add-on--an enhancement to our already existing
systems.

To illustrate, let me say a few words about this
planning method.   A key element in the process is the
creation of an Idealized Design of your business or
organization.   The design is completed under the
presumption that the existing organization was
destroyed last night.   The presumption provides the
freedom to think about what "should be" without the
encumbrance of having to describe how to do it (that is
left to later phases of planning).   Simultaneously,
another group identifies all of the problems and impediments in today's system.   The ideal design should
solve all of the identified problems in the current
operating environment; if not, the design is faulty.

In our Quality Improvement Process we concentrate
on removing problems.   In the new planning process,
solving problems is not the key element, it is only a
part.   Implementation of the design provides the added
benefit of being able to influence the future within
which you will be operating.   This change in focus is
different than the original approach yet it builds on
our past experience, and its implementation is viewed
as a logical evolutionary step.

We are currently engaged in the task of installing
this interactive planning method at our corporate level
of management.   Other levels of the organization will
use the new planning system as an enhancement to their
quality networks.   The phrase "Quality Through
Planning" captures the essence of this undertaking.

The new planning process also builds on the new
survey technology from Texas A&M.   All organizations
are being requested to provide, as part of their annual
planning objectives, "an assessment of the gap that
exists between customer expectation and perceived
levels of service," and to provide an explanation of
how they expect to go about reducing these gaps.

We are happy to say that we have had the
opportunity to participate more closely with the
continued research being conducted by the Texas A&M

group. We are hopeful that the research will provide
additional insight on the internal factors that
contribute to the service quality gap and provide a way
to show how modifications in management behavior can
influence the quality of service delivered.

The researchers expect to provide these insights
by building upon their initial findings. The original
studies identified reasons why there would be a gap
between the customer perception of delivery and his
expectations. These reasons were associated with the
delivery of the service and focused on areas over which
management has control. Great value was placed on the
identification of these problem areas. Yet the mere
acknowledgment of their existence raises major
questions about the nature and importance of the contributing factors. The identification of these factors
provides the means for helping managers do a better job
in the delivery of quality service. We see this as a
very   worthwhile   venture,   deserving   our   full
participation and backing.

ASPIRATIONS FOR IMPROVED QUALITY

As we look to the future, we'd like to describe it
in terms of our fondest aspirations.

1.   We would hope to gain significant insight into our
     customers needs and to continually provide an
     enhanced level of service that would even exceed
     their expectation levels.

2.   We hope this can be accomplished as a result of
     improving the skills of our management personnel,
     and most   important, by improving the skills of
     the people who have the immediate responsibility of
     servicing a customer.

3.   We would like to see customer contact people having
     broader range of capabilities so they can respond
     more completely to customer needs.

4.   We would like to see this develop to the point
     where the only limitation on the personal growth of
     Met Life employees is self imposed.

35

Computer Integrated Manufacturing: Allen-Bradley's Experience

J. Tracy O'Rourke

A whole new philosophy of business must be adopted by companies considering Computer Integrated Manufacturing. They must broaden their thinking to strategic dimensions and they must be ready for bold action.

Some manufacturers already know that. They have felt the sharp sting of international competition. They have discovered that their competitors are on the next continent, as well as in the next town.

TRENDS IN MANUFACTURING

American manufacturers have probably felt more pain than any other group. An article in The Economist recently reported that America's share of world trade shrank from 21 percent to 14 percent in the past quarter century. In that time, the U.S. trade balance tumbled from a healthy $5 billion surplus to a $150 billion deficit.

And it gets worse. In the past five years alone, America's trade balance in manufactured goods fell from an $11 billion surplus to a $32 billion deficit. America's manufacturing export volume tumbled 32 percent and every billion dollars of lost exports costs an estimated 25,000 American jobs.

Efforts by politicians and others to save blue-collar jobs have probably slowed U.S. manufacturers' progress in becoming more competitive and could move us toward the position that British manufacturers find themselves in. In a recent article, Peter Drucker wrote:

> The British example indicates a new and critical economic equation: a country, an industry, or a company that puts the preservation of blue-collar jobs ahead of international competitiveness will soon have neither production nor jobs.

Despite the dismal statistics, manufacturing is still very big business in the United States. Manufacturing's contribution to the gross national product has remained a constant 20 to 25 percent since 1947. The wealth is there, even if the jobs aren't.

There are going to be a lot fewer people operating our factories of the future and those who remain are going to be Very Important People--VIPs in a very real sense because they will have significantly more responsibility than the factory worker of today. The trend is already evident. The Wall Street Journal recently reported that since late 1982, when the recession bottomed out, manufacturing jobs have increased by less than 6 percent while manufacturing output as measured by the Federal Reserve Board index has increased by nearly 30 percent. During the past two years output has continued to rise while the number of manufacturing jobs has dropped.

MEETING GLOBAL COMPETITIVE CHALLENGE WITH CIM

U.S. manufacturers increasingly recognize that today the only way they can meet global competitors head on--and win--is with CIM.

So what's causing all this change? Peter Drucker recently observed that economic dynamics have shifted us from a national economy to a global one. Only a few decades ago the world's economies were based on natural resources. A nation's climate, waterways, and minerals created the competitive advantages that defined its manufacturing specializations.

Then, in the 1960s, we began to see the emergence of a created competitive advantage: lower labor costs. Developing countries with large labor pools attracted manufacturers from more industrialized countries who sought the advantage of lower labor costs. These developing nations shaped their own manufacturing specializations. They gradually built capital, and developed infrastructures of transportation and communication. Today they can put pressure on the more advanced economies. Thus, the international playing field has been leveled considerably.

Natural resources have decoupled from national boundaries, direct labor is decoupling from production, and capital and technology have combined to form a new driving economic force. Today, any country can buy raw materials such as oil and copper on the world market-- and they'll pay approximately the same price as anyone else. So, natural resources are no longer a differentiator. Neither are the wages of a nation's labor force, because direct labor is an insignificant cost factor in advanced manufacturing.

These changes are forcing advanced nations to compete with three resources: capital, technology, and know-how--the kind of technical "know-how" found only in large pools of educated professionals. Properly blended, this capital, technology, and people form a new competitive advantage and a new natural resource: Computer Integrated Manufacturing, or CIM.

Some view CIM with fear, seeing it as an unnatural, disruptive force. But it's far from that. CIM isn't revolutionary--CIM is evolutionary.

It's the most recent development in the quest to produce better quality products at a lower cost with less human involvement. That quest began 200 years ago, with the Industrial Revolution--an upheaval that thrust farm workers into city factories, and changed a barter-based economy into a currency-based one.

Mechanization followed, harnessing steam generation to power machines. Electrification gave birth to modern manufacturing. Then came computer control, making automation possible.

WHAT IS CIM?

Computer Integrated Manufacturing is the next logical step in this evolution--a step that requires no great leap of faith. CIM is an advanced business philosophy that unifies a company's administration, engineering, and manufacturing.

To manufacturers inundated by the tides of global competition, CIM is an attractive strategy for survival, growth, and prosperity. With CIM, they can achieve the levels of quality, cost, and service they need to become competent global competitors.

If this is true, why aren't all manufacturers rushing into some form of CIM? Some have tried that. The path traveled by new technology is rarely smooth, so some who have plunged into CIM or some aspect of it, maybe without a good plan in place, have had a rough trip.

But with a few years of experience behind us, we are fortunately seeing a new realism and renewed faith in the future of CIM technology on the part of U.S. manufacturers. We may occasionally see negative head-lines but the demands of global competition will over-come whatever queasiness exists regarding CIM technologies.

FINANCIAL JUSTIFICATION OF CIM

Unfortunately, once a manufacturer concludes that CIM is the path to survival, he faces the wilderness of financial justification. Even though manufacturing and accounting experts have contributed much to knowledge and theories about CIM justification, the process remains intimidating.

Traditional accounting methods which have served industry for decades lose relevance when applied to manufacturing systems that are flexible and reusable. Some of CIM's most valued benefits are just not quantifiable by traditional accounting methods, nor are there proven new ways of evaluating them.

It's easy to see the problems one encounters when trying to justify CIM to a board of directors that has always dealt with traditional accounting methods and is unfamiliar with CIM's advantages. At this point, the justification of CIM remains as much a strategic issue as a financial one. And for many manufacturers who are getting their first glimpse of global competition, the strategic issue can be interpreted as: "Do I want to stay in this business or don't I?"

CIM DEFINITION

Before going further, we need to define CIM. Computer Integrated Manufacturing integrates the "factors of production" to organize every event that occurs in a manufacturing business, from receipt of a customer's order to delivery of the product. The ultimate goal is to integrate the production processes, the material, sales, marketing, purchasing, administra- tion and engineering information flows into a single, closed-loop, controlled system.

CIM is a whole new philosophy of business. It is a whole new way of thinking that requires new strategies, new management techniques, and new manufacturing dynamics. "True CIM" means the complete integration of a company's office, engineering, and manufacturing.

In many firms, engineering is already experiencing tremendous productivity gains through computer-aided design (CAD) and computer-aided engineering (CAE). A similar process has begun in the office, where more than two million desk top computers have stopped the slide of productivity. But it's not overall producti- vity, it's individual productivity, because generally these work stations aren't integrated with larger data bases.

Manufacturing is just now turning to a variety of computerized controls. And in these early stages, that's creating islands of automation on the factory floor.

The world-class company of the future will integrate all three areas--the office, engineering, and manufacturing--into a closed-loop system for the entire production cycle. This closed-loop CIM system can provide quantum improvements in a number of areas.

CIM BENEFITS

CIM offers improved Human Productivity. A manufacturer's ultimate goal should be to make every action in his facility add value to the product. But look at any company and you'll see human beings at every level performing tasks that don't add value-- tasks that are unproductive. For instance, it makes no sense for people to read instruments, write down the readings, then keyboard them into a computer, when that can be done on an automated basis. Computerizing that work, and integrating it into a system, can result in higher productivity and lower cost. It frees workers for more productive tasks and shifts management's emphasis from supervising people to supervising machines.

CIM can improve Capital Resource Productivity. CIM makes better use of capital resources, helping us use fewer "things"--fewer machines, lower inventory, and less space--to achieve greater output at lower cost.

CIM can improve Quality. Quality is today's competitive edge, and the customer sets the standard for it. To paraphrase author John Guaspari, the customer can't define quality, he just knows it when he sees it. You'll know it when he sees it too, because quality will reward you with higher profits and larger market shares.

CIM helps you afford external quality--customer satisfaction, at a lower cost--because it gives you higher internal quality--the elimination of waste from the design, engineering, and production cycle.

CIM gives Economies of Scale through Economies of Scope. No doubt about it, a CIM investment is a major one. Clearly, you need enough units flowing through your system to justify the investment.

But how can you make money on production runs as small as a lot size of one? Unlike traditional manufacturing, which is based on large volumes of identical products, CIM produces many products with many product variations. In a minute's time, a CIM assembly line can serve a multitude of needs--not just one. It can produce a lot size of one, or a lot size of thousands--one after the other. It offers greater volume through greater variety, letting you spread your investment over many different products, with the total number of units creating the economy of scale for profitability.

CIM offers manufacturers a Rapid Response to the Marketplace. CIM flexibility gives you the ability to be first into the marketplace, to gain significant market share before competitors can catch up. It lets you respond rapidly to a changing marketplace with greater customer service, and with improved product development cycles, the flexibility to meet even the smallest demands, and the shortest possible delivery time.

If you're the best in the world at these things, you'll be a winner. You'll have the highest human productivity, the lowest allocation of capital resources per unit, the lowest waste, the best quality, and the fastest response to changing markets. All of that will make you the low-cost producer. And if your marketing people do their job right, you'll be the market leader too. Unfortunately, there's not a company today that's doing all these things.

CIM BARRIER: MANAGEMENT'S LACK OF VISION/COMPUTER TRAINING

Considering the benefits, a decision to invest in Computer Integrated Manufacturing should be an easy one. The necessary computers and communications technologies are available and proven in operation. In spite of this, many manufacturers don't feel threatened enough to take CIM seriously. They've seen what's happened in the automotive, steel, and consumer electronics industries, but they don't see themselves in a race for survival.

Let's look at the major barriers to Computer Integrated Manufacturing.

Offsetting America's lead in advanced technologies is the limited vision of many executives. On one hand is the "successful" CEO who sees little need to tinker with what's worked in the past--even though new world competitors are nibbling at his market share. At the other end of the spectrum is the shortsighted CEO whose vision extends only as far as the next quarterly report. He's as cost-conscious as they come--just don't rock the status quo by suggesting any major changes in the way his firm designs and produces his product.

I think it's appropriate here to quote from that widely read writer, "Author Unknown":

The irrational pattern of human behavior repeats itself again and again, individually and collectively. Even when our old forms are failing miserably, even when they cannot handle the

problems of the day, they are fiercely defended;
those who challenge them are derided.

But, the problem doesn't just rest with the
company's CEO.   Too often manufacturing management has
no idea of the competitive situation.   No one has ever
asked them to examine their cost drivers.   They're too
busy getting product out the door anyway.   Survival?
That's someone else's responsibility, not theirs!

Management educated in the 1950s and 1960s present
another level of complexity.   Most of these people were
schooled long before today's technologies emerged.   Too
few are computer literate or have sufficient respect
for the power of this basic tool.   And that problem is
really deeper than the management level.   In one major
U.S. company with 3,500 electrical engineers, only 500
understand programmable controllers.   Those 3,000 have
never been trained on them even though the PLC has been
a part of American industry since 1970.

Management at all levels has been trained to
propose incremental measures rather than those that
offer quantum improvements.   That's because smaller
projects usually only require the approval of plant
managers, while larger ones have to work their way up--
sometimes all the way up to the board of directors.   So
we've been inching our way to survival and prosperity,
scaling projects to levels of authority.   Unfortunately, we've fallen behind the rest of the world, and
now, we have to race ahead to catch up.   But we can't
race ahead with incremental steps.   We have to leap
forward with quantum improvements.

CIM BARRIER: TRADITIONAL INVESTMENT MODELS

Industry's accounting tools pose serious problems
too.   Traditionally, return on investment and internal
rate of return have been used to evaluate capital
projects.   While these methods work well to evaluate
the financial implications of an individual machine,
they're not appropriate for analyzing all the benefits
of a CIM installation.   That's the problem!
Traditional financial justification procedures based on
internal rates of return, or short payback periods, are
perhaps the greatest single barrier to adoption of new
manufacturing technologies by U.S. industries.

Let's examine the inadequacies of the current
investment model.   Traditional cost models just weren't
designed for today's Computer Integrated Manufacturing.
A classic case in the metal working industry is the use
of full absorption accounting, where direct labor is
generally used to absorb overhead charges.   But today,
direct labor accounts for less than 10 percent of the

sales value of a product, and CIM can lower that
percentage even further. So, obviously, measuring
costs based on labor's absorption of overhead is
questionable when so little direct labor is involved.
What is needed is a new accounting methodology, one
that distinguishes between those items that add value,
and those that only add cost.

A second challenge is to measure those benefits
that are difficult to quantify. Traditional accounting
models like return on investment and internal rate of
return focus on the time value of money, and give
little thought to the strategic opportunities and
threats that technological advances present. It's
automatically assumed that money on hand is worth more
than money promised in the future. A cost deferred is
preferable to a cost incurred.

This kind of thinking puts many companies into a
"Mediocrity Mode." Using these models, an investment
is made in hopes of realizing significant return in the
form of cash flow in subsequent years. The projected
annual returns are discounted to reflect the time value
of money. At the end of the justification period the
investment has some residual value. Traditionally, an
investment is justifiable if the sum of the discounted
cash flows plus the residual value exceeds the original
investment.   Too often, this process encourages
accountants to focus on the tactics of discount rates
and residual values, instead of taking a broader, more
strategic view.

Let's take a critical look at these accounting
models and examine their shortcomings.   They're
tactical instead of strategic. These models are short-
term, project-oriented tactical evaluation tools, not
strategic planning tools.   To survive, grow, and
prosper, we have to adopt new business strategies for
manufacturing companies. But with current financial
models, we'll never get there. Short-term, incremental
measures rarely lead to quantum advances. They require
short payoff periods, frequently three to five years,
This, despite the fact that CIM is designed for longer
life. Many CIM investments won't pay for themselves
within that time frame. They're not supposed to.
They're long-term strategic measures. In fact, most
CIM systems will probably outlive the products they're
designed to produce. But a CIM system's reusability
will let it shift easily to other production.

In other words, a CIM investment won't reach
terminal value until long after the traditionally
accepted three- to five-year periods--and even that
value may equal or exceed that of the original investment. They use inappropriate costs of capital. Discount rates for traditional models are usually based on
rates for five-year treasury notes, thirty-year bond

rates. But these are all inappropriate for long-term planning--they're merely snapshots in a moment of time. Just five years ago, interest rates on thirty-year bonds were two to three times higher than today's rates. It's easy to imagine how many long-term projects those rates killed or postponed.

For CIM planning, what you need is a blended, long-term, composite view of rates, based on historical and projected trends. They favor low-risk cash-producing investments. But many projects of strategic importance won't produce positive cash for five years or more. Unfortunately, these models make no provision for today's cash user becoming tomorrow's cash provider.

They ignore intangibles that could enhance revenue. How will more consistent quality improve your market share? How will the flexibility to make product changes enhance your competitive situation? How will reduced economic order quantities and shorter delivery times affect customer satisfaction? These intangibles can have real value in today's marketplace. Unfortunately, they have no value in today's models. The challenge is for management to see these intangibles as revenue enhancers and assign them value. This will require subjective judgement, but management is well skilled at that. Many of its decisions are based on experience and intuition as well as fact.

CIM BARRIER: LACK OF INTEGRATED APPROACH

After examining the shortcomings of these traditional models, it's clear that management must rethink its guidelines for justification procedures. Of even greater importance is management's responsibility to develop a strategic approach on CIM.

CIM is a major philosophical change for all of us. Until recently, we considered the office, engineering, and manufacturing separately. Even today, too many people think of CIM in terms of the factory floor alone. But for CIM to succeed, we must integrate all three business environments--office, engineering, and manufacturing--welding them into a single, smooth-running machine. To achieve that, work produced in all areas and at all levels of your company must be done as efficiently as possible, with the highest quality, at the lowest possible cost.

CIM must involve everything and everyone--from the chairman of the board to the custodian. In other words, CIM is a whole new way of doing business. It requires new management techniques, new strategies, and new organizational structures to accommodate new manufacturing dynamics.

CIM BARRIER: LACK OF STRATEGIC ORIENTATION

CIM also requires new ways of looking at
investments. For years, management made long-term
financial projections based on the assumption that
things would go on as they always had.

This worked well from World War II through the
late 1970s. But it's a different world now. It's no
longer safe to measure capital expenditure investments
against a status quo alternative of not making the
investment.

Who today feels comfortable assuming a
continuation of current market share, current selling
price, and current costs? Instead, one should consider
the prospects of declining cash flows, shrinking market
share, and smaller profit margins. The status quo
means little to hungry worldwide competitors. No doubt
about it, successful investments must yield returns
greater than the cost of the capital invested. But
enlightened management must evaluate CIM investments in
terms of long-term objectives. In other words, management's viewpoint must be changed: it should start with
a strategy, not an investment.

Unfortunately, the IRR and NPV models are project
or tactically oriented--not strategically oriented. As
such, they're not relevant to today's struggle for
survival. To survive, manufacturers have to look
beyond tactics. They can't move from tactic to tactic
until a strategy emerges. That's why CIM automation
architecture has to drive CIM accounting architecture.

Manufacturers must take strategic long-term views
of their future and where they want to be. They must
initiate strategic planning processes, which will lead
to new business strategies throughout their companies.

Many firms have fought their way through the
justification barriers and implemented CIM with
surprising results. Recent studies by Frost & Sullivan
and the Yankee Group show that improvements of 50 to
over 75 percent are the norm for those who dare to
implement Computer Integrated Manufacturing.

Allen-Bradley has been involved in numerous CIM
projects throughout the world, but the one we know best
is our very own: the World Contacter Assembly Facility
in Milwaukee. Some years ago, we saw that the merging
international standards for motor controls would
eventually threaten one of our core businesses. To
remain competitive, we not only needed a new product,
we needed one that could be sold profitably anywhere in
the world. Off-shore sourcing was examined and quickly
discarded. The engineering skills were already in
place in Milwaukee--and they were unmatched anywhere in
the world.

The solution was a new product, and a new production line, developed by design and production engineers working as a team. Other experts were assigned to the project from Quality Assurance, MIS, Marketing, and Purchasing. The objectives were to develop a world-class product and a production facility with the following goals:

High volume--600/hr.

125 product variations in lot sizes of one

Products built-to-order in 24 hours

Located within existing plant

Competitively insignificant labor costs

So, we built our World Contacter Facility as part of a strategic plan to capture a higher share of the world market for electrical contacters. The result? We met all our objectives.

Today we have a totally integrated CIM facility whose relative cost per unit is 60 percent that of machine assisted labor. And this system has already demonstrated a flexibility to respond to market demands. Instead of the original 125 product variations, it now offers 600 variations...in lot sizes as small as one. And relative return on assets is five times what it would have been with traditional manufacturing.

Today, we're producing consistently high quality contacters with a rejection rate of just twenty units per million, and at a lower per unit cost than anyone else in the world. And that's not all. We're the global market leader. And with this CIM facility, we intend to stay that way.

RULES FOR IMPLEMENTING CIM

So how do you implement CIM? What are the rules? Based on our global experience with Computer Integrated Manufacturing, here are some do's and don't's:

- Don't start the trip without a roadmap. You must have long-term objectives and these should be supported by long-range plans.

- Remember, what's tailor-made fits best. For new products, design the product and the process simultaneously. Don't make one a slave to the other.

- Set your sights high--make zero defects and a lot
  size of   one your goal. If survival's your target
  you're aiming too   low. Shoot for profit--thrive,
  don't survive. Raise your   sights and make quality
  a competitive weapon. Seek the   flexibility to give
  your facility a longer life than the   product it's
  producing. And remember, CIM's economies of   scope
  make economies of scale possible. If you can produce
  profitably to a lot size of one, imagine how
  profitable   you'll be on high-volume runs.

- Make your system great, then automate. Automating a
  system   that's out of whack can only give you
  automated confusion. Bad manufacturing practices and
  unstable processes shouldn't   be automated. After
  all, you wouldn't put your company's   general ledger
  system under computer control if you couldn't
  balance it by hand.

- Only automate things that add value. In existing
  plants, separate your costs into those that add value
  and those that   add cost. Then, get rid of the
  latter and automate the   former.

- Don't focus only on reducing direct labor. Direct
  labor   accounts for less than 10 percent of the
  sales dollar for   most products produced in the
  United States. You can achieve bigger   savings by
  reducing indirect labor, trimming inventory, and
  lowering the cost of quality.

- Plan from the top down. Implement from the bottom
  up. Draw   up a blueprint then start building a
  solid foundation. Don't   buy hardware until you
  have a blueprint. Plan first and   implement second.
  Make sure you have the right building   blocks in
  place--things you can ultimately integrate.

- Eat the elephant one bite at a time. For existing
  plants,   automating step by step is fine. Don't try
  to do more than   you're prepared to. But remember,
  there's a difference   between   step-by-step
  automation and piecemeal automation.   Everyone must
  follow the same plan.

- Don't implement without trained people. CIM requires
  an   innovative and highly productive organization.
  Each employee   is a critical link in the system.
  These individuals must be   properly trained,
  motivated, and rewarded.

SUMMARY

The global economy is here. The need for CIM is clear. In a short span of time resources have decoupled from national boundaries, labor is decoupling from production, capital has become the driving force of trade.

Today, competition comes from all directions. Markets demand faster response. And windows of opportunity slam shut before many realize they've opened.

Yesterday's ways of doing business are passe. Yesterday's investment models are irrelevant. We can't profit by them--we can't even survive by them. Deferring a CIM investment until returns meet accepted norms could be a decision to drop out as a principal player--or worse.

Those who find it impossible to take action must have been in the mind of Russell Baker of the International Herald Tribune when he said:

> It's unAmerican to manufacture things. Asians
> manufacture things. Americans acquire, divest,
> merge and file for bankruptcy.

It's a brave new world. And it requires bravery, not fear. You can survive on fear. But you can't very well thrive on it. CIM offers you a way to thrive. It is a way to fuse all the elements of a manufacturing company into a single, smooth-running machine, making you the leader in quality, cost, and service.

Getting to that point will take some time, because CIM requires a company to change the way it thinks and works. But that's OK. Because you don't have to buy the whole store at once. You can think big and start small. And that's the point. It's time to start drawing up a long-term CIM strategy, and time to start down the road to survival, growth, and prosperity.

36

Emerging Technologies: New Materials and Optoelectronics at DuPont

Alexander MacLachlan

My invitation to speak at this seminar indicated the topics of the seminar had been chosen based on a special report in <u>Fortune</u> entitled "The High Tech Race." A race implies that some entrants surge ahead while others lag behind and perhaps even drop out. This is very definitely the case in technologies.

EMERGENCE/OBSOLESCENCE OF TECHNOLOGIES

Some examples of products and technologies that are very familiar to many of you but have now literally dropped out of use will bring this point home. How many of you used the slide rule? I know I used one frequently in my college and early professional days. However, the advent of inexpensive consumer electronics in the form of the pocket calculator has literally made the slide rule obsolete. Most of us recall when copies of any correspondence or report were made with numerous layers of messy carbon paper. Chester Carlson's invention of electrophotography and the subsequent development of the Xerox or dry copying machine have likewise made carbon paper almost obsolete. Similarly, although it's an item that was widely used here at the Utah State Dairy Store and elsewhere, when did anyone here last see a glass milk bottle?

These examples illustrate the process whereby existing technologies dominate the market today but are being replaced or superseded by <u>emerging technologies</u> that have been demonstrated and have a small but growing market share at the present time. Likewise, some <u>new or developing technologies</u> of today will become the <u>emerging technologies</u> of the future, while others will fail to mature and capture a significant share of the market.

The challenge to us in the "High Tech Race" is to identify and competitively utilize those emerging technologies that will have a significant impact on the

products we use and the way we live in the future.
There are a number of such emerging technologies in the
field of advanced materials and optoelectronics; I have
chosen these four examples to talk about this afternoon
that are of special interest to my company.

1.   Plastic automotive body parts

2.   Barrier plastics in food packaging

3.   Optical storage technology

4.   Lightwave transmission of voice and data signals

PLASTIC AUTOMOTIVE BODY PARTS

     Prior to the mid-1970s the use of plastics in
exterior automotive body parts was generally limited to
relatively small parts and trim.   Then Ford and General
Electric initiated a program to upgrade the bumper
systems on several Ford models by converting to
injection-molded engineering thermoplastics.   Shortly
thereafter, Pontiac introduced the Fiero, which used
engineering plastics for all major exterior body parts.
The success of the Fiero proved to be a major turning
point that led the automotive industry to believe that
an all-plastic vehicle could be a viable and desirable
option.   Since then the pace of change has quickened
and today's automotive industry is broadly initiating
programs to replace sheet metal with plastic in a
variety of body parts.
     There are several advantages to plastic body parts
for both the manufacturer and the customer.   Injection
molding of thermoplastics is a cheaper method of forming automotive body parts than is stamping of sheet
metal.   The use of plastics also permits rapid styling
changes, and of course, reduces the weight of the
automobile.   Advantages to the customer include a
longer car life than with steel because of better
corrosion resistance and dent resistance.
     Developing plastics with the necessary properties
for use in the demanding application of automotive body
parts is no easy task.   These properties include fast
molding and set times, dimensional stability in use,
long-term durability, and, of course, toughness or
impact resistance.
     Historically, polymer based businesses grew
through the discovery of entirely new generic polymers
that offered exciting new properties.   Today, blending
and alloying existing polymers is a major technology
for developing new compositions with unique properties.
In a multiphase polymer blend, the two or more

components exist in discrete microscopic domains. In automotive plastics, a rubbery phase is dispersed throughout a harder or somewhat more brittle polymer to absorb the energy of impact and improve toughness, while the more brittle phase provides the required dimensional stability and surface finish. At DuPont, we have developed a whole family of polymer blends for use in a variety of automotive body parts.

How far will the substitution of plastics for metals in automotive body parts go? Past trends and projections indicate that by the end of this century our automobiles will be over 30 percent plastic. As an indication of this trend, I invite you to look around the parking lot as you leave the seminar and notice how many plastic bumpers you see on new model cars.

BARRIER PLASTICS IN FOOD PACKAGING

The second emerging technology in advanced materials that I would like to discuss is the use of barrier plastics in food packaging. Since the introduction of cellophane in 1924, plastics have played an important role in packaging technology. This has been especially true in flexible packaging. However, until recently, the all-plastic package couldn't approach the criteria for an ideal rigid container: that it be as strong as steel, as inert and impermeable as glass, as tough and formable as plastics, and as inexpensive as paper. However, the advent of new plastic resins and processing technologies combined with rising costs for the energy-intensive production of glass and metals has set the stage for major growth in rigid plastic packaging.

To be suitable, the packaging material must contain the product and keep the environment from degrading the product's quality over a shelf life ranging from thirty days to over two years. For example, oxygen permeable containers allow oxygen to seep in, robbing food of flavor and freshness and accelerating spoilage.

As was the case in automotive body parts, no one polymer today can supply all the necessary barrier properties, structural needs, and ability to be processed under a variety of conditions such as heat sterilization or retorting. DuPont has, however, developed a family of resins that can be fabricated into single layer or multi-layer packages that will meet the needs of a specific application. A cross section through a typical multi-layer package might consist of three to six layers to provide suitable barrier and structural properties and meet processing needs. We have developed a family of barrier resins and have

introduced them in the market for both food and other
packaging uses.   We utilize different materials for
oxygen, aroma or odor, and hydrocarbon barriers in a
variety of applications.   Many products are being marketed in packages utilizing our barrier resins by a
variety of manufacturers.   Currently about 19 percent
of primary food packaging is plastics but market
analysts expect this to increase to over 40 percent by
the turn of the century.

There are, of course, numerous other emerging
technologies and products in advanced materials.
Battelle Memorial Institute recently published a report
from which I have abstracted which shows the major
present and future applications of engineering
plastics.   (See Figure 36.1)

Figure 36.1   **ENGINEERING THERMOPLASTICS**

MARKET AREA	ATTRACTIVE PROPERTIES	APPLICATION
Electronics	Electrical properties	Chip carriers
	Dimensional stability	Mounting, housings
Food packaging	Barrier properties, heat resistance	Films, containers
Automotive body parts	Lightweight, toughness, durability	Vertical panels, doors fenders, hoods, and truck lids
Membranes	Permeability	Gas separation, chemical waste processing
Construction	Impact resistance, transparency	Plumbing, floors, ceilings

You will recognize two of the five market areas as
ones I have just spoken about; my company also has
major programs in the other areas.

Another important area I have not discussed is
advanced ceramics, which is finding new applications
because of the resistance to heat, wear, and chemical
attack.   I have also not mentioned fibers or fiber
reinforced composites; they too are a major factor in
emerging advanced materials technology with broad
applications in aircraft, automotive body parts, and
sporting goods.

OPTICAL STORAGE TECHNOLOGY

I would like to turn now to some emerging technologies in electro-optics, the first being optical storage technology. The amazing focusing power of the laser has enabled the technology to be developed for the storage of very large amounts of information or data in an extremely compact and easily accessible form. Many of you who are music lovers are familiar with this technology through the purchase and use of music recordings on compact discs, or CDs, which are rapidly replacing records and cassette tapes in the music industry.

The same technology can be used to store a variety of information in the same compact and easily accessible fashion. For example a fourteen-inch optical data disc can store up to 400,000 pages of text or 500 chest x-rays.

What permits this compact data storage? Molded from plastic, the discs contain a series of microscopic "pits" and "lands" or flat areas. As the reading laser scans over the disc, each "land" generates a bright spot of reflected light while each "pit" generates a dark spot. This occurs because of interference between the light reflected from the flat surface immediately adjacent to the top of the "pit" and that reflected from the bottom of the "pit" which is a quarter of a wave-length of light deep. A photo detector reads these bright and dark spots as a stream of binary digits into which any type of information can be encoded. The focusing power of the laser permits the "pits" and "lands" to be extremely small. Each "pit" is no larger than a bacterium. On a CD, some 2 billion of them, laid down in a continuous spiral nearly three miles long, fit on a disc less than five inches in diameter. That's compact information storage!

The DuPont company, in a joint venture with N. V. Phillips, is producing not only optical discs with information already impressed or stored on them, but also larger fourteen-inch data discs on which the customer can enter or engrave his own information with the use of a laser. This "write once" capability should be of real value for permanent storage of information such as tax and census data. Even more exciting is the probability that advances in optical storage technology will put erasable discs on the market in a few years or less. This will allow customers to erase and rerecord information at will, as a computer floppy disc or magnetic tape now does. As you can see, optical storage technology is going to have a major impact on the "information age" we are living in.

LIGHTWAVE TRANSMISSION OF VOICE/DATA SIGNALS

The final emerging technology I would like to
cover is the use of electro-optics in telecommunications. Prior to the early 1980s all telecommunications
systems were essentially electrical. But the development of very high speed light-emitting devices and the
perfection of techniques for manufacturing ultrapure
glass for fiber optic transmission lines have made the
transmission of voice and data signals via lightwaves
both practical and increasingly common. Today virtually all new long-distance telephone lines and many
local area data transmission networks in offices and
businesses use lightwave systems. In fact, the DuPont
company is installing one of the world's largest local
area systems at our facilities in the Wilmington,
Delaware, area.

Lightwave or electro-optical transmission systems
have several advantages over electrical systems. The
laser and fiber optic system have the capability to
handle and transmit immense volumes of data at any
given time. This is due to several factors including
the low dispersion of the signal as it travels down the
fiber, the ability of the laser to emit very short
pulses of light, and the fact that the fiber optic line
can carry a number of different frequencies of signal
at the same time. This information-carrying capacity
is referred to as the band-width of the system. Some
experts believe the bandwidth of a lightwave system is
almost limitless. One fiber optic line can carry as
much information as thousands of copper wires such as
those in an electrical cable. Another advantage is
that because of the low losses in the fiber optic
cable, the number of expensive repeater or
amplification stations is reduced.

The basic components of an electro-optic
telecommunications system consist of a laser signal
transmitter, a fiber optic transmission line, and a
photo detector receiver. Because of our experience in
materials development and our engineering and manufacturing expertise, our company has recently formed a
joint venture with British Telecommunications to
develop, manufacture, and market electro-optic
telecommunication components.

A number of enhancements to potentially expand the
application of light wave communication systems are
under way, including the development of practical high-
speed light switching devices. Light signals today are
controlled electronically, much as electronic signals
could only be switched mechanically in the days before
the vacuum tube. At DuPont we are investigating the
behavior of a variety of non-linear optical materials
that could be used in switching and controlling light

signals and might one day play a role in expanding the use of lightwave communications.

SUMMARY

I hope this brief tour through some of the emerging technologies in advanced materials and electro-optics has both stimulated and informed you. I would like to close on the same note as I opened, that being the special report in Fortune on the "High Tech Race." All of us--as individuals, companies, schools, or even countries--are involved in this race and, as I hope this talk has demonstrated, the consequences and rewards of such a race can be enormous. However, to compete in any type of endeavor requires appropriate resources, preparation, and commitment. Let me mention some of what I believe are the necessary ingredients for us to compete in the "High Tech Race."

First, we must have a continuing supply of well-educated researchers and scientists from our universities and colleges. Second, we must have a clear understanding and appreciation of the value and contribution that technology can play in all our lives. Third, this will, I believe, lead to a commitment on the part of our society to continue to compete in a high tech race rather than opt for a role on the side-lines as a so-called service economy. Finally, of course, we need government policies and economic prac-tices that will enhance rather than fetter the capabi-lity of America's businesses to compete and prosper in the high tech race. Perhaps seminars such as the one we are participating in today will help foster and further these ideas.

37

Dow Chemical's Quality and Productivity Improvement

Wayne R. Pero

Most of my career with Dow Chemical has been spent in manufacturing. I have worked in several of our plants in the United States and have spent a couple of years overseas. Even more recently, I was fortunate to have the experience of managing our Denver sales office. In that job, I became very aware of the need to pay close attention to all customer requirements, not just product quality. I saw firsthand how difficult it can be to make a sale when up against a tough competitor.

You might say I've been around the block once or twice. I'm old enough to have watched the U.S.A. go from the enviable leader in quality and productivity to a nation struggling for survival in industry after industry. I am pleased, though, that I am young enough to, hopefully, see this country regain that leadership position in quality and productivity--because I'd like to see my kids and their kids enjoy the same standard of living and opportunities that my generation has enjoyed.

RESPONDING TO WORLD COMPETITION AT DOW

About the only way this is going to happen, though, is if companies--and individuals like you and me--make quality and productivity a priority. Everyone--at every level in American businesses and in American schools--must actively and enthusiastically pursue quality and productivity improvement. Being world competitive must be a strategic intent--in other words, a first consideration in everything we do, not just a passing fancy.

Dow has that strategic intent, and we're making the changes that quality consciousness demands. The changes have not come easy, but the reward to this change has been finding a training program that helps the organization change.

But, before I tell you about our successful approach to quality and productivity training, I'd first like to tell you a little bit about Dow. It is truly a multinational company.

- One-half of our sales are in the United States,

- One-half of our new capital is spent in the United States, and

- 53 percent of our people are based in the United States.

About half of our products are what we call basic chemicals--products like chlorine, caustic solvents, and plastics such as polystyrene, polycarbonate, and polyethylene. These products are generally sold in high volume and, more than likely, are upgraded or used in the upgrading of other products before you and I see them on the shelf or in the store window.

The other half of our products are specialties. These products are usually sold in smaller volume on a performance basis for specific end uses. For example, ion exchange resins are used in water purification, or to remove metal from water. You'd find epoxy resins in can coatings, in micro chips, and in circuit boards. Latexes are used in paper coating, and in carpet backing and adhesives; and METHOCEL* cellulose ethers are used as thickeners for a variety of food products such as ice creams.

There is still another group of specialty products that go directly to the shelf for you and me to buy. These include pharmaceutical products like NICORETTE* and SELDANE* antihistamine and consumer products such as SARAN WRAP* and HANDI-WRAP* plastic films and ZIPLOC* plastic bags. We also produce herbicide and insecticide products for specialty agricultural uses.

Our products are extensive and varied. Worldwide, we employ 51,300 people to develop, manufacture, and service them. Incidentally, we hire an average of 400 to 600 new college graduates per year in the United States.

Over the years Dow has recognized the importance of good people management. The fact that we have paid close attention to this most important resource in the past has made it easier to put a process in place for continual improvement.

- We have good working relationships between managers and employees, and between hourly and salaried personnel.

- We try hard to create a work place that allows total involvement, as well as the opportunity for everyone to contribute and receive recognition for good performance.

- There is a real place for the individual and the freedom for this individual to feel part of the management process.

     Although we are mostly a salaried organization, we have good relations with bargained-for employees and the several unions who represent these employees. Teamwork is a common thread.

     I make these special points because good human relations is the key to any quality and productivity management process. Managing people to create the desire for continuous improvement is the core of any quality and productivity program and must be the focus of any learning program.

     As with most companies, Dow recognized the need to change in the early 1980s. We realized then that all the things that made us successful in the 1960s and 1970s were not going to work in the 1980s and 1990s. World competition was demanding a change. That change would take a major training effort and a full-time commitment by management to lead the change.

     The U.S. Area Operating Board started this change with the announcement of a Quality Performance steering committee in late 1983, a move that led to the creation of my job. This steering committee created a road map that laid out for the organization what the management process was, exactly. This steering committee decided that the process must be customer driven, it must involve everyone in the organization, and it would require continual improvement through the elimination of waste. As you will see, the process is built around those three points.

ELEMENTS OF QUALITY IMPROVEMENT PROCESS

     With that background on Dow I would like now to describe the key elements of Dow's Quality Performance Improvement Process that our training is built around. I will include the steps we've taken to implement this management process.

     First, we have found that this management process always works. It is making our businesses better--a whole lot better. It is making so much difference that we can see why those companies that do it will survive and those that don't will not.

The quality improvement process is simple. So simple, in fact, we have, over the past decade, overlooked the obvious--the focus. The process starts with the customer. It is, and needs to be, customer driven. That is why we exist. This applies to companies, to teams, to functions, and for individuals. We learn daily through surveys and meetings with customers that we do not always clearly understand which requirements are most important to them. And, the closer we get to the customer, the more we find areas where we can improve.

The second element of the quality performance improvement process is having a good working relationship with the suppliers. Suppliers should have well-defined requirements and should be held accountable for their performance based on those requirements. Suppliers should be selected on performance, not just on price.

The third element of the quality performance improvement process is to focus on the process. Continual improvement comes from working on the process, not the people. The reason this is a management process is because only the managers can fix the process. The workers work in the process and know where the waste is, but only the managers can change the processes to eliminate the waste. Unfortunately, we have found that the managers do not see the waste as clearly as the workers. The higher up the line the managers are, the further they are from the waste and the less they see it. Managers see waste at 3 to 5 percent, while the workers will say 30 to 50 percent. The manager's job in the new process is to break down the barriers so the workers will openly talk about the waste without fear of reprisal or loss of jobs.

In the new process, the focus is identifying, through the use of statistical tools, which processes are in control and which ones are not. Then, everyone in the organization works on reducing the variability to bring the processes into control. Everyone works on centering the output of the processes so the aim point meets the customer requirements all the time. This may sound simple, but we have found most of our processes are not in control statistically. We have even found that some of our more sophisticated manufacturing processes under computer control can be out of statistical process control. For sure, most of our non-manufacturing processes do not have the predictability that we want. In fact, we see most of our improvement opportunities in the non-manufacturing areas such as accounting, order entry, invoicing, and planning, to name a few.

The fourth element of the quality performance improvement process is the need to measure. These measurements, to be effective, should measure how well the customer requirements are being met, how well the suppliers are meeting our requirements, and how well the key processes are running. Are they in control? Do they have the right aim point? As basic as these measurements are, I will bet that most companies are the way Dow used to be. We had tons of measures but few that answered these very important questions.

TRAINING PROGRAMS

Our challenge was to develop a series of training programs that would help our managers and employees work in this new culture. Our first training challenge was to launch an effective program that would help us focus on quality and the customer. We used the "Towards Excellence" program, put together by Zenger-Miller for Tom Peters, the co-author of In Search of Excellence. "Towards Excellence" draws out the fact that most, if not all, successful companies have an obsession for quality--in the form of quality products or services. The customer is where it all starts--and ends.

The "Towards Excellence" program was very successful, as it helped break down barriers between the different functions. As we focus more on the customer, and our thinking becomes more oriented to our business groups, we continue to build trust between the functions and even within functions, because there is more sensitivity to internal customers as well.

Our next challenge was to develop our own training to make quality and productivity a way of life in Dow. We looked closely at all the quality gurus--Deming, Juran, Crosby. We also visited other companies to see what was working for them. The conclusion of all this research was that, because it was a management process, each company had to tailor their own training to fit their own personality. So, that we did.

The product of this research and internal development was the Quality Performance Workshop. It is not only the content, but also the approach to training that has made this workshop successful. The Quality Performance Workshop was originally designed for our commercially oriented management teams that are made up of research and development, manufacturing, technical service and development, and sales personnel. The workshop has since been modified for use within the various functions, including the administrative functions.

The structure of the workshop incorporates the key elements of any formula for training. First, training has to bring about dissatisfaction with the way things

are being done.   Then it has to provide a vision of
what the new comfort zone is to look like.

And, finally, the training has to help the
participants take the first step in reaching that new
plateau.   In other words, application has to be an
integral part of the process.   These three elements
constitute the formula for training: Dissatisfaction +
Vision + Application = Change.   Throughout our Quality
Performance Workshop, we have built in these three
elements--dissatisfaction, vision, and application--to
create the change in attitude that is necessary for
quality and productivity improvement.

Early on, we learned that training in numbers
often does not work because a critical mass is not
established in the work place.   So, we made it imperative that only natural work groups--in other words,
teams with a common purpose of business, job, or functional responsibility--would participate in the workshops.   For instance, with our commercial teams, we
request that the business team be part of the same
workshop in which their market management teams are
participating.

We also recognized that it is critical that
managers, as well as employees, be equally involved in
the training.   Since the manager is part of the natural
work group, the manager is automatically part of the
training.

MODULE FOR QUALITY PERFORMANCE WORKSHOP

So, what is this Quality Performance Workshop?
It's an extensive two- or three-day seminar.   The
length depends on the exposure the teams have
previously had to the key elements of the quality
management process.

The workshop is designed to systematically
progress through these six basic training modules:

1.   Introduction or What's-In-It-For-Me (WIIFM) Module

2.   Mission Statement

3.   Visions of the Future

4.   Cost of Waste

5.   Simple Problem-Solving Tools

6.   Measurements

In each module, the teams will accomplish specific
goals and they'll learn from each other.   The skills

taught in the Quality Performance Workshop are ongoing, dynamic, and intended to change thinking and actions.

A key to the success of this training is the fact that less than 15 percent of workshop time is spent in lecture. Ten percent is allowed for question-and-answer sessions, while the participants spend the rest of the time--75 percent--making decisions and working on their very real problems.

Time is allotted for team presentations at the end. In these presentations, the teams summarize what they have accomplished in each module. They also outline how they, as a team, are going to use the quality management process in the future to address the real problems they face.

In the first module, the WIIFM Module, the training concentrates on creating dissatisfaction and shaping the vision. Through team workshop exercises, lectures, and video presentations, the group is introduced to the quality improvement process, which they spend time examining. As the work groups diagram and understand the quality improvement process, they become more comfortable with the need to change. They see that what they have been doing is not necessarily wrong, and they begin to see the huge potential for improvements; they begin to imagine a better system.

In the second module, the teams spend time preparing a mission statement that applies to their own specific work situation. They identify who they are, who their customers are, and what their customers' needs are. They learn that a common mission is important to the team's success.

In the next module the teams, through group interaction, practice divergent and convergent thinking and, through this interaction, create visions of the future. What will it look like when they are successful? It is in this module, the third, that they create what we call process quality statements.

In the fourth module, the teams learn more about the cost of waste. They discover that waste has been building up for years and that we've gotten used to tolerating waste in all forms. This waste exists in the form of waste of time, in capital, materials, and lost business and it can range from 15 percent to greater than 50 percent of cost of sales. The teams start out thinking of waste in the 5 to 10 percent range, but by the end of the module, they learn how to identify the things that are barriers to the realization of their visions.

Teams are taught the skills and techniques in the fifth module that will enable them to take the complicated problems and break them down into solvable units. They learn to use simple tools and techniques. Fishbone diagrams, block flow charts, run charts, and histograms are just a few of the powerful tools that the groups learn to use to identify and solve their specific problems.

Through simple problem-solving techniques the teams learn where to work on waste, thus bringing the element of application into the equation. As part of the training, they again practice divergent and convergent thinking, learning how it provides a powerful mix of creativity and conformity in the team environment. Statistical process control techniques--SPC--are also introduced in the context of problem solving, and teams begin to be comfortable with these methods of pinpointing and controlling variability in processes.

As the final step of the Quality Performance Workshop, the work groups learn the performance measurement techniques that will enable them to keep track of where they are and where they are going. Measurements, they learn, are what convert a program into a process, and measurements are needed to identify processes or parts of processes that need improvement. Work groups are shown how to use measurement techniques such as indexing to chart progress toward their own particular goals.

The final stage of the workshop agenda--as I mentioned earlier--is for the work groups to summarize what they have accomplished in each module. They conclude their presentations with an outline of ways they are going to apply the concepts and techniques to their job functions in the future.

RESULTS FROM WORKSHOP TRAINING

That should give you a pretty good idea of the way our quality and productivity training works. In a nutshell: we foster dissatisfaction, create a vision of a better way, and encourage application of the concepts, tools, and techniques that will result in change for the better. And, we strive to get everyone urgently and relentlessly pursuing quality and productivity improvements. However, we don't stop there. We do it over and over again--continual improvement--and we never stop looking for ways to be better.

Is it working? Well, to date, over 80 percent of the commercial teams have gone through the workshop. In all cases it has had an impact on how the teams are approaching their businesses. For example, one business team, and the market management teams reporting to it, credit savings of $17 million in the first year to the implementation of workshop techniques. Another business team credits the workshop with significant cost reductions to the business, and, even more importantly, the development of different market strategies.

The process does work. One team in Georgia worked with a customer to reduce product variability to well below previously accepted industry standards.   And, they're shooting for even further reductions. One of our customers, an industry leader, says we revolutionized their industry in less than six months. In another business, one that we should know how to run because we've been in it over forty years, we found ways to save $2 million in the first year.

We know we will have work yet to do to establish the complete cultural change we're aiming for, but to date, we're pleased with the progress we've made with our quality and productivity training.   Building a quality culture across America is going to take the efforts of every one of us--you and me included.   For America to be world competitive and recognized as the quality and productivity leader in the 1990s will require a return to the work ethic our parents and their parents practiced. It will require a national strategic intent to build quality into our products and to be the best at what we do. It will take management and workers working together, not against each other. It will take industry, government, and academic cooperation, not confrontation.

I sense a real awakening in America. I know you recognize your part as evidenced by your attendance at this conference. I know Dow recognizes what it has to do and knows there is nothing more important than our total commitment as a nation to the need to be the global quality and productivity leader.

Pillsbury's Productivity Through Motivation and Participation

Jim Shadler

Assets make things possible, but people make things happen. The success and continued growth for ten years in our Profit Improvement Program is due to the participation and involvement of people, more people every year.

Our program, which we call PIP, was formally organized in 1977. The objective was to formalize ongoing cost reduction efforts to realize greater savings, improved quality and productivity. Productivity has always been a part of our culture. We had a lot of sensitivity in our manufacturing operations when we said we wanted a productivity program because people thought they were doing a pretty good job. They had been, but the program formalized our efforts.

As far as profit impact is concerned, the P & L data below clearly demonstrate the value of reducing cost of goods.

	NOW	INCREASE SALES	PIP COST REDUCTION
SALES	$ 100	$ 120	$ 100
COST OF GOODS	60	72	54
GROSS MARGIN	40	48	46
SALES AND ADVERTISING	20	24	20
CONTRIBUTION	20	24	26
FIXED	10	10	10
OPERATING INCOME	10	14	16
CORPORATE FEES	2	2	2
PROFIT BEFORE TAX	8	12	14
TAX	4	6	7
NET PROFIT	4	6	7

It shows that sales of $100, less cost of goods of
60 percent, eventually provides a net profit of $4.
The next column reflects a sales increase of 20 percent
along with a corresponding increase in cost of goods
which provides a profit increase to $6. But in the
competitive environment that we all live in, it is not
always easy to increase sales or increase prices, but
you have the challenge of doing something to your
bottom line. In this event, the cost of goods area is
the one you really want to attack and the one that can
have a significant impact on your bottom line.
Obviously you would like to do both, goods reduction,
which when you drop to the bottom line shows you have
been able to increase your profit to $7. When a 10
percent reduction in cost of goods gives an improved
profit level over a 20 percent increase of sales, the
value of a PIP program is readily evident.

Top down endorsement is extremely critical which I
think you hear about in every successful program. It
starts with the chairman and president and goes through
the group areas into the business unit. Everybody has
to support the program. We are fortunate that we have
total support throughout our system and everyone feels
strongly about the program and the contributions that
are made.

KEYS TO COST CONTROL

One of our primary principles is to examine the
basic cause of cost. This involves three key areas.
The first one is that perfection is no barrier to
change. We are always searching for improvement, and
the emphasis that you currently hear in the Japanese
programs where they talk of continuous improvement
falls right in line. There are many synergies between
the things that we have been doing for ten years and
many of the new things that you hear about today. It
is not that the Japanese are the only ones that have
ever done them. I think the American industry has done
a pretty good job, but we have been slow to talk about
some of the things that we have accomplished.

Second, the savings potential is the full existing
cost--not just part of the cost but the full existing
cost. We feel that every dollar of cost should contribute its fair share to profit. All the dollars we
spend should be making money for us. If not, we should
take a look at why we are spending those dollars.

Third, we never consider any item of cost
necessary. No matter how good something may be--no
matter how great it may have been in the past, no
matter how tied to it people may be as we face the

future--we do not consider any item of cost as being necessary. We try not to overlook any opportunity.

ELEMENTS OF PIP

The main elements of our program include the following: The coordinator is extremely key. We have coordinators at our group level, at our business unit level, and in our various manufacturing facilities. We have gone from the forty-six that were mentioned in the introduction to now over sixty so that each of these facilities has a coordinator. And I might add that these coordinators come out of the existing complement. We take those people that we feel have growth potential, and their reward and recognition is promotion.

An information or cost accounting system is extremely critical. You have to have a means to track your progress.

Productivity MBOs--everybody in the system has a productivity MBO. We start at the group level, go down to the business unit, go into departments, and then to individuals. Everybody is tied to the same overall productivity objective, but they are individualized so that you know specifically what your role is and what your challenge is to help achieve the total objective.

We feel that PIP is a focal point for idea exchange. When you have over sixty facilities, a number of business units, and have individual programs within all of those different areas, the key thing is that you have communication between all of those various areas so good ideas can be shared. What might be good for Business Unit A might also be terrific for Business Unit C. The coordinator organization is the one that moves between operations and shares the information.

In terms of what is a project--a project must be deliberate. By that I mean, you must take some deliberate action to cause something to happen. If you look at the commodity markets where the price of wheat went down $1 and is worth $1 million, that is not considered a PIP. That happens no matter what you do. But, if you were to take flour out of the formula and you saved some money, that is a deliberate action. That becomes a PIP.

PIP ideas must be measurable. Our finance area is charged with the responsibility of measurement and reporting results. The reason for that is that when you have a variety of different departments many times you have groups that work together on a given project. You may have the distribution area, the R & D area, and the manufacturing area all working on a project. Let's

say it is a $1 project.   Each of those three areas
feels that they are contributing a $1.   They would tell
you that it would be $3 but obviously this is not so.
The finance people take all the input--give credit
where credit is due, but then put on the bottom line
$1.

A major directional shift occurred after about two
years.   At the start of the program, we had a top-down
driven and supported program which is absolutely
crucial.

But what happened after about two years, we
changed from a top-down driven to a top-down supported
and a bottom-up driven program.   People really drive
the program, and the reason why we have had a successful program for ten years is because of the
participation level of the people in the system.

I look at productivity programs a little bit like
I look at a sky rocket.   Over the first couple of years
that sky rocket takes off and goes up--straight up,
then it bursts into stars, bangles and lights and the
people are going ooh and aah.   And then people seem to
run out of interest, run out of ideas, and the sky
rocket slowly starts burning out, going down, and at
the end of year five, it is gone.   Year six, somebody
in the headquarters office is sitting behind his desk
saying, "Gee, why don't we have a productivity program.
Let's hire a consultant, let's get started," and you go
through the whole thing again.   Fortunately, we have
been able to avoid that kind of sky rocket effect.   We
have been on a steady track up for ten years.   The
reason is because of the people.

The source of the ideas really comes from the
people in the field.   When you start the program, the
people in the headquarters environment have big ideas.
They are terrific and you certainly want to capture
those, but at the same time you kind of run out of
those big ideas after a while and that is the point
where the sky rocket is either going to keep going or
starts coming down.   This is where the change to a
bottom-up driven program is important--where you get a
lot of small ideas.   For example, we have one business
unit that in the period of time it has been in the
program, and this particular unit has been in the
program for seven years, has never once had an idea or
project that has been in the top ten.   And when I speak
of top ten, I am talking about projects that are $3
million, $2million, $100 million, or whatever it is,
but in rank order.   They have never been in that ten
biggest project category.   But they have had twice the
participation of the next closest business unit and for
seven straight years they have contributed at a level
that leads our business units.   It just proves that
participation is absolutely critical.

The real keys to our success are employee participation and involvement and the more the better. Today we have sixty times the participation level that we had the first year of the program.

Communications must be continuous to avoid misunderstanding and to provide visibility. People want to know what is going on. If you do not tell them, they lose interest.

Training is vital since as more people participate it is extremely important that you train your people. Help them search out new and different ways to find ideas. And our latest effort, which I will talk about more, a little bit later, is the Statistical Process Control program where we are teaching basic statistics and statistical process control to all of our people. They are learning the value and the use of control charts to identify opportunity areas.

As I mentioned numerous times, people are absolutely critical to our program and as we say, assets make things possible--we have to provide the assets, but the people are the ones that really make things happen.

PRODUCTIVITY FORMULA

Here is a simple formula that we feel very strongly about: Quality plus safety plus quality-of-life equals productivity.

As far as quality is concerned it is our absolute #1 priority and under no circumstances will it ever be compromised. No project will ever be accepted in our PIP program unless quality is either maintained or, preferably, enhanced. A business unit makes the call on what its specifications are, what their requirements are, but the overall policy of the company is that quality must be maintained. We feel that there is a direct tie between quality and performance. If you do it right the first time and you do not have to redo things, you don't have a problem with your productivity.

We feel the same thing about safety--both product safety and human safety--where it must be maintained or enhanced. We recognize that some projects may not have an immediate dollar impact that you can measure as far as human safety is concerned, but we feel that any time we can improve safety in our system we help improve productivity. We want to keep our good workers on the job. We do not want them hurt or off the job with substitutes who have to take the time to learn what is going on.

Quality-of-job-life is extremely important. There are two key areas--facility and environment. I have

always said the better a facility looks, the better it
runs. It has always been the case in our organization.
We do some little things like making sure the fronts of
our facilities are landscaped. Trimming of hedges,
neatly trimmed lawns, and a little paint does not cost
a great deal of money and does not require capital.
Once you move into our plants there is an attractive
reception area, the offices, they are well kept, and
the same holds true when you move into the manufacturing area. The important thing is that the people
should feel proud of the place where they work. They
spend a third of their lives there so we want them to
feel proud of their plant.

We also feel that the atmosphere in which our
people work is extremely important. We like to make
sure that our people are part of what is going on. We
want them to have pride in the operation. We want them
to have the feeling that they are helping to improve
productivity. The fact that we involve the people and
that they take pride in the operation gives everyone a
positive attitude. Those positive attitudes lead
directly to productivity improvement.

We have economics 1A lectures where we meet with
all of our people and explain to them what is going on
in the marketplace, what it takes to be a low-cost
supplier, and what it means to be a high-share-low-
cost-high-volume operator. We relate share with volume
so everyone understands that the better our share, the
higher the volume demand and the more production that
will run in the plant. That is why we try to get our
people involved in the planning process so they can
work to reduce costs and hopefully increase jobs.

When I visit a plant, I do not go in and ask, "How
are we doing on productivity?" The first thing I do is
take a look at the operation, walk through, visit with
people, and the first question I ask is, "How is your
quality?" I also ask about our safety and visit with
the union officials in organized locations or talk with
the group leaders where we are not organized. I do not
have to ask about productivity, it is obvious.

PIP SUCCESSES AT FOUR PLANTS

I would like to discuss four examples of plant
programs. The first plant is a seasonal operation, the
second an old facility, the third a recent acquisition,
and the fourth a new plant. You will note similarities
as I go through this, but the important thing is that
each location is free to adopt whatever program they
want and call it whatever they want. They all know
they are part of the PIP effort and that they have the
objective of contributing a certain number of dollar

savings. But in terms of how they implement, they have complete freedom to do whatever it is they want to do.

Plant A--Seasonal Operation. This was a canning plant. It was on a "hit list" that was targeted for closure. They were new to the productivity challenge. It was a newly acquired business for us and they had no program.

When I went to the plant for the first time, everyone was sitting with their chins on their chests because someone had said they were about doomed for closure. All I said to them was don't wait to get "shot." You do not volunteer for the firing squad. What you do is stand up and be counted. Demonstrate what you can do. If no one feels you, no one is going to think about you, and obviously it is going to happen. I said, "Get your chin up off of your chest, and let's take a look at the things you can do for yourself. Then, maybe, there are some things others can do to help you, but let's get going."

So what did they do? Sure, a survival program helps insure commitment. Anytime you are about to drown, you are going to do a lot of things and you are going to be awfully committed to get to shore. A plantwide program was initiated in our fiscal year 1981 with nine projects and 10 percent participation. Our 1987 goal is 100 projects and 90 percent participation (in just six years). There was considerable involvement by the plant manager whose recent reward for the job that he did was a promotion. There have also been strong PIP coordinators. Earlier I said they are high potential people. There have been six in this period of time and all six of these individuals have been promoted to significantly better jobs. With this reward and recognition, people are clamoring to be coordinators in our system.

As far as awareness is considered, we have initial plant manager meetings where the plant manager describes the program. Following this there are a variety of awareness activities--bulletin boards, communications meetings, newsletters, and new employee orientation. New employee orientation is very interesting. When you have a canning plant, you have regular employees and seasonal workers. We have to hire approximately three times the number of people we have working on any period of time because of turnover. So if we have a canning plant with 1,500 people, we will hire 4,500 people during the period of about four to five months. We train every one of those people. We train them in human safety, product safety, and PIP.

Communication is critical to continued interest. Instead of once a month posting on how the plant did, it is a daily update. Anytime there is a new project, it is posted immediately. We also have presentations to and by our senior management group. We go into the plants and see what is going on. We also report on how the company is doing. But what is really impressive is that there is a period of time when we go out into the plants and to the work stations. Employees at their work stations have tagged items in blue, red, and green. Those in blue are safety items, those in red are quality items, and those in green are profit improvement items. They take the time to say this is the way it was, this is the way it is, and this is what I have done. For five or ten minutes that production employee has probably two or three vice presidents of the company occupied by telling them exactly what he has done. Can you imagine how that individual feels? The pride that that person has in his work place. And it is just terrific to see the kinds of things that they do.

As far as promotional programs, there are a number of things done to generate interest and make sure that we recognize and reward people. Awards range from recognition of every idea to annual awards. Plant A has now become one of the most productive plants in our system. It is no longer on the "hit list." It has won our group "outstanding facility award" twice and this past year we have decided to turn this plant from a seasonal plant into a year-round operation, which is a tremendous change for the people who would normally work for four months and then go back to farm work, house work, or whatever it might be. There is much excitement, and the plant is getting visibility throughout the system.

Plant B--Old Facility. It was about to move onto a target "hit list." They developed what they call a SPOKE program and they are now in an expansion phase. SPOKE is Safety, Product Quality, Opportunity, Keep the Promise (a quality assurance program), and Exchange of Ideas. They developed a mission statement that said they wanted to be the best facility in the company. And they absolutely believe it. You say to anybody in that plant "What do you think about your facility?" They reply, "We may not be best but we are going to be best."

In this location we have quality circles. Again depending on the plant and plant management, they can do what they want. Our facilitator/coordinator in this location is an advisor to the group leaders. The group leader is selected from six to ten volunteers that are in each of the various groups. Training has been provided in a number of areas: brainstorming, cause/effect diagram, criteria testing, pareto principle, flow charts, planning charts, fishbone diagrams. You have to help the individuals understand how to find opportunity.

The results have been impressive. They have won "facility of the year" award, and their annual contributions are among the top in the company.

Plant C--Recent Acquisition. We feel that our PIP program is a great vehicle for new plants to become immediate members of our "family." PIP is a great opportunity for them to issue a challenge to the system. This particular location said, "We may be the newest member, but we are going to be the best."

PIP became a great rallying point for this location. Plant management got on board with the initial ideas coming out of the plant management team. A newly appointed coordinator began developing the plant program, and initial savings were $2 million.

We made an acquisition within the last year and the plant identified first-year savings of $3.6 million. So when you have brand new plants in the system, sending that challenge to the existing plants that have been around twenty, thirty, forty, fifty years, it creates a lot of interest and activity in the system.

Plant D--Newly Constructed Plant. It had a participative management organization and had the challenge of being the low-cost plant in its business unit. The total plant operates with participative management. It does not have first line supervision. Instead, work groups are responsible for various areas in the plant. Since they started off at a higher cost level, they were given the challenge to become the least-cost operation in their business unit.

They were in the limelight because a lot of people were looking at them because of their management structure. They initially got a coordinator and set up a task force. The task force identified gaps, "Here is where our standard or optimum is, here is where we are, here is the gap, what are the programs to fill the gap." They set up their tracking system, communications, and awards. They prepared a brochure for all of the employees that says, "Here is the profit improvement program, here is the eligibility and how it is administered, here is the reporting process, the plant role, goals and recognition program." The result, at this plant, is that they are now the least-cost plant in their business unit.

INGREDIENTS FOR SUCCESSFUL PIP

Programs are individualized by plant, but they all include: (1) selection of high potential coordinators who are absolutely the best people we can find; (2) high visibility of the program throughout the entire organization; (3) clear objectives so that everyone understands what they are doing; (4) emphasis is on communications since it is extremely important to let people know what is going on; (5) encourage everyone to participate (no idea is too small); (6) becomes a part of management MBOs; (7) receives strong support throughout organization; and (8) provides recognition and rewards.

Recognition is vital. It is just human nature that no matter who you are, chairman of the board, vice president, or employee in the plant, everybody likes to have a pat on the back. People like to be recognized for what they do. As far as awards and recognition is concerned, we have facility awards, business unit awards, and group awards. The plants and business units have a variety of awards that are used to give recognition to individuals, e.g., jackets, shirts, plaques, and banners.

But the big event is our annual awards program where we recognize people throughout our U.S. Foods organization. This is our "academy awards" or our Deming award. For those of you who are familiar with the Japanese industry and the Deming award you know that it is a major event. It takes a lot of effort to win that particular award. I am not saying ours is the same but to our people it seems like it. We recognize a key individual, team, key facility or department, facility idea initiator, and staff idea initiator. At our awards banquet, those attending include the president of the company, general managers, and senior officers, responsible managers, and, obviously, the participants. It is a tremendous motivator. Every time a person is recognized and called up to receive their award, they say, "I am coming back next year." When they go back to their plant, they get tremendous recognition. Imagine you are somebody running a machine or a maintenance man in one of the plants, and you have the president of the company and executive officers all listening to your accomplishment. It is a tremendous motivator.

Again the successful ingredients are that programs are individualized by plant or function, all programs emphasize participation, objectives are set by each location, results are given visibility, management support is strong, and recognition and awards are given. Our tenth anniversary of our program is this year, and this will be an all-time record year for us.

NEW TOOLS

Even though we have had excellent results and we think everything is just terrific, we recognize that what has been great in the past is not necessarily

going to be great in the future. We feel that new
tools are needed, and when I say new tools, we have
what we now call a total quality commitment in our
system.

There is a culture change under way to reinforce
our commitment to quality and productivity throughout
the organization. And when I say total organization is
challenged, I am talking about sales, marketing, finance, operations, and not just the plants. The challenge is to eliminate waste and to add value. It is
just terrific. I just love this new culture change.
When people come to me and say we would like to hire
the following people, and we would like to do this kind
of program, and we would like to do all of these neat
things, my response is, "Does it add value and does
that change things in a hurry?" If people cannot
demonstrate that it adds value, we are not going to do
it. That is the challenge throughout the entire
system.

In the plants our strategic direction is to
eliminate waste and defects. Our vehicle is going to
be JIT--Just in Time Manufacturing--and the challenge
of continuous improvement is the name of the game. JIT
involves the total pipeline from raw material supply,
to vendors, to customers. Single source supply is not
new to us. We have had a single source supplier to the
tune of millions of dollars and as part of an
association with them for thirty-five years. So it is
not exactly new.

We will go through considerable training of our
people--especially in SPC, Statistical Process Control.
We feel that the new tools are needed for new ideas
because major opportunities are not always apparent.
In that regard, people are really for the first time
beginning to understand "the process." A good example
is rework. I know a lot of people have looked at this
one. Instead of figuring out a whole lot of ways in
which to improve the handling of rework, we have just
said we are not going to have any rework. We are going
to work on eliminating rework, and we are going to do
it right the first time. This sets up a new world of
opportunities. It is a whole new different philosophy
for our people.

We feel that our ten years of increasing results
have been just terrific, and we are very pleased. We
know we cannot just continue to paddle water and do
what we have been doing for ten years. We have to
change and in the next ten we are going to deal with
improved quality, improved service (extremely important
as it relates to our customers), and we are going to
have a program of significant improvement through total
quality commitment.

We intend "To Be the Best Food Company in the World." Our values articulate what we really feel and obviously tie directly to the theme of my presentation. First, people make the difference. Second, our quality and productivity improvement is a direct result of the job our employees have done. Quality is essential--a top priority in everything we do--and at no time will we ever compromise on quality. Every employee understands and supports this value. And third, excellence must be a way of life. This is a guiding value for everyone in our organization and top of mind in everything we do.

We are obviously pleased with our program and the results that we have achieved. We are now looking forward to implementing a whole new effort, TQC, into our system and realizing continuing, growing improvement in our quality and productivity.

39

Boeing's Quality Strategy: A Continuing Evolution

John R. Black

I want to give you an overview of the quality improvement process within the Boeing companies--where it started, how it grew, and where we are today. The process has not always been a smooth one--for we are involved in a major cultural transition. What we are changing to today, however, is not entirely different from the way things were back when Bill Boeing made his first two airplanes in 1916.

He and Navy Commander Conrad Westervelt built a single-engine seaplane, which was called the B & W Number 1, using their initials. Those planes measured only 27 1/2 feet long and had a range of 320 miles. That's tiny compared to today's commercial airliners. In fact, the tail of the 747 is more than twice as high as the B & Ws were long, and the 747 can travel up to 7,000 miles without refueling.

Those first B & Ws were assembled in two locations--the wings and floats at a shipyard on Puget Sound's Elliot Bay, and the fuselage in a hangar on Seattle's Lake Union. When the first plane was completed in June 1916, Bill Boeing test flew it himself. He didn't delegate the job; he was committed to the belief that his people had done quality work. And he was willing to stake his life on it. By 1928, his company had become one of the nation's largest aircraft builders.

It was the capability and ruggedness of the Boeing B-17 Flying Fortress that helped us win World War II. 12,726 B-17s were built--more than any other multi-engine aircraft in history, with Boeing alone producing 6,981 of them. The rest were produced by other manufacturers, following the Boeing plans.

Today, over half of the commercial jet airliners in the free world were built by Boeing. The 707 ushered in the jet age, more 727s have been sold than any other jet airliner, the 737 was 1986's best-selling jet airliner, and the 747 is the flagship of any

airline--with greater passenger capacity and longer
range than any other airliner.

CHANGING CORPORATE CULTURE

Our formal efforts to change our culture started
in 1980, with the 757 Productivity Program and the
birth of quality circles in several of the Boeing
companies.   Perhaps I should tell you a little about
Boeing at this point.   There are currently six main
Boeing companies, including Boeing Commercial Airplane
Company (which I represent, and which also includes
Boeing of Canada, formerly DeHavilland).   BCAC
currently has in excess of 40,000 employees.
The others are Boeing Military Airplane Company,
Boeing Aerospace Company, Boeing Computer Services,
Boeing Electronics Company, and Boeing Vertol(which
makes helicopters). A seventh company, Boeing Services
Division, was created in January 1987 to bring together
activities that support all of the other companies--
such as graphics, medical, security, etc.   Total Boeing
employment is 128,000 worldwide, with 82,000 in the
greater Seattle area.
Our goal is to deliver products that meet our
commitment to excellence--to make Boeing the recognized
standard for the quality, after-sales support, and
technical and economic performance of those products.
Quality circles began for the most part primarily
in manufacturing areas such as the Commercial Airplane
Company's giant fabrication division in Auburn,
Washington; the Military Aircraft Company's plant in
Wichita, Kansas; and the Aerospace Company's manufacturing facilities at the Kent Space Center in
Washington State.   And since Boeing Computer Services
maintains support personnel at all of the different
company locations, they were also early in implementing
circles at Boeing.
Before I describe how the quality circle process
grew at Boeing, I'd like to tell you about a comprehensive attempt we made to change the company culture that
began in 1980 and ended in 1984.   Established in March
1980, the 757 Productivity Program was far more than
just another attempt to encourage our people to work
smarter and faster.   The program focused on developing
a comprehensive and cohesive strategy for implementing
employee involvement.
Our first priority on 757 was defined as employee
involvement (EI)--an umbrella term that involves the
people and the organization in improving productivity
and quality of worklife.   EI emphasizes changing the
management style from "telling" to "listening," and
minimizing adversarial relationships.   It involves the

employee in group decision making, problem solving, and goal setting (team efforts), and vertical communication (primarily upward).

Employee involvement is a process and a discipline for accomplishing an evolutionary change from an existing management structure, system, and style to one that encourages and facilitates less management structure, more participative systems, and maximum openness and communication.

An executive council was established to provide overall guidance and decision making. The management team became the steering committee, responsible for the plan. Department teams became coordinators, and employee teams became active participants in identifying problems and recommending solutions.

We conducted employee surveys at the request of department managers. Over 450 operational factors, both positive and negative, were identified as appearing to influence individual and group productivity. The results were reported confidentially to department managers for appropriate action. Employees were interviewed both individually and through job reaction questionnaires.

A unique aspect of this process was the formation of a Quality of Work Life Committee consisting of employees and managers. The committee addressed survey results, developed recommendations, and presented them to the organization's steering committee for decisions and implementation.

However, when the 757 and 777 projects were merged in 1984, the 757 Productivity Program encountered considerable resistance from managers who were suddenly exposed to the cultural change process. They did not understand what it could do for them. They saw it as unnecessary. Without their support, the 757 Productivity Program was discontinued. And as a result, the momentum to move to a more participative management style within BCAC slowed.

BIRTH OF QUALITY CIRCLES

At the Boeing companies, the quality circle process had been initiated in 1980, with the first three circles. This process has been highly successful and continues to be. By the end of 1986, there were 846 circles active throughout the Boeing companies, and the process is still growing rapidly. Of these circles, 7,901 employees were involved in the process at the end of the year, averaging 9.3 members per circle.

These circle members are not just locating and tackling problems in a haphazard manner, they are all trained, either by their facilitator or by their

training group, in the use of systematic problem-
solving tools. 6,478 circle members and 669 leaders
were trained in 1986.

Of the projects worked, half of them provided cost
savings, while the other half resulted in improvements
that are highly significant changes in an organization's culture. Support at the top isn't enough.
Managers throughout the structure, and especially
middle managers, must be committed to the process for
real cultural changes to be produced. When we created
the BCAC Quality Improvement Center in 1986, we made
sure to educate all managers in the company in the new
philosophy as the first order of business.

Improving quality was also made a special priority
at Boeing Aerospace Company in 1983. However, back
then, it was treated as a motivational program. Just
run out the banners, print brochures, make awards, and
suddenly the front line troops will do all the improvement for us--so the theory went. They called it The
Pride in Performance Program and got the graphics guys
busy doing a four-color brochure to tell our people and
our customers what wonderful things we were doing.

There was a lot of truth to that. We were doing a
great job at the time. We were told over and over
again that we were at the top of the aerospace industry
for quality. So we figured we didn't have a lot to
do--just tweak the system a little and get the word
out. We put the Pride in Performance posters up, sent
the brochures out, and waited for the results to come
in.

QUALITY IMPROVEMENT CHARACTERISTIC: MANAGEMENT
COMMITMENTS

In 1984, Bill Selby, who had been Director of
Manufacturing, became Director of Operations--the man
in charge of the Pride In Performance Program. He was
convinced that there had to be more to quality
improvement than we were doing at the time. He said
our first priority was to find out what the experts
were saying and what other companies were doing.

Bill and I went to hear Juran, Deming, Crosby,
Conway--and anyone else who seemed to have a handle on
what quality improvement was all about. We visited
Tektronix, ARMCO, Motorola, McDonnell Douglas,
Honeywell, Hewlett-Packard, and IBM--and we were
impressed by what we saw. And so we brought in J. M.
Juran, Al Gunderson, and finally Bill Conway to present
seminars to our senior management on what quality
improvement was all about.

The outcome was that the experts and other
companies were saying if we were really going to do

quality improvement, it would take a lot more than just launching into another productivity program or expanding the circle process. Even the word "program" suggests something that's only going to be around for awhile. Quality improvement is a process--one that must become a part of the way we do business every day.

All the experts were saying that management commitment to the process is essential if we really wanted it to work. In fact, the estimates were that 85 percent of benefits to be gained by quality improvement relied on managers working together to change the systems within which their people work. As a result, we decided that our strategy would be to train managers first. We didn't want to get people fired up about the process, only to have them be stopped by their managers or by middle management, who saw the process as a potential threat. We wanted to have them on board first!

At every company we looked at, senior management-- usually the CEO or president--was the spark plug behind the quality effort. It wasn't something senior management paid lip service to and then delegated to somebody else. It was something they believed in and led, personally.

QUALITY IMPROVEMENT CHARACTERISTIC: COMMITMENT TO CHANGE

A second common characteristic is a commitment to changing the management style. Managers can no longer sit in their offices and issue edicts. The management role has changed to that of coaching and supporting his or her people to do their very best. It involves a change in the organizational culture.

What is really going on in an organization is not always very obvious. A lot of the time these unwritten rules are only known by people who have been there for some time and they may be quite different from what is said.

What we must do is to change this culture not only to adopt the new priorities, but also make them clear to all members of the organization and follow through on them.

In order to change the culture, we must change the structure--which means changing our ways. Management must:

- Lead by example.

- Listen--employees are the experts about their own jobs.

- Push responsibility and authority levels down.

- Foster teamwork.

- Provide continuous personal appreciation, recognition, and reward.

- Provide the resources that allow people to excel.

We are accomplishing cultural change through the following five-step process:

1. Identifying the norms that currently guide executive behaviors and attitudes.

2. Identifying the behaviors necessary to make the organization successful for today and tomorrow--not just what worked yesterday.

3. Developing a list of the new norms that will move the organization forward.

4. Identifying the culture gaps--the difference between desired norms and actual norms.

5. Developing and putting in place an action plan to implement the new cultural norms. You have to develop agreements that these new norms will indeed replace the old ones and that this transition will be monitored and enforced.

QUALITY IMPROVEMENT CHARACTERISTIC: TRAINING

A third characteristic we discovered was that everyone in the company must receive comprehensive training in both the need for continuous improvement and in the tools to achieve it. Because only management can change the systems within the company, it is essential to train management at the outset of the process. This is the approach to continuous improvement that was taken at BAC, and it is the approach to improvement we have taken at BCAC.

When I became Director of the newly formed Quality Improvement Center at BCAC in February 1986, I saw our first task as training all management to know why it is vital that we pursue continuous improvement and how to do it, from the very start. Only in that way are we making sure we have a true top-down approach. BCAC is a sizeable company, and so this has been no small task. We have more than 43,000 managers plus about 100 senior executives. And once we train them, we have more than 36,000 other employees to bring into the new system.

We have developed a two-day Managing Quality Seminar that we will continue teaching to groups of 100 to 250 managers until all BCAC managers are trained. The seminar begins with senior management stating the need for achieving a cultural change in management. Divisional directors share what is happening in their areas. And on the second day of the seminar, speakers are brought in from other major corporations to tell our people what they are doing. Already we have heard from IBM, McDonnell Douglas, Ford, Chrysler, Hewlett-Packard, Harley Davidson, ALCOA, and many others; and we videotape their presentations to use with our improvement teams.

We are also bringing in Bill Conway, president of Conway Quality Inc., to help all our managers understand the urgency of working to make this cultural transition--starting today! This year we are hosting ten meetings with Bill Conway on the University of Washington campus--for groups of 400 managers plus university students from the business school.

We train team facilitators, team leaders, and trainers. Those trainers work in line organizations to expand the training of quality improvement into all areas of the company. We have designed and taught classes in process control methods, process improvement, and are developing classes in other statistical methods for process improvement.

We believe that the implementation of statistical methods for process control is essential to provide accurate data for decision making and to give us adequate feedback to ensure that we meet our improvement goals. Already some of our divisions are setting up their own courses to expand this training to all their people. We plan to have all of our managers attend the two-day Managing Quality Seminar and the Conway Seminar by the end of 1987.

IMPLEMENTATION STRUCTURE/STRATEGY

To do all this and to make it work requires a structured approach. All of the six major Boeing companies have formed Executive Councils to oversee the process throughout their companies. We have formed steering committees and quality councils, from the president's level on down to make sure that the improvement process is increasingly integrated throughout all of our business systems.

Our companies have also formed Quality Improvement Centers to guide and assist the process. At BCAC, we have a permanent staff of ten people, and a group of twelve functional representatives from major areas of the company. Identified as people who have the

potential to become the leaders of tomorrow, these reps
come to the center for a one-year rotation, after which
they return to their parent organizations, where they
continue their efforts to implement the quality
improvement process. While at the center, they provide
direct contact, consultation, and support for their
organizations' improvement efforts.

To make sure that improvement objectives are being
met--to find the opportunities for improvement with the
biggest payoffs, and to accurately track what we are
accomplishing--it takes data. A group of top managers
in BCAC has formed a team to develop the criteria for
measuring the quality of all BCAC products and services. Meanwhile, the new Quality Assurance Division
has spearheaded a renewed process of talking to our
customers about their quality needs and problems. By
fully identifying customer requirements and the highest
impact customer quality problems, we can focus our
improvement efforts to deal with the most important
quality issues our customers have.

Within BCAC, we have a six-year strategy to
achieve the transition to a total quality improvement
process. We saw 1986 as our year of commitment, in
which we spent much of our time educating management on
the need to implement the quality improvement process
and to get their commitment to be supportive of the
process and personally involved in it. An example of
that commitment is the creation of a number of pilot
quality improvement teams, who receive training in QI
methods and are given a major quality problem to
resolve--a problem that has a significant effect on a
number of organizations.

We call 1987 our year of implementation, as we
expand the number of quality improvement teams, develop
measurement systems to locate and quantify waste, identify major impediments to continuous improvement, and
initiate a supplier improvement process. We see 1988
as the year for transition and transformation, with
inspection becoming more a responsibility of everyone
and less relying solely on traditional methods. We are
working to get a supplier improvement process in place,
statistical management implemented, quality planning
integrated into the business planning cycle, and union
participation.

We anticipate significant results to be achieved
in 1989, with all employees trained in the use of
statistical tools, a pilot project on self-regulating
work teams in place, and layers of management reduced.
In 1990, we foresee that we will be using only suppliers that have their processes under statistical
control. And at the end of the six-year process in
1991, we anticipate that we will have achieved reduced
inspection, restricted hiring, self-regulating work

teams, BCAC business processes restructured to minimize waste, and reduced levels of management.

This process of continuous improvement, to which we are committed, is not one that can suddenly be grafted into a company. Every organization--in our company, in your company--must make it their own. Top management leadership MUST be provided, and ALL management must be brought on board. Only when that happens, when all the people are committed and the process is locked in for the long term, will it achieve the breakthrough that it is capable of providing for you and for us--the key to economic success in the future.

40

Measuring White-Collar Productivity

William Bradshaw

During the 1970s factory productivity increased 85 percent while white-collar productivity increased only 4 percent. Historically indirect/white-collar labor has been neglected and all the emphasis has been placed on controlling direct labor because of its direct relationship to the product cost. Over the years technological developments have caused a dramatic shift in work force demographics toward indirect labor and service jobs. Just a few short years ago a typical manufacturing corporation's work force was comprised mostly of direct labor workers. Today indirect labor makes up well over 50 percent of a manufacturing company's work force and consumes about 70 percent of its payroll. Both numbers vary among different type companies but both numbers are expected to continue to increase dramatically in the years ahead. In fact, it has been predicted that by the year 1990 nearly 90 percent of all employed North Americans will be working in either white-collar or service-oriented occupations.

The old saying, "People do best what the boss checks," proves true here. The boss has been checking the direct workers and they have been doing well. We have not been checking the indirect workers, we have not been measuring their output. Does this mean that if we start checking the white-collar workers their productivity will improve?

MYTHS ABOUT MEASURING WHITE-COLLAR PRODUCTIVITY

The great nemesis to improving white-collar productivity has been the inability to quantify the end results of the white-collar employee. There are many myths about performance measurement that stand in the way of developing useful measurement programs. Until those myths are destroyed no significant measures can be developed. Here are some of the most common.

MYTH #1:   You can't measure "creativity" or "judgement" and, therefore, a meaningful measure of what we do can't be developed.

The first half of this statement is true. Personality characteristics such as creativity and good judgement are extremely difficult to measure. Yet few, if any, businesses exist for the purpose of providing "creativity" or "good judgement." Almost all businesses exist solely for the purpose of providing a product or service to customers. While "creativity" and "good judgement" may be invaluable in providing the product or service, they are a means to an end. Performance measures should be designed to report changes in the product or service provided, particularly as these changes relate to the satisfaction of the customer.

MYTH #2:   It isn't possible to develop a measure that will truly reflect the performance of this group.

Correct!   No single measure is likely, in and of itself, to be satisfactory. All measures are only indicators of performance. Usually it is necessary to have several indicators which, taken together, can provide meaningful information to management. The search for a single, ideal measure will inevitably result in no measure being developed. We suggest multiple measures for each important outcome. Each measure, in and of itself, is deficient. But taken together, the various measures do serve as good indicators of overall performance. Obviously, too many measures are just as bad as too few. Thus, it is important to seek a proper balance.

MYTH #3:   This measure is misleading. It fails to consider those things beyond our control.

This myth fails to recognize that all measures are only useful to indicate changes in level of performance. Rarely is it possible to develop a measure that weighs out all extraneous factors. A measure cannot be expected to explain a change, but only to indicate when a change occurs. At times, any measure will bring attention to a change in performance that is the natural by-product of circumstances beyond anyone's control. An attempt to charge every measure with the responsibility of reporting changes and explaining where they occur dooms the measurement program to failure at the start.

MYTH #4:   We need a more sophisticated measure.

Wrong!   While complicated measures are elegant and intellectually stimulating for some people, most of us find them frustrating and confusing. It is important to keep measures as simple as possible, particularly when they are first introduced. The real test of a good measure is whether the non-mathematically or nonstatistically inclined manager can easily understand the measure.

MYTH #5: It is easy to measure when the job is to manufacture "widgets." Then you have something to count. But our job is to provide a service. There is nothing to count.

Obviously, it is easier to measure performance when there is a product that can be physically counted and measured. Yet, it is incorrect to say that service delivery results in no countable by-products. Schedules are met or not met. Complaints from customers are received or not received. Costs are controlled or not controlled. There are many "results" of providing service that are countable and/or serve as indicators of the level of service provided. We all know that the provision of a service involves doing something. That is why we pay people. And, whenever people do something, there is an action, event, or behavior that can be counted.

MYTH #6: Since all measures are corruptible, it is a waste of time to develop them. Somebody will always find a way to "beat the system."

It is true that no measure is sufficiently pure that no one can find a way to corrupt the numbers to his own benefit. Yet, it is usually possible to make the effort required to corrupt the measure more trouble than the effort required to focus on real improvement in performance. For example, it is only prudent and reasonable to develop measures from raw data that are the least subject to manipulation by the person or group being measured.

MYTH #7: It's just not possible to measure what we do. There are "undefined" or "intangible" aspects of our job that cannot be measured.

Almost every job has components that are difficult, if not impossible, to measure without resorting to considerable trouble and expense. These "intangibles" or non-measurable factors should be given appropriate consideration in evaluating total performance. This does not mean that the measurable components should be ignored, or that measures should not be developed where it is possible to do so. As much as 20 percent of any job is probably not subject to measurement or, if subject to measurement, the cost would far outweigh the benefits. However, there is considerable merit in measuring the remaining 80 percent.

MYTH #8: We are so busy in our day to day activities, we cannot take time out to establish a measurement program.

We never have enough time, and you certainly don't have to stop what you're doing to establish measures.

Who knows how much time may be saved in the long run?
When productivity improves, resources formerly consumed
to produce the unwanted or unnecessary can be captured,
saved, or reapplied in some other fashion.

USING THE PRODUCTIVITY OBJECTIVES MATRIX

    Now that we have dispelled all the myths and sold
management on the necessity of measurement, the next
step is to choose a method that will work for you.
There are several methodologies to get you to where you
want to go.   There are several software packages
available; some PC based, others are mainframe
packages, and some are strictly manual efforts.
    The "package" is not a "measurement" package.   It
is a total program of productivity improvement activities. Measurement is, as it should be, only a portion
of the entire program which must include the activities
such as:

- Clarifying organizational objectives

- Examining internal outputs, inputs, and activities

- Identifying key departmental customers

- Understanding customer expectations

Measurement is not an end product, it is only a means
to an end.
    Dr. James L. Rigg, Director of the Oregon
Productivity Center, Oregon State University, is the
author of the Productivity Objectives Matrix.   It was
first recognized as a solution to the measurement of
knowledge worker productivity from an article in the
National Productivity Report Newsletter.
    The Objectives Matrix system is used to combine
the monthly results of key productivity indicators
(ratios) to yield a single number, the Index. This is
accomplished by plotting monthly results on the
Objectives Matrix and obtaining the corresponding performance score. The performance score is multiplied by
the weighting factor to obtain the weighted scores.
The combined total of weighted scores yields the
Productivity Index.
    The method begins with establishing criteria or
productivity ratios on a functional/departmental basis,

as shown in box 1, Figure 40.1.   I have used the
following criteria:

CRITERIA	MEANING
1. ASAP Rating	Assuring Survival by Aiming for Perfection (Measure of customer service between 0 and 100 developed through customer survey)
2. $\dfrac{\text{Tot CI Achieved}}{\text{Tot. Dept. Budget}}$	Total Cost Improvement Dollars divided by Total Department Budget
3. $\dfrac{\text{Tgt Dates Set}}{\text{Tgt Dates Met}}$	Target Dates Set divided by Target Dates Met
4. $\dfrac{\text{Hrs Spt on Proj Work}}{\text{Hrs Avl for Proj Work}}$	Hours Spent on Project Work divided by Hours Available for Project Work
5. $\dfrac{\text{Empl on Measure}}{\text{Empl Avl for Meas}}$	Employees Working Against Established Productivity Measures divided by Employees Available for Measuring

This allows a department to be measured against its own
objectives.   Care must be taken when determining
criteria to ensure that the criteria are specific,
quantifiable, and consistent.   Data results must be
easily obtainable.   Criteria should involve short-term
objectives that are compatible with long-term plans.

Development of productivity criteria should
involve the interaction of all levels within a functional area.   Who knows what is going on better than
the people doing it?   Management should not try to set
criteria in isolation from the workers.   This ensures
that criteria requirements for all levels are represented on the Objectives Matrix.   The process causes
management to clarify goals and objectives, focus on
issues that are most central to departmental/divisional
productivity, and it increases the visibility of information relative to white-collar productivity and therefore increases productivity awareness or the "Hawthorne
effect."

Four to six criteria should be all that is
necessary to measure the productivity of a department.

Figure 40.1.   Productivity Objectives Matrix- Criteria

1	ASAP RATING	TOT CI ACHIEVED / TOT DEPT BUDGET	TGT DATES SET / TGT DATES MET	HRS SPT ON PROJ WORK / HRS AVL FOR PROJ WORK	EMPL ON MEASURE / EMPL AVL FOR MEAS					
	#	RATIO	%	%	%					PERFORMANCE
	35	25-1	100	95	90					- - 10
	33	20-1	99.5	94	80					- - - 9
	30	15-1	99	93	75					- - - 8
3	29	14-1	98	90	70					- - - 7
	28	13-1	96	85	65					- - - 6
	27	12-1	94	80	60					- - - 5
	26	11-1	92	75	55					- - - 4
2	25	10-1	90	70	50					- - - 3
	20	9-1	85	65	40					- - - 2
4	18	8-1	80	60	30					- - - 1
	17	5-1	75	55	25					- - - 0

DEPT.
MONTH

SCORES

									SCORE
5	25	15	15	20	25				WEIGHT
									VALUE =

JAN	FEB	MAR	APR	MAY	JUN	JUL	AUG	SEP	OCT	NOV	DEC

Although there are no strict rules regarding the kind of criteria selected, two areas should be represented, EFFICIENCY and EFFECTIVENESS.

After the criteria have been established, management through the use of existing records, historical data, and subjective estimations assesses current performance levels, as shown in box 2, Figure 40.1. These levels may be adjusted during the first three to six months of the reporting period. After the norm is established, results are entered on the level corresponding to a score of three on the Objectives Matrix.

A scale is now developed from the norm (line 3) to the criteria objective (line 10). Each level on the scale represents mini objectives or hurdles. As an organization moves up the criteria scale the value of the productivity index increases, as shown in box 3, Figure 40.1. Occasional declines in productivity are accounted for and intermittent values are inserted in the squares below level three, as shown in box 4, Figure 40.1.

The final step in assembling the Objectives Matrix is the weighting of criteria, as shown in box 5, Figure 40.1. The sum of these weights equals 100 and can be distributed in any informative fashion in increments of five. Weighting must be carefully applied and distributed appropriately. Weighting leads to a prioritization of objectives and a clarification of mission or direction (box 5).

The first three to six months of Objectives Matrix reporting are used to monitor the accuracy of the matrix and make adjustments where necessary. Fluctuations along criteria scales must be accounted for. Criteria data are audited to ensure they are being reported accurately and current performance levels are adjusted; weightings may be changed when management discovers that current performance levels are greater/lower than anticipated. It is during this period that criteria may be added or deleted.

The Objectives Matrix is still an extremely useful tool during the initial monitoring period. Management is directed by the Objectives Matrix to audit information sources, reprioritize objectives, change the weightings of contradictory objectives. In essence, management is using the matrix to make decisions, which increases their control over the organization.

Once all adjustments have been made, the objectives, scale, and weightings must be finalized. A base is established and productivity levels are measured against base to determine net increase or decrease in productivity, as shown on Figure 40.2. It is imperative that the matrix not be changed since this

Figure 40.2. Productivity Objectives Matrix – Index

ASAP RATING	TOT CI ACHIEVED / TOT DEPT BUDGET	TGT DATES SET / TGT DATES MET	HRS SPT ON PROJ WORK / HRS AVL FOR PROJ WORK	EMPL ON MEASURE / EMPL AVL FOR MEAS					DEPT. / MONTH
#	RATIO	%	%	%					PERFORMANCE
35	25-1	100	95	90					- - 10
33	20-1	99.5	94	80					- - - 9
30	15-1	99	93	75					- - - 8
29	14-1	98	90	70					- - - 7
28	13-1	96	85	65					- - - 6
27	**12-1**	94	80	60					- - - 5 SCORES
26	11-1	92	75	**55**					- - - 4
25	10-1	**90**	**70**	50					- - - 3
20	9-1	85	65	40					- - - 2
18	8-1	80	60	30					- - - 1
17	5-1	75	55	25					- - - 0
4	**5**	**3**	**3**	**4**					SCORE
25	15	15	20	25					WEIGHT
100	75	45	60	100					VALUE = 380

JAN	FEB	MAR	APR	MAY	JUN	JUL	AUG	SEP	OCT	NOV	DEC
380											

changes the relationships and nullifies comparisons between periods.

The Objectives Matrix can provide management with a powerful tool. It presents productivity data in a meaningful form of productivity ratios, and the matrix is easily readable and quickly identifies problem areas. Construction and implementation of the matrix forces management to prioritize objectives and clarify mission and direction. Most important, the Objectives Matrix measures the productivity of knowledge workers on their own terms in an accurate and meaningful manner.

41

Managing Business:
Today and Tomorrow

Richard A. Jacobs

GENERAL PRINCIPLES FOR COMPETITIVE SURVIVAL

Every organization has two tasks: operating today's business and creating tomorrow's business. The requirements and demands of these two tasks are always in conflict. In today's business, companies are concerned with performance against current competition. People are faced with limited or infrequent change. They minimize risk. They respond immediately to problems. They pay close attention to costs. They place a premium on experience.

In tomorrow's business, people are faced with significant change. They accept uncertainty. They concentrate on development. They focus on opportunity. They reward for creativity.

In today's business, winners <u>execute</u>. They focus on excellence in performance rather than on administrative tasks. In tomorrow's business, winners <u>innovate</u>. They reward for risks that pay off.

If there are two businesses in every company, there are three kinds of people. <u>Wealth producers</u> design, make, sell, and distribute a company's products and services. There can not be too many of them. <u>Wealth maintainers</u>, comprised of management and support staff, help wealth producers be more effective in their jobs. You need a few of them. The remaining kind of people are <u>wealth dissipators</u>. Often among the most capable people of any organization, they work energetically and independently to solve problems, establish controls, and create opportunities. However, when their activities are out of sync with the current goals of the organization...when the solved problems are trivial, the cost of controls are greater than the benefits and the opportunities spurious...they keep things from happening. They squander the efforts of the wealth producers.

In a time when the lifeblood of this country depends on improved productivity, the executives who

understand and manage these principles will win the
global, economic battle of competitive survival.   I
plan to build on these themes.   I'll share with you new
research on productivity.   Then I'll build on conclusions from this research to suggest ways for seizing a
productivity advantage.

NINE FINDINGS ON ACHIEVING SUPERIOR PRODUCTIVITY

    In November 1984, A. T. Kearney published Seeking
and Destroying the Wealth Dissipators.   Despite
predictions to the contrary, our research suggested
that, for most companies, staffs of executives, managers, professionals, and other white-collar support
people were out of control.   They were out of control
in size, cost, and most important, contribution.   Our
analysis could not have been more right.   In 1985 and
1986, 68 percent of the largest U.S. companies cut
staff.   As a strategy this downsizing is not working as
executives hoped.   Most believe their competitive position has not improved.   They have had second thoughts
as customer complaints increased, product introductions
lagged, and payoffs from automation and process
improvement dwindled.
    But downsizing is the only game in town.   Seventy-
seven percent of these same companies plan further cuts
by 1990.
    For most companies seeking the pot of gold at the
end of the profit and loss statement, downsizing is not
the answer.   The truth is they make a mess of it.   They
don't understand how they got into economic trouble.
They impose across the board staff cuts for expediency
and avoid the hard work of determining where the fat
really is.   They fail to understand the specifics of
problems and opportunities so they can't focus their
efforts.   They don't know how to find lasting solutions.   These companies get caught in a vicious cycle.
Cut to improve profits, patch problems but increase
costs, cut again but deeper.
    Some companies are avoiding this trap.   They're
re-examining the profit equation (profits = revenues -
cost) to understand what controls costs and what builds
revenues.   New research suggests these better methods
work.
    One consequence of the "Wealth Dissipator" report
was the tremendous surge of interest in benchmarking, a
modern, sophisticated approach to comparative/compe-
titive analysis.   More than two hundred host companies
have obtained objective answers to their questions and
used this information to forge leading productivity
programs.   Certain data and findings from these benchmark studies transcend industry and company situations.

We drew nine conclusions of significance about successful companies that provide some very broad answers to
the question: "How do I compare?"   (Successful or
winning companies in this context refers to those who,
over the past five years, have consistently outgained
others in their industry in sales, earnings, and shareholder value compared to the resources expended to
achieve these gains.)

1. Successful companies continue to operate with
    between one half and one fewer staff individuals
    per $1 million of revenues. That equates to a cost
    equal to an average company's net profit.   Down-
    sizing in itself will not make a company success-
    ful.   However, successful companies have more
    effective and more efficient staffs.   They work on
    the right things and they do these right things
    right.   This helps them to focus on the right way
    to build revenues and control costs.

2. Organizations are becoming flatter and increasing
    spans of control.   The number of direct, salaried
    individuals reporting to a manager has increased to
    4.0 people, up 11 percent compared to 1983.   How-
    ever, leaders have a wide margin, averaging spans
    of control of 5.2 individuals.
         There are many factors affecting span and each
    situation needs to be evaluated on its merits.
    Nevertheless, leaders found that by increasing each
    manager's span by about one person, it was possible
    to reduce total staff costs by about 10 percent in
    an organization of 1,000.   In addition, one or two
    levels were taken out of the organization, thereby
    speeding responsiveness and improving quality and
    service.

3. Corporate staffs tend to follow one of four
    organizational philosophies:

    - Holding companies treat their divisions as a
      portfolio of investments and exercise virtually
      no control over current operations.

    - Autonomous value is a decentralized philosophy
      in which central staff functions exist at the
      pleasure of the divisions.   The divisions fund
      central staffs typically for economy or
      expertise.

    - Central value is a mixed philosophy in which the
      headquarters decides which functions to
      centralize, their role and authority.   Divisions
      must use these functions.   A matrix management

organization is one example of a central value philosophy.

- Functional companies centralize most activities except plant and sales operations. Even the upper management of these activities may be centralized.

No larger corporation is "pure" in its philosophy. However, the more diverse the businesses of the company, the more likely it will use a decentralized style. By the same token, most large functional companies have a dominant business comprising the majority of their revenues.

At the end of 1986, 71 percent of all large companies were operating in some form of decentralization, an increase of 8 percent in only three years. It tends to reflect the entrepreneurial spirit of current American management.

Decentralizing does not assure success. On the other hand, the more "mixed" the philosophy, e.g., concurrent strong central functions and strong (willed) profit center divisions, the more likely the company will not achieve superior performance.

4. Corporate staffs are being shrunk and only partially replaced in the field/division. In theory, a functional structure should contain the fewest number of staff individuals. In practice, every company in our data base who moved staff activities from corporate to division, in the last three years, reduced their total staff employment. It is a clear indictment that the large, centrally dominant corporations have permitted their staffs to lose touch with what adds value to the business.

In total, more successful companies have about 10 percent fewer corporate/central staff, regardless of philosophy.

5. Most organizations have trouble with central value and matrix structures. It is not impossible for central value companies to achieve superior performance, but it requires a more sophisticated and effective management. As an example, we compared engineering costs of companies operating with a central value mode with companies using other philosophies. The central value engineering costs were 3 percent to 10 percent higher compared to those companies who operated in either a fully centralized or decentralized mode.

Successful companies operating in a central value mode typically take one of two directions.

Each approach has advantages and disadvantages.   In
one there is a dominant central function which
"makes the rules."   Procter & Gamble is a successful company operating in this manner.   Typically,
companies operating like this can execute well
provided there is a stability in the business.
However, when innovation, change and responsiveness
are important, problems develop.

In the second type of direction, companies
"cross the matrix" (interface and make decisions)
low in the organization.   This works well for
change but increases business risk.   At a recent
client meeting of a company using this "cross the
matrix" approach, the president told us of a recent
occurrence at a trade show.   An executive of a
competitor company told our client, "Instituting a
consignment policy has set off a new trade war!"

It took the president a week to find out if
this statement were true.   Unfortunately, it was.
A critical business decision was made by marketing
managers well down in the organization.

6. <u>Downsizing has affected administrative support
    personnel far more than managerial exempt</u>.   The
    death of middle managers is greatly exaggerated.
    Due in part to popular salary classification and
    administration systems, managers faced with a
    directive to downsize will act first on secre-
    tarial, clerical and other support personnel.   Such
    salary-driving systems are also working against
    redeployment.   Consequently, some parts of many
    companies   are   still   over-staffed   while,
    concurrently, important development areas suffer.

    These downsizing changes also are causing
    cultural shocks for many managers.   Some are
    benefiting through broader responsibility from
    eliminated organization layers and wider spans of
    authority.   On the other hand, they see their
    career progress slowing, with fewer opportunities
    for which to strive.

    These are the lucky managers.   More likely,
    they are in the same position, working harder and
    underutilizing their skills by doing tasks previ-
    ously performed by non-exempt staff.   They too have
    limited career opportunities.

    The real winners are contract firms.   In an
    effort to reduce fixed costs and guard against
    economic downturn, companies are turning to con-
    tract personnel in record numbers.

    Companies continue to use clerical and
    administrative temporaries, albeit far better
    trained and computer literate.   The significant
    shift is in the management and professional arena.

Some companies, as an example, have virtually
eliminated important functions such as engineering,
treasury, benefits administration, advertising,
public relations, systems and data processing.
Consultants are being used increasingly for
specialty management or professional functions
particularly in logistics, marketing research,
telecommunications, and automation. Yet others
have changed the mission of their company to one of
design, marketing and sales; choosing to have
others make their products and maintain their facilities, with better quality and lower costs than
when done in-house.

7. <u>Successful companies, regardless of organizational
    philosophy, are making cultural changes</u>. Downsi-
    zing won't work unless change takes place in the
    values and beliefs of the organization. For exam-
    ple, most companies believe it is good to push
    decision making down in the organization. However,
    when people are punished for making a mistake in
    management judgement, they will again push deci-
    sions back up. A manager may be "forced" to cut
    staff, but when rewarded for the size of staff,
    they'll hire back at first opportunity.
         The nature of cultural shifts is obviously
    different from company to company. However, change
    is more evolutionary than revolutionary. For many
    successful   companies   there   is   a   shift,
    nevertheless, from:

    - Competitive quality and service to superior
      quality and service.

    - Paternalistic to people sensitive.

    - Avoiding risks to taking calculated risks.

    - Making decision by group consensus in meetings
      to having accountable individuals make
      decisions.

    - Insisting on uniform management styles to
      permitting different management styles.

    - Rewarding for seniority, administrative skills
      and size of staff to rewarding for business
      contribution.

8. Leaders of successful companies are beginning now
    to deal with "tomorrow's" areas of concern.   Two
    questions are dominant:

    - What is the right organization and operation for
      competing in a global marketplace?   For most,
      the fact that they must compete globally is
      recognized as a fundamental for survival.   Whe-
      ther they should organize on a geographic, pro-
      duct, market or other basis is less clear.
      Whether they should make or source their product
      and components seems far more important than
      further vertical integration.   Whether joint
      ventures are advantageous and for how long is an
      uncertainty.

    - How   should   the   technology   and   product
      development   process be organized   and   managed?
      Product life cycles will continue to shrink.
      Surviving companies will be under immense pres-
      sure to meet competitive innovation, satisfy
      changing market segments and differentiate their
      own offerings.   No area seems to rank as high in
      executive dissatisfaction as the return and
      responsiveness from their technological invest-
      ment.   Consequently, executives are exploring
      new organization structures and practices to
      correct the increased failings of current func-
      tional   approaches   and   project   management
      systems.

9. Companies stay successful "today" by understanding
    what it takes to win in their industry.   Success
    breeds failure.   When companies grow they tend to
    forget the basics that spawned their success.
    Problems are solved with brute (people) force.
    Staffs grow and strive for functional and position
    power rather than business excellence.   It is not
    surprising the failing rate of "excellence" and
    "enterprise of the year" companies is so great.
        The successful companies who attain and
    sustain their position seem to understand and focus
    on the few things needed to compete.   Then they set
    goals and continually differentiate themselves from
    their competitors.   Leadership, rather than just
    professional management, seems to make these goals
    clear to the organization, integrates effort to a
    common purpose and drives achievement.

APPLE AS A ROLE MODEL

Picking a role model for success is as risky as a star athlete having his/her picture on the cover of Sports Illustrated. Nevertheless, Apple Computer, Inc. demonstrates many of the positive attributes of these benchmark findings.

Founded in a garage, Steve Jobs and Stephen Wozniac built one industry, personal computers; and fostered a second, packaged software. The company grew to nearly a billion dollars in sales in a few years after its founding. Its success was centered on the educational market. The Apple II and IIe were the technological models for the industry. Jobs, as CEO, was the classic entrepreneur; autocratic, demanding, inspirational and undisciplined. His style drove Wozniac from the business. Jobs brought in professional managers. He decentralized the company into product oriented division to ward off IBM's growth in the business market. He added staff and pursued new technologies in the Lisa (predecessor to the Macintosh). But growth and preeminent position disappeared as quickly as it had come.

The bad news was Jobs had lost sight of what had made Apple successful. The good news was he hired John Sculley to build order out of chaos.

Sculley brought more than professional management to Apple. He brought leadership. Recognizing two conflicting cultures would doom Apple, he took his argument to the board of directors. The board chose Sculley, Jobs left.

Sculley downsized initially to trim some of the fat from growth. He then established clear goals and integrated the executives toward common business objectives. He helped individuals understand their jobs and for what they were responsible. This permitted meaningful delegation of decision making. He eliminated the divisions and adopted a functional philosophy. This permitted further economies and, with wider spans of control, helped flatten the organization and make it more responsive. He instituted procedures and discipline where impulsive spontaneity had existed previously.

Sculley did more than control costs. He redeployed talent to technological innovation with a focus on desk-top publishing. This helped Apple gain a niche so strong as to begin to attract the elusive but important business market. Technological leapfrogging has placed Apple in a leadership position again. Revenues and shareholder value are growing. Sculley has Apple back in the winner's column.

LESSONS LEARNED FROM SUCCESSFUL COMPANIES

These results achieved by Apple and the more successful companies in our benchmark data base are impressive: sales, earnings, and shareholders growth are <u>consistently</u> outpacing others in their industry. However, their leaders aren't satisfied.   They are relentlessly fine-tuning current practices while simultaneously working now on the important issues of the future.   What lessons can be learned from these successful companies?   I'd like to share several things to reflect upon as you evaluate your own productivity efforts:

- Why do companies undertake a productivity program?

- What attributes seem to be common to successful programs?

- Why aren't the results right, when I'm doing all the right things?

There seems to be a variety of reasons why companies started and sustained their programs.   However, the reason tended to cluster into two groups, today and tomorrow.   <u>Today</u> dealt with competitive and off-shore cost squeezing.   Not all companies were in economic trouble, but an economic change was evident. "We're not growing as fast as before; returns are still better than the industry but not like the 'old days'," were typical executive remarks.   Even the higher tech IBM or Hewlett-Packard matures.

The <u>tomorrow</u> reasons focused about a need to do different and a general discomfort (not necessarily a problem) with investment.   "New products take forever. I'm not sure I'm getting what I should from research and engineering, MIS or automation," were popular comments.

Some company's reasons fell on both sides of the aisle.   Nevertheless, executives believe a certain level of pain has to be experienced to get real results from any program.   A forest products CEO told us, "Everyone gives lip service to productivity programs. But when you are ridin' high no one really cares.   It isn't until it hurts enough at my level that <u>I'm</u> going to make sure it's a top priority for everyone."

If the reasons for undertaking the programs are right, am I following all the right rules and formulas for success?   Alas, there is no formula for organizational effectiveness and productivity.

We doubt if there will ever be.   No two companies are enough alike to apply a universal technique for success taken from a management cookbook.

However, examining the more successful companies uncovered a number of common attributes. They all have:

- <u>Productivity programs spanning administrative and operational areas</u>. These programs have been in place for a number of years. While techniques have changed, the program's continuity has not faltered.

- <u>Top management committed to the program</u>. They, in turn, have personally involved the organization in it.

- <u>Leaders who know where they are going</u>. Specifically, they have a strategic plan for their company and objectives aimed at getting there. They clearly communicate their objectives to the organization and integrate functional efforts toward these objectives.

- <u>Different ways to integrate efforts</u>. Some use hierarchical MBOs, others use benchmark targets; some have total quality programs, still others tailored reporting systems. What seems to be common is that individuals down in the organization seem to understand what counts in their job, know how it helps achieve company goals, and work at ways to link improvement in their area toward these goals.

Often, executives will comment as they leave a briefing I've given on organizational effectiveness. They'll say kindly it was time well spent and they were delighted to learn that they are doing all the right things, i.e., following the attributes. But the remarks of some don't register. It is almost like the old cliche, "If you are so smart, why aren't you so rich?" Their companies just aren't getting the right performance results. Of all the analysis on what makes for success, I believe there is more to learn from these companies than virtually any other.

When all the rules are followed and the right results aren't there, it's usually because <u>significant</u> change has failed to take place in structure, process or focus.

What is significant <u>structural</u> change? We know "ten percent across-the-board" doesn't work, although profits are helped for a time. In a 10 percent cut the organization continues to do the same things the same way as in the past. People just work harder for a while. But there is a limit to this. We also know when companies make arbitrary cuts in excess of 20 percent, disaster is likely to follow. Development virtually stops, service and quality deteriorate and customers are driven away.

In the programs that work, there _is_ important change likely to affect in excess of 10 percent of the organization...but it isn't just arbitrary cut.   It usually includes significant _redeployment_.   Some departments may disappear, others double in size, still others change not at all.   Redeployment rather than head-chopping is the key.

To change _process_ requires significant change.   By process we mean change to the culture and decision-making practices of the organization.

For Chrysler, Brunswick, Apple Computer, or Firestone, it meant changing CEOs.   At Xerox, DEC, and DuPont, it took bold leaders who were willing to change the practices that had helped their past personal success.   They discarded important products whose time had passed, abolished businesses they had hungered to lead at one point, early-retired   friends and permitted differentiation in a culture that had historic pride in sameness.

It takes time for culture to change.   For the organization that historically has "killed risk takers," for example, no mere memo will produce results.   It will take consistent leadership action to back their words; appropriate motivational reward; and elimination of traditional committees that make decisions by avoiding decision making.

Introducing significant _focus_ change is the most difficult concept for executives to grasp.   The principles seem straightforward.   Every organization has limited resources.   Consequently, have the organization focus or concentrate its resources on the one, two, or three things that are most important for current and future success.   Doing a few right things right is far better than squandering resources on ten, fifteen, or twenty continually changing objectives.

Practicing the principle is something else.   At the core of failure is an inability to manage the conflict inherent in today's business and tomorrow's business.

Remember our opening principles?   Today's business is making, selling, and distributing existing products to existing customers.   Tomorrow's business is involved with change...developing new markets, designing new products and investing in technology and science to change processes.   It is developing tomorrow's competitive advantage.

In most organizations the vast majority of management, technical and professional individuals have some responsibility for both today and tomorrow. Regardless of management's desire, today's business problems always take priority.   It isn't until one looks back over the past year or two or three and

notices the absence of significant change that the
seriousness of the situation hits home.

What can be done? It isn't as simple as "add more
people." Even if resources were unlimited, which they
never are, there are always more demands than people
available.   However, it is possible to simulate the
creation of resources by improving the efficiency of
investment.   The productivity leaders are doing the
following:

1. Of all the work done in the business, determine
    what activities are necessary.   (What activities
    would you pay for as a sole proprietor of your
    business?)   Of these necessary activities, under-
    stand what are done for today's business and what
    activities are done for tomorrow's.

2. Keep only enough resources in today's business for
    doing the job.   Continually strive to do better
    with fewer.

3. Put what resources you can afford into tomorrow's
    business.   Integrate executives to focus on a few
    right projects; don't permit them to work in their
    own islands of effort.   Improve their focus with
    clear, well-understood priorities of effort.

4. Most important, organizationally separate today's
    activities from tomorrow's to the extent practical.
    The most efficient organization is the one having
    the fewest individuals with responsibility for both
    today's and tomorrow's business.

5. Use functional structures for today's business; use
    project structures for tomorrow's business.   Func-
    tional structures are most efficient; project
    structures are best for change.

6. Use different processes (i.e., systems and
    practices) for today's business versus tomorrow's.
    Differential management practices are essential for
    tomorrow's change.   Stability of practices is best
    for today's operation.

The list can go on. But why is gaining focus the
most difficult to accomplish? Perhaps, it is because
companies are in business to be themselves, not change
themselves.   Without focus, executives lose perspective.   They strive for functional and administrative
excellence and lose sight of the corporate goals and
what produces results. As Walt Kelly's <u>Pogo</u> said, "We
have met the enemy and they is us."

New perspectives are needed. Even the basic theme of this year's Partners Productivity Seminar must be challenged.

Ask not how quality, science or technology can improve productivity. Rather, ask how can productivity improve the business. Only then consider if quality, science or technology are among alternative solutions. Finally, and only then, ask if they are the best alternatives.

SUMMARY

We examined nine benchmark findings that showed how some companies were able to achieve superior productivity and consistently outpace their competitors in sales, earnings and shareholder growth.

We found that companies undertake and sustain organizational effectiveness/productivity programs because of today's competition or tomorrow's need for differentiation. However, until the pain is severe or important enough to the CEO, any program will have limited results.

Certain attributes go with successful programs: involvement and commitment, clear strategies, integrated efforts, and strong leadership. But the mere presence of these attributes does not always assure success. Significant change must take place: in the structure and deployment of the organization, in its processes, cultures, and practices, and in its ability to focus. While an organization can't create resources, it can get far greater leverage from its available resources by understanding and managing the conflict in today's business versus tomorrow's business.

Part V

MANAGERIAL STRATEGY

42

Guidelines for Action
Y. K. Shetty and Vernon M. Buehler

Throughout the American business scene, the drive for greater competitiveness has become a major issue. At the top of many legislative and corporate agendas now rests the determination to find the best means to restore U.S. competitiveness. Our national agenda should include measures such as encouraging research and development and capital investment, eliminating government constraints in the global marketplace, and improving education and training. To be fully effective, the private sector must improve its competitiveness through improved productivity growth, quality products and service, innovation, and technology. It is management--not the government--that makes a company productive and competitive. Making companies productive and competitive is the ultimate responsibility of the manager.

This concluding section provides general guidelines based on the experience of companies that have excellent reputations for productivity, quality, and innovation. These rich experiences are drawn from a wide range of companies, industries, technologies, and programs. They suggest that improving productivity, quality, and innovation can greatly enhance the competitive advantage of a company and its profitability. To be sure, there is no complete consensus on how to organize, plan, and implement efforts aimed at improving competitive advantage through productivity enhancement, quality improvement, and heightened innovation. There are differences in strategies, program content, and the choice of techniques and methods. Despite these differences, there are certain commonalities among the companies. Over the years, we have studied over 200 companies with outstanding reputations for success in enhancing productivity, quality, and innovation. These common elements show up consistently irrespective of the company, industry, markets, and technological setting.

SOME GUIDELINES

- Firm commitment of top management is essential for successful efforts in improving productivity, quality, and innovation. This commitment has to be translated into company philosophy, goals, strategies, policies, and culture--that guide all employees.

- Recognize that the human resource component of the company is the key to the ultimate success of improvement efforts. Employees and managers at all levels should be convinced that productivity, quality, and innovation are essential for survival, growth, and prosperity in an intensely competitive global market.

- Strategic focus on productivity, quality, and innovation is essential for enduring results. Management's concern for productivity, quality, and innovation must be consistent with the other strategic objectives and integrated into the regular business strategy and operations. A fundamental criterion for providing strategic focus is that each issue--productivity, quality, and innovation--be consistent with each other. In excellent companies, these elements send similar messages and reinforce each other.

- Productivity, quality, and innovation are closely linked and are critical for gaining maximum competitive advantage. Quality enhances productivity through reduced costs. A low--cost producer can use its cost advantage to increase profit margins and/or lower prices for improved profits. Quality also affects a firm's sales. The reputation for higher quality decreases the elasticity of demand and provides opportunities for companies to charge higher prices and earn higher profit margins. Innovation is closely linked with productivity and quality. Innovation, particularly technological innovation, influences productivity and quality in many ways, including conserving resources, increasing output, and improving product performance and reliability.

- Information and manufacturing technologies play a key role in improving productivity and quality. These technological innovations provide numerous opportunities for businesses to improve their efficiency and flexibility. Some of these opportunities include office information systems, factory automation, computer integrated manufacturing, and artificial intelligence. In order to get the full potential of these technologies, they must be managed strategically from all stages extending from the generation of new ideas to their commercial application in manufacturing processes or a new generation of products.

- A variety of approaches and techniques have the potential for improving productivity, quality, and technological innovation. Approaches for productivity and quality improvements have been suggested by authorities such as Deming, Juran, Crosby, and others. Techniques to improve quality and productivity include statistical quality control, quality circles, office automation, zero-defects, computer-aided design and manufacturing, productivity incentives, and combinations thereof. No single approach or technique is best for all companies. The approaches adopted and the techniques selected must be tailored to a company's situation. Also, there are opportunities for combining a number of approaches and techniques that meet the unique requirements of a company in maximizing benefits.

- Benefits of productivity and quality improvements and technological innovation are optimum when there is a close relationship between functional departments and when there is team effort. Productivity and quality problems cut across functional boundaries, and efforts for improvement should concern all elements of a company. Removing barriers between different functional specialists, providing education, and offering incentive for teamwork are essential for desired results. Such teamwork not only improves quality and productivity but increases the competitive position by shortening the time for product development and commercializing new ideas.

- Productivity, quality, and innovation initiative must be managed. Improvement must be planned, organized, monitored, controlled, and continuously revitalized. Productivity, quality, and innovation must have goals and standards. Systems should be designed to measure progress against planned improvements in these areas, and to ensure that actions are taken when results deviate from plans. For enduring results, all these aspects--goals standards, measurements, and corrective action--must be fully integrated into the total activities of a company.

Taken together, these general guidelines cluster around three major areas: (1) certain essential preconditions for effectiveness, (2) a variety of approaches and techniques, and (3) a systematic program for planning, implementing, and controlling the improvement initiatives.

CONCLUDING NOTE

Productivity, quality, and innovation are the pillars of competitive advantage. That much is generally agreed upon by thoughtful management scholars and practicing managers. Strategic focus on the three components offers the best way for companies to improve their competitive situation and become profitable. Given the imperative role of these in gaining competitive advantage, American business must have a sense of urgency about its need to enhance productivity, improve quality, and encourage innovation in order to remain competitive.

Bibliography

Abernathy, William J. The Productivity Dilemma: Road
    Block to Innovation in the Automotive Industry.
    Baltimore: Johns Hopkins University Press, 1978.
Abernathy, William J., B. Clark, and A. M. Kantrow.
    Industrial Renaissance: Producing a Competitive
    Future for America. New York: Basic Books, 1983.
Abernathy, William J., and Robert H. Hayes. "Managing
    Our Way to Economic Decline." Harvard Business Re-
    view (July/August 1980): 68-81.
Adam, E. E., Jr., J. C. Hershauer, and W. A. Ruch.
    Productivity and Quality--Measurement as a Basis
    for Improvement. Englewood Cliffs, N. J. :
    Prentice-Hall, 1981.
Adams, Harold W. "Solutions as Problems: The Case of
    Productivity." Public Productivity Review (Septem-
    ber 1975): 36-43.
Allen, David, and Victor Levine. Nurturing Advanced
    Technology Enterprises: Emerging Issues in State
    and Local Economic Development Policy. Westport,
    Conn.: Praeger, 1986.
Alston, Jon P. "Three Principles of Japanese Manage-
    ment." Personnel Journal (September 1983): 758-63.
Alvesson, Mats. Consensus, Control, Critique: Para-
    digms in Research on the Relationship Between
    Technology, Organization and Work. London: Gower,
    1987.
Anderson, John C., and Robert G. Schroeder. "Getting
    Results from Your MRP System." Business Horizons
    (May-June 1984): 57-64.
Anderson, John C., Roger G. Schroeder, and Gary D.
    Scudder. "White Collar Productivity Measurement."
    Management Decision (Winter 1986): 3-8.
Andrew, Charles G., and George A. Johnson. "The Crucial
    Importance of Production and Operations Manage-
    ment." Academy of Management Review (January
    1962): 143-47.
Ardolini, C., and J. Hohenstein. "Measuring Producti-
    vity in the Federal Government." Monthly Labor Re-

<u>view</u> (November 1974): 13-20.

Argote, Linda, Paul S. Goodman, and David Schkade. "The Human Side of Robotics: How Workers React to a Robot." <u>Sloan Management Review</u> (Spring 1983): 31-41.

Baillie, Allan S. "The Deming Approach: Being Better Than the Best." <u>Advanced Management Journal</u> (Autumn 1986): 15-24.

Baily, Martin N., and Alok K. Chakrabarti. <u>Innovation and the Productivity Crisis</u>. Washington, D.C.: Brookings Institute, 1987.

Bain, David. <u>The Productivity Prescription: The Manager's Guide to Improving Productivity and Profits</u>. New York: McGraw-Hill, 1986.

Bakewell, K. G. B. <u>How to Find Out: Management and Productivity</u>. New York: Pergamon Press, 1970.

Bastone, Eric, and Stephen Gourlay. <u>Unions, Unemployment, & Innovation</u>. New York: Basil Blackwell, 1986.

Basu, A. P. <u>Reliability and Quality Control</u>. New York: Elsevier, 1986.

Baumgardner, Mary. "Productivity Improvement for Office Systems." <u>Journal of Systems Management</u> (August 1981): 12-15.

Bewley, W. W. "America's Productivity Decline: Fact or Fiction?" <u>Financial Executive</u> (April 1982): 31-35.

Blair, John D., and Carlton J. Whitehead. "Can Quality Circles Survive in the United States?" <u>Business Horizons</u> (September/October 1984): 17-23.

Blau, G., and M. Rosow. <u>Trends in Product Quality and Worker Attitude: Highlights of the Literature</u>. Studies in Productivity No. 3, Scarsdale, N.Y.: Work in America Institute, 1978.

Bloom, G. F. "Productivity: Weak Link in Our Economy." <u>Harvard Business Review</u> (January/February 1971): 4-14.

Bobbe, Richard A., and Robert H. Schaffer. "Productivity Improvement: Manage It or Bust It?" <u>Business Horizons</u> (March/April 1983): 62-69.

Bohlarder, G. W. "Implementing Quality-of-Work Programs." <u>MSU Business Topics</u> (Spring 1979): 33-40.

Botkin, James, et al. <u>The Innovators: Rediscovering America's Creative Energy</u>. Philadelphia: University of Pennsylvania Press, 1986.

Bradford, David, and Allan Cohen. <u>Managing for Excellence</u>. New York: John Wiley, 1984.

Brief, Authur P., ed. <u>Research on Productivity: Multidisciplinary Approach</u>. New York: Praeger, 1984.

Brockner, Joel, and Ted Hess. "Self-Esteem and Task Performance in Quality Circles." <u>Academy of Management Journal</u> (September 1986): 617-23.

Buehler, V. M., and Y. K. Shetty, eds. <u>Proceedings: Managing Productivity Enhancement: Company Exper-</u>

iences. Logan: College of Business, Utah State
University, 1979.
---. Productivity Improvement: Case Studies of Pro-
ven Practice. New York: AMACOM, American Manage-
ment Associations, 1981.
Buffa, Elwood S. "Making American Manufacturing Compe-
titive." California Management Review (Spring
1984): 29-46.
---. Meeting the Competitive Challenge: Manufacturing
Strategy of U.S. Companies. New York: Dow Jones-
Irwin, 1984.
Buijs, J. "Strategic Planning and Product Innovation--
Some Systematic Approaches." Long Range Planning
(October 1979): 23-24.
Burch, E. E. "Productivity: Its Meaning and Measure-
ment." Atlanta Economic Review (May/June 1974):
43-47.
Burck, C. G. "Working Smarter." Fortune (July 15,
1981): 68-73.
---. "What Happens When Workers Manage Themselves."
Fortune (July 27, 1981): 62-65.
Burgelman, Robert, and Leonard R. Sayles. Inside Cor-
porate Innovation. New York: The Free Press,
1985.
Burnham, D. C. Productivity Improvement. New York:
Columbia University Press, 1973.
Bylinsky, Gene. "America's Best-Managed Factories."
Fortune (May 28, 1984): 16-24.
Callon, Michel, et al. Mapping the Dynamics of Sci-
ence and Technology. Dobb's Ferry, N. Y.: Sheri-
dan House, 1986.
Capannella, J., and F. J. Corcoran. "Principles of
Quality Costs." Quality Progress (April 1983): 17-
21.
Caron, Paul F., and Stanley J. Haddock. "Developing a
Quality Assurance Program." Internal Auditor (De-
cember 1986): 37-42.
Case, Kenneth E., and Lynn L. Jones. Profit Through
Quality and Quality Assurance Programs for Manu-
facturers. Norcross, Ga.: American Institute of
Industrial Engineers, 1978.
Chinloy, Peter. Labor Productivity. Cambridge, Mass.:
Ballinger, 1981.
Christopher, William F. Productivity Measurement Hand-
book. Cambridge, Mass.: Productivity Inc., 1983.
Clark, K. B. "Impact of Unionization on Productivity:
A Case Study." Industrial & Labor Relation Review
(July 1980): 451-69.
Cole, Robert E. Work, Mobility, and Participation: A
Comparative Study of American and Japanese Indus-
try. Berkeley: University of California, 1979.
Cole, Robert E. "Will Quality Control Circles Work in
the U.S.?" Quality Progress (July 1980): 30-33.

Cole, Robert E., "Learning from the Japanese: Prospects and Pitfalls." Management Review (September 1980): 22-28.

Cole, Robert E. "The Japanese Lesson in Quality." Technology Review (July 1981): 29-34.

Cole, Robert E., and D. S. Tachiki. "Forging Institutional Links: Making Quality Circles Work in the U.S." National Productivity Review (Autumn 1984): 417-29.

Collier, D. A. "The Service Sector Revolution: The Automation of Services." Long Range Planning (December 1983): 10-20.

Committee for Economic Development. Productivity Policy: Key to the Nation's Economic Future. New York: Committee for Economic Development, 1983.

Connell, G. W. "Quality at the Source: The First Step in Just-in-Time Production." Quality Progress (November 1984): 44-45.

Cook, M. H. "Quality Circles--They Really Work, But." Training and Development Journal (January 1982): 4-6.

Corn, Joseph J. Imagining Tomorrow: History, Technology, and the American Future. Cambridge, Mass.: MIT Press, 1986.

Craig, C. E., and R. Clark Harris. "Total Productivity Measurement at the Firm Level." Sloan Management Review (Spring 1973): 13-29.

Crandall, N. F., and L. M. Wooton. "Developmental Strategies of Organizational Productivity." California Management Review (Winter 1978): 37-46.

Crosby, Philip B. Quality Is Free: The Art of Making Quality Certain. New York: New American Library, 1980.

---. Quality Without Tears: The Art of Hassle-Free Management. New York: McGraw-Hill, 1984.

Crystal, G. S. "Motivating for the Future: The Long-Term Incentive Plan." Financial Executive (October 1971): 48-50.

Cummings, L. L. "Strategies for Improving Human Productivity." The Personnel Administrator (June 1975): 40-44.

Cummings, Thomas G., and Edmond S. Molloy. Improving Productivity and the Quality of Work Life. New York: Praeger, 1977.

Dale, B. G., and J. Lees. "Quality Circles: From Introduction To Integration." Long Range Planning (February 1987): 78-84.

Davis, L., and A. Chern, eds. The Quality of Working Life. New York: The Free Press, 1975.

Davis, L. E., and James C. Taylor, eds. Design of Jobs. Santa Monica, Calif.: Goodyear Publishing, 1979.

Deal, Terrence E., and Allan A. Kennedy. Corporate Cultures: The Rites and Rituals of Corporate Life.

Reading, Mass.: Addison-Wesley, 1982.

———. "Culture: A New Look Through Old Lenses." Journal of Applied Behavioral Science 19 (1983): 498-
505.

Dearden, J. "How To Make Incentive Plans Work" Harvard
Business Review (July/August 1972): 117-24.

Delbecq, Andre L., and Peter K. Mills. "Managerial
Practices that Enhance Innovation." Organizational
Dynamics (Summer 1985): 24-34.

Deming, W. Edwards. "Improvement of Quality and Productivity Through Action Management." National Productivity Review (Winter 1981-1982): 12-22.

———. Quality, Productivity, and Competitive Position.
Cambridge: MIT Press, 1982.

———. Out of the Crisis Cambridge: Massachusetts Institute of Technology Center for Advanced Engineering Study, 1986.

Dermer, Jerry. Competitiveness Through Technology:
What Business Needs From Government. Lexington,
Mass.: Lexington Books, 1986.

Devanna, M. A., C. Fombrun, and N. Tichy. "Human Resource Management: A Strategic Approach." Organizational Dynamics (Winter 1981): 51-67.

Dewar, C. The Quality Circle Handbook. Red Bluff,
Calif.: Quality Circle Institute, 1980.

Drucker, Peter F. "Entrepreneurial Strategies." California Management Review (Winter 1985): 9.

———. "The Discipline of Innovation." Harvard Business
Review (May-June 1985): 67-72.

———. Innovation and Entrepreneurship: Practice and
Principles. New York: Harper and Row, 1985.

Edwards, S. A., and M. W. McCarrey. "Measuring Performance of Researchers." Research Management (January 1973): 34-41.

Eilon, S., B. Gold, and J. Soesan. Applied Productivity
Analysis for Industry. New York: Pergamon Press,
1976.

Eldrige, Lawrence A., and Charles A. Aubrey, II.
"Stressing Quality—The Path to Productivity."
Magazine of Bank Administration (June 1983): 20-
24.

Fabricant, Solomon. A Primer on Productivity. New York:
Random House, 1971.

Feigenbaum, A. V. Quality Control. New York: McGraw-
Hill, 1951.

———. Total Quality Control: Engineering and Management. New York: McGraw-Hill, 1961.

Fein, Mitchell. Rational Approaches to Raising Productivity. Norcross, Ga.: American Institute of
Industrial Engineers, 1974.

———. "Improving Productivity by Improving Productivity
Sharing." The Conference Board Record (July 1976):
44-49.

Ferris, G. R., and J. A. Wagner, III. "Quality Circles
    in the United States: A Conceptual Reevaluation."
    Journal of Applied Behavioral Science 21 (1985):
    155-67.
Fitch, Thomas P. "Putting the Emphasis on Quality."
    United States Banker (May 1984): 28-32.
Fitzgerald, L., and J. Murphy. Installing Quality Cir-
    cles: A Strategy Approach. San Diego: University
    Associates, 1982.
Flynn, Patricia M. Technological Change: Production
    Life Cycles & Human Resource Planning. Cambridge,
    Mass.: Ballinger, 1987.
Follini, J. R. "Production Certifies the Quality of Its
    Work." Industrial Engineering (November 1971): 10-
    17.
Fombrun, C. J., Noel M. Tichy, and Mary Anne Devanne.
    Strategic Human Resource Management. New York:
    John Wiley, 1984.
Foote, George H. "Performance Shares Revitalize Execu-
    tive Stock Plans." Harvard Business Review (Novem-
    ber/December 1973): 121-30.
Ford, Robert N. "Job Enrichment Lessons for AT&T."
    Harvard Business Review (January/February 1973):
    96-106.
Forrester, J. W. "Innovation and the Economic Long
    Wave." Management Review (June 1979): 16-24.
Foulkes, Fred K., and Jeffrey L. Hirsch. "People Make
    Robots Work." Harvard Business Review (Janu-
    ary/February 1984): 94-102.
Freeman, Christopher. Design, Innovation & Long Cycles
    in Economic Development. New York: St. Martin's
    Press, 1986.
Frost, Carl F., John H. Wakely, and Robert A. Ruh. The
    Scanlon Plan for Organization Development: Identi
    ty, Participation, and Equity. East Lansing: Mi-
    chigan State University Press, 1974.
Fuchs, Victor R., ed. Production and Productivity in
    the Service Industries. New York: Columbia Uni-
    versity Press, 1969.
Gadon, Herman. "Making Sense of Quality of Work Life
    Programs." Business Horizons (January/February
    1984): 42-44.
Gainsburgh, Martin R. "Productivity, Inflation, and
    Economic Growth." Michigan Business Review (Janu-
    ary 1971): 15-21.
Gale, Bradley T. "Can More Capital Buy Higher Producti-
    vity?" Harvard Business Review (July/August 1980):
    78-86.
Garvin, David A. "Quality on the Line." Harvard Busi-
    ness Review (September/October 1983): 65-75.
---. "What Does 'Product Quality' Really Mean?" Sloan
    Management Review (Fall 1984): 25-43.

---. "Quality Problems, Policies, and Attitudes in the United States and Japan: An Exploratory Study." _Academy of Management Journal_ (December 1986): 653-73.

---. _Managing Quality_. New York: The Free Press, 1987.

Geare, A. J. "Productivity from Scanlon-Type Plans." _Academy of Management Review_ (July 1976): 99-107.

Gerstein, Mark S. _The Technology Connection_. Reading, Mass.: Addison-Wesley, 1987.

Gilder, George. _The Spirit of Enterprise_. New York: Simon & Schuster, 1984.

Gitlow, Howard S., and Shelly J. Gitlow. _The Deming Guide to Quality and Competitive Position_. Englewood Cliffs, N.J.: Prentice-Hall, 1987.

Gitlow, Howard S., and Paul T. Herts. "Product Defects and Productivity." _Harvard Business Review_ (September/October 1983): 131-41.

Glaser, Edward M. _Productivity Gains Through Worklife Improvement_. New York: Harcourt Brace Jovanovich, 1976.

---. "Productivity Gains Through Worklife Improvement." _Personnel_ (January 1980): 71-77.

Gold, Bela. "CAM Sets New Rules for Production." _Harvard Business Review_ (November/December 1982): 88-94.

Goldberg, Joel A. _A Manager's Guide to Productivity Improvement_. New York: Praeger, 1986.

Gray, D. O., and T. Solomon. _Technological Innovation: Strategies for a New Partnership_. New York: Elsevier, 1986.

Grayson, C. Jackson, Jr. "Productivity's Impact on Our Economic Future." _The Personnel Administrator_ (June 1975): 20-24.

Greenberg, Leon. _A Practical Guide to Productivity Measurement_. Rockville, Md.: BNA Book, 1973.

Griffith, Gary. _Quality Technician's Handbook_. New York: Wiley, 1986.

Grootings, Peter. _Technology and Work: East West Comparisons_. Wolfeboro, N.H.: Longwood, 1986.

Grove, Andrew S. _High Output Management_. New York: Random House, 1983.

Gryna, Frank M., Jr. _Quality Circles: A Team Approach to Problem Solving_. New York: AMACOM, American Management Association, 1981.

Guaspari, John. _Theory Why: In Which the Boss Solves the Riddle of Quality_. New York: AMACOM, 1986.

---. "The Role of Human Resources in 'Selling' Quality Improvement to Employees." _Management Review_ (March 1987): 20-25.

Hall, Peter. _Technology, Innovation and Economic Policy_. New York: St. Martin's, 1986.

Hallett, J. J. "Productivity and Quality: The Never Ending Quest." Personnel Administrator (October 1986): 22.

Hanley, J. "Our Experience with Quality Circles." Quality Progress (February 1980): 22-24.

Harrison, Jared F. "Why Won't They Do What I Want Them to Do?" Improving Performance and Productivity. Reading, Mass.: Addison-Wesley, 1978.

Hayes, Glenn E. "Quality and Productivity--The Education Gap." Quality (October 1982): 50-51.

Hayes, Robert H. "Why Japanese Factories Work." Harvard Business Review (July/August 1981): 57-66.

Hayes, Robert H., and Kim B. Clark. "Why Some Factories Are More Productive than Others." Harvard Business Review (September/October 1986): 66-74.

Hayes, Robert H., and R. W. Schmenner. "How Should You Organize for Manufacturing?" Harvard Business Review (January/February 1978): 105-18.

Hayes, Robert H., and Steven C. Wheelwright. "Link Manufacturing Process and Product Life Cycles." Harvard Business Review (January/February 1979): 133-40.

---. Restoring Our Competitive Edge: Competing Through Manufacturing. New York: John Wiley, 1984.

Heaton, Herbert. Productivity in Service Organization. New York: McGraw-Hill, 1977.

Hershauer, J. C., and W. A. Ruch. "A Worker Productivity Model and Its Use at Lincoln Electric." Interfaces (May 1978): 80-89.

Hershfield, David C. "Barriers to Increased Labor Productivity." The Conference Board Record (July 1976): 38-41.

Hickey, James J. Employee Productivity: How to Improve and Measure Your Company's Performance. Stratford, Conn.: Institute for the Advancement of Scientific Management and Control, 1974.

Hill, Chris T., and James Utterback, eds. Technological Innovation for a Dynamic Economy. New York: Pergamon Press, 1979.

Hinrichs, John R. Practical Management for Productivity. New York: Van Nostrand Reinhold, 1978.

---. "Avoid the 'Quick Fix' Approach to Productivity Problems." Personnel Adminstrator (July 1983): 39-43.

Holzer, Marc, ed. Productivity in Public Organizations. New York: Kennikat, 1976.

Hornbruch, F. W., Jr. Raising Productivity. New York: McGraw-Hill, 1977.

Horwitch, M. Technology in the Modern Corporation: A Strategic Perspective. New York: Pergamon, 1986.

Hostage, G. M. "Quality Control in a Service Business." Harvard Business Review (September/October 1975)

98-118.

Hutchins, Dave. "Quality Is Everybody's Business." Management Decision (Winter 1986): 3-7.

Hyer, N. L., and V. Wennerlov. "Group Technology and Productivity." Harvard Business Review (July/August 1984): 140-49.

Hykes, Dennis, and Colin Herskey. "Cultivating Entrepreneurism in Smokestack Industries." Management Review (March 1985): 38.

Iacocca, Lee, and William Novak. Iacocca: An Autobiography. New York: Bantam Books, 1984.

Ingle, Sud. Quality Circles Master Guide: Increasing Productivity with People Power. Englewood Cliffs, N.J.: Prentice-Hall, 1982.

Ishikawa, Kaoru. Guide to Quality Control. Tokyo: Asian Productivity Organization, 1972.

Jacobs, Herman S., with Katherine Jillson. Executive Productivity. New York: American Management Associations, 1974.

Jehing, J. J. "Profit Sharing, Motivation, and Productivity." Personnel Administration (March/April 1970): 17-21.

Judson, A. S. "New Strategies to Improve Productivity." Technology Review (July/August 1976): 61-67.

---. "The Awkward Truth About Productivity." Harvard Business Review (September/October 1982): 93-97.

Juran, Joseph M. Quality Control Handbook. New York: McGraw-Hill, 1974.

---. "Japanese and Western Quality: A Contrast in Methods and Results." Management Review (November 1978): 20-28, 39-45.

---. "Japanese and Western Quality--A Contrast." Quality Progress (December 1978): 10-17.

---. "Product Quality--A Prescription for the West." Management Review (June 1981): 8-14.

Kanter, Rosabeth Moss. The Change Master: How People and Companies Succeed Through Innovation in the New Corporate Era. New York: Simon & Schuster, 1983.

---. "Frontiers for Strategic Human Resource Planning and Management." Human Resource Management (Spring/Summer 1983): 9-21.

Katzell, M. E. Productivity: The Measure and the Myth. New York: AMACOM, American Management Association, 1975.

Kelly, Charles M., and James M. Norman. "The Fusion Process for Productivity Improvement." National Productivity Review (Spring 1983): 164-72.

Kendrick, John W. Understanding Productivity: An Introduction to the Dynamics of Productivity Change. Baltimore: Johns Hopkins University Press, 1978.

---. International Comparisons of Productivity and Causes of the Slowdown. Cambridge, Mass.: Bal-

linger, 1984.
---. Improving Company Productivity: Handbook with
    Case Studies. Baltimore: Johns Hopkins University
    Press, 1986.
Kendrick, John W., and Daniel Creamer. Measuring Compa-
    ny Productivity rev. ed. New York: Conference
    Board, 1975.
Kendrick, John W., and E. S. Grossman. Productivity in
    the United States: Trends and Cycles. Baltimore:
    Johns Hopkins University Press, 1980.
Kohl, R., et al. "Can America Meet Foreign Competition?
    A Treatise on Productivity." Journal of Small
    Business Management (January 1982): 56-58.
Kolmin, F. W. "Measuring Productivity and Efficiency."
    Management Accounting (November 1973): 22-24.
Konz, S. "Quality Circles: Japanese Success Story."
    Industrial Engineering (October 1979): 24-37.
Kreitner, R. "Identifying and Managing the Basics of
    Individual Productivity." Arizona Business (May
    1976): 3-8.
Landau, Ralph, and Dale Jorgenson. Technology and Econ-
    omic Policy. Cambridge, Mass.: Ballinger, 1986.
Langdon, Richard, and Roy Rothwell. Design and Innova-
    tion: Policy and Management. New York: St.
    Martin's, 1986.
Langevin, Roger G. Quality Control in the Service In-
    dustries. New York: AMACOM, American Management
    Association, 1977.
Lawler, Edward E., III. "Human Resource Productivity in
    the 80s." New Management (Spring 1983): 46-49.
Lawler, Edward E., III, and John A. Drexler, Jr. The
    Corporate Entrepreneur. Los Angeles: Center for
    Effective Organizations, University of Southern
    California, Graduate School of Business, 1980.
Lawler, Edward E., III, and Gerald E. Ledford, Jr.
    "Productivity and Quality of Work Life." National
    Productivity Review (Winter 1981-1982): 23-36.
Lawler, Edward E., III, and S. A. Mohrman.   "Quality
    Circles After the Fad." Harvard Business Review
    (January/February 1985): 65-71.
Lawrence, Colin, and Robert Shay. Technological Innova-
    tion, Regulation, and the Monetary Economy. Cam-
    bridge, Mass.: Ballinger, 1986.
Lefton, Robert E., V. R. Bussotta, and Manuel Sherberg.
    Improving Productivity Through People Skills. Cam-
    bridge, Mass.: Ballinger, 1981.
Lele, Milind M., and Uday S. Karmarker. "Good Product
    Support Is Smart Marketing." Harvard Business
    Review (November/December 1983): 124-31.
Leonard, Frank S., and W. Earl Sasser. "The Incline of
    Quality." Harvard Business Review (September/Octo-
    ber 1982): 163-71.

Lesko, Matthew. Lesko's New Tech Sourcebook. New York: Harper and Row, 1986.

Levering, Robert, Milton Moskowitz, and Michail Katz. The 100 Best Companies to Work for in America. Reading, Mass.: Addison-Wesley, 1984.

Limpercht, Joseph A., and Robert H. Hayes. "Germany's World-Class Manufacturers." Harvard Business Review (November/December 1982): 137-45.

Lokiec, Mitchell. Productivity and Incentives. Columbia, S.C.: Bobbins Publications, 1977.

Lubar, Robert. "Rediscovering the Factory." Fortune (July 13, 1981): 52-64.

Luke, High D. Automation for Productivity. Huntington, N.Y.: Krieger, 1972.

MacKinnon, Neil. "Launching a Drive for Quality Excellence." Quality Progress (May 1985): 46-50.

Maidique, M. A. "Entrepreneurs, Champions, and Technological Innovation." Sloan Management Review (Winter 1980): 59-76.

Main, Jeremy. "Ford's Drive for Quality." Fortune (April 18, 1983): 62-70.

Maital, Shlomo, and Noah M. Meltz. Lagging Productivity Growth. Cambridge, Mass.: Ballinger, 1980.

Mali, Paul. Improving Total Productivity. New York: John Wiley, 1978.

Mammone, J. L. "Productivity Measurement: A Conceptual Overview." Management Accounting (May 1980): 36-42.

Martin, James. Technology's Crucible: An Exploration of the Explosive Impact of Technology on Society during the Next Four Decades. New York: Prentice-Hall, 1986.

McBryde, Vernon E. "In Today's Market, Quality is Best Focal Point for Upper Management." Industrial Engineering (July 1986): 51-56.

McConnell, C. "Why Is U.S. Productivity Slowing Down?" Harvard Business Review (March/April 1984): 102-11.

Mehl, Wayne. "Strategic Management of Operations: A Top Management Perspective." Operations Management Review (Fall 1983): 29-36.

Mensch, Gerhard O. Stalemate in Technology. Cambridge, Mass.: Ballinger, 1979.

Metzger, Bert L. Increasing Productivity Through Sharing. Evanston, Ill.: Profit Sharing Research Foundation, 1980.

Miles, Raymond E., and Charles C. Snow. "Designing Strategic Human Resources Systems." Organizational Dynamics (Summer 1984): 36-52.

Miller, Donald B. "How to Improve the Performance and Productivity of the Knowledge Workers." Organizational Dynamics (Winter 1977): 62-80.

Mills, Stephen, and Roger William. <u>Public Acceptance of</u>
    <u>New Technologies</u>. New York: The Free Press, 1986.
Monden, Y. "Adaptable Kanban System Helps Toyota Main-
    tain Production." <u>Industrial Engineering</u> (May
    1981): 29-46.
Mooney, M. <u>Productivity Management</u>. Research Bulletin
    No. 127. New York: Conference Board, 1982.
Mudel, M. E., ed. <u>Productivity: A Series for Indus-</u>
    <u>trial Engineers</u>. Norcross, Ga.:   American Insti-
    tute for Industrial Engineers, 1977.
Munchus, G., III. "Employer-Employee Based Quality
    Circles in Japan: Human Resource Policy Implica-
    tions for American Firms." <u>Academy of Management</u>
    <u>Review</u> 2 (April 1983): 255-61.
Murphy, John W., and John T. Pardeck. <u>Technology and</u>
    <u>Human Productivity: Challenges for the Future</u>.
    Westport, Conn.:   Greenwood, 1986.
Naisbitt, John. <u>Megatrends: Ten New Directions Trans-</u>
    <u>forming Our Lives</u>. New York: Warner, 1982.
---. <u>Reinventing the Corporation</u> New York: Warner,
    1985.
Nassr, M. A. "Productivity Growth Through Work Measure-
    ment." <u>Defense Management Journal</u> (April 1977):
    16-20.
National Center for Productivity and Quality of Working
    Life. <u>A Plant-Wide Productivity Plan in Action:</u>
    <u>Three Years of Experience with the Scalon Plan</u>.
    Washington, D.C.: 1975.
---. <u>Improving Productivity: A Description of Select</u>
    <u>Company Programs, Series 1</u>. Washington, D.C.:   De-
    cember 1975.
---. <u>Improving Productivity Through Industry and Com-</u>
    <u>pany Measurement</u>. Washington, D.C.:   October 1976.
National Commission of Productivity and Work Quality. <u>A</u>
    <u>National Policy for Productivity Improvement</u>.
    Washington, D.C.:   1975.
National Research Council Staff. <u>Scientific Interfaces</u>
    <u>and Technological Applications</u>. Washington, D.C.:
    National Academy Press, 1986.
Naumann, A. "The Importance of Productivity." <u>Quality</u>
    <u>Progress</u> (June 1980): 18-26.
Ouchi, William. <u>Theory Z: How American Business Can</u>
    <u>Meet the Japanese Challenge</u>. Reading, Mass.:   Ad-
    dison-Wesley, 1981.
---. <u>The M-Form Society</u>. Reading, Mass.: Addison-Wes-
    ley, 1984.
Packer, Michael B. "Measuring the Intangible in Produc-
    tivity." <u>Technology Review</u> (February/March 1983):
    48-57.
Papacosta, Pangratios. <u>The Splendid Voyage: An Intro-</u>
    <u>duction to New Sciences and Technologies</u>. Engle-
    wood Cliffs, N.J.:   Prentice-Hall, 1986.

Pascale, Richard Tanner, and Anthony G. Athos. The Art of Japanese Management: Applications for American Executives. New York: Simon & Schuster, 1981.

Peeples, Donald E. "Measuring for Productivity." Datamation (May 1978): 222-28.

Peloquin, J. J. "Training: The Key to Productivity." Training and Development Journal (February 1980): 49-52.

Peters, Thomas J., and Nancy Austin. A Passion for Excellence. New York: Random House, 1985.

Peters, Thomas J., and Robert H. Waterman, Jr. In Search of Excellence: Lessons from America's Best-Run Companies. New York: Harper and Row, 1982.

Pierre, Andrew J. The Technology Gap. New York: New York University Press, 1986.

Pinchot, Gifford, III. Intrapreneuring. New York: Harper and Row, 1985.

---. "Innovation Through Intrapreneuring." Research Management (March/April 1987): 14-19.

Pipp, Frank J. "Management Commitment to Quality: Xerox Corp." Quality Progress (August 1983): 12-17.

Pitt, Hy. "A Modern Strategy for Process Improvement." Quality Progress (May 1985): 22-28.

Porter, Michael E. Competitive Strategy. New York: The Free Press, 1980.

Quinn, James Brian. "Technological Innovation, Entrepreneurship, and Strategy." Sloan Management Review (Spring 1979): 19-30.

Ramquist, Judith. "Labor-Management Cooperation: The Scanlon Plan at Work." Sloan Management Review (Spring 1982): 49-55.

Randall, R. "Job Enrichment Savings at Travelers." Management Accounting (January 1973): 68-72.

Ray, G. F. "Innovation as the Source of Long Term Economic Growth." Long Range Planning (April 1980): 9-19.

Reddy, Jack, and Abe Berger. "Three Essentials of Product Quality." Harvard Business Review (July/August 1983): 153-59.

Rees, A. "Improving the Concepts and Techniques of Productivity Measurement." Monthly Labor Review (September 1979): 23-27.

Regan, John F. Even More Productivity: Expanding Effectiveness & Efficiency in Plant. Philadelphia: Swansea Press, 1987.

Reich, Robert B. The New American Frontier. New York: Times Books, 1983.

---. "The Next American Frontier." The Atlantic Monthly (March 1983) 43-58.

Richardson, John H. "Manpower and Material--Overlooked Elements of Productivity." Production Engineering (August 1983): 30-31.

428   Bibliography

Roberts, Edward B., "Generating Effective Corporate
    Innovation." Technology Review (October/November):
    1977.
---. Generating Technological Innovation. New York:
    Oxford University Press, 1987.
Rockart, John F. "An Approach to Productivity in Two
    Knowledge-Based Industries." Sloan Management Re-
    view (Fall 1973): 23-33.
Rogers, F. G. "Buck." The IBM Way. New York: Harper and
    Row, 1986.
Rolland, I., and R. Janson. "Total Involvement as a
    Productivity Strategy." California Management Re-
    view (Winter 1981): 40-48.
Ross, Joel E. Productivity, People and Profits. Reston,
    Va.: Reston Publishing, 1981.
Ross, Joel E., and Lawrence A. Klatt. "Quality: The
    Competitive Edge." Management Decision (Winter
    1986): 12-17.
Ross, Joel E., and William C. Ross. Japanese Quality
    Circles and Productivity. Reston, Va.:   Reston
    Publishing, 1982.
Ross, Joel E., and Y. K. Shetty. "Making Quality a
    Fundamental Part of Strategy." Long Range Planning
    (February 1985): 53-58.
Ross, R. L., and G. M. Jones. "Approach to Increased
    Productivity: The Scanlon Plan." Financial Execu-
    tive (February 1972): 23-29.
Roy, Robin, and David Wield. Product Design and Tech-
    nological Innovation. Phildelphia: Taylor and
    Francis, 1986.
Ruch, William A., E. E. Adam, Jr., and J. C. Her-
    schauer. "Developing Quality Measures for Bank
    Operations." Bank Administration (July 1979): 47-
    52.
Rumberger, Russell, and Gerald Burke. The Impact of
    Technology on Work and Education. Philadelphia:
    Taylor and Francis, 1986.
Rutigliano, Anthony J. "An Interview with Peter
    Drucker: Managing the New." Management Review
    (January 1986): 38-42.
Ryan, John. "The Productivity/Quality Connection--Plug-
    ging in at Westinghouse Electric." Quality Prog-
    ress (December 1983): 26-29.
Schaffer, R. H. "Productivity Improvement Strategy:
    Make Success the Building Block." Management Re-
    view (August 1981): 46-52.
Schainblatt, Alfred H. "How Companies Measure the Pro-
    ductivity of Engineers and Scientists." Research
    Management (May 1982): 10-18.
Schlesinger, Leonard A., and Barry Oshry. "Quality of
    Work Life and the Manager: Muddle in the Middle."
    Organizational Dynamics (Summer 1984): 4-19.

Schmenner, Roger W. "Every Factory Has a Life Cycle." Harvard Business Review (March/April 1984): 121-27.

Schonberger, Richard J. Japanese Manufacturing Techniques: Nine Hidden Lessons in Simplicity. New York: The Free Press, 1982.

---. "Just-in-Time Production: The Quality Dividend." Quality Progress (October 1984): 22-24.

---. World Class Manufacturing. New York: The Free Press, 1986.

Schonberger, Richard J., and James P. Gilbert. "Just-in-Time Purchasing: A Challenge for U.S. Industry." California Management Review (Fall 1983): 54-68.

Sepehri, Mehran, ed. Quest for Quality: Managing the Total System. Technology Park, Atlanta: Industrial Engineering and Management Press, 1987.

Shaw, John C. The Quality-Productivity Connection in Service Sector Management. New York: Van Nostrand Reinhold, 1978.

Shaw, John C., and Ram Capoor. "Quality and Productivity: Mutually Exclusive or Interdependent in Service Organizations?" Management Review (March 1979): 25-28, 37-39.

Sherman, George. "The Scanlon Concept: Its Capabilities for Productivity Improvement." The Personnel Administrator (July 1976): 17-20.

Sherman, H. D. "Improving the Productivity of Service Businesses." Sloan Management Review (Spring 1984): 11-23.

Sherman, Stratford P. "Eight Big Masters of Innovation." Fortune (October 15, 1984): 66-84.

Shetty, Y. K. "Key Elements of Productivity Improvement Programs." Business Horizons (March/April 1982): 15-22.

---. "Management's Role in Declining Productivity." California Management Review (Fall 1982): 33-47.

---. "Managerial Strategies for Improving Productivity." Industrial Management (November/December 1984): 24-28.

---. "Corporate Responses to the Productivity Challenge." National Productivity Review (Winter 1984-1985): 7-14.

---. "Quality, Productivity, and Profit Performance: Learning from Research and Practice." National Productivity Review (Spring 1986): 166-73.

---. "Product Quality and Competitive Strategy." Business Horizons (May-June 1987): 46-52.

Shetty, Y. K., and Joe Barrett. Productivity: A Resource Guide. Roy, Utah: Barrett Management Services, 1981.

Shetty, Y. K., and Vernon M. Buehler. Quality and Productivity Improvements: U.S. and Foreign Compa-

nies' Experiences. Chicago: Manufacturing Productivity Center, 1983.

---. eds. Productivity and Quality Through People:
Practices of Well-Managed Companies. Westport,
Conn.: Quorum Books, 1985.

---. Quality, Productivity and Innovation. New York:
Elsevier, 1987.

Shetty, Y. K., and Joel E. Ross. "Quality and Its
Management in Service Businesses." Industrial Management (November/December 1985): 7-12.

Sibson, Robert E. Increasing Employee Productivity. New
York: American Management Association, 1976.

Siegel, Irving H. Company Productivity: Measurement
for Improvement. Kalamazoo, Mich.: W. E. Upjohn
Institute for Employment Research, 1980.

Sinha, Madhav H., and W. O. Willborn. Essentials of
Quality Assurance Management. New York: John
Wiley, 1986.

Skinner, W. "The Productivity Paradox." McKinsey Quarterly (Winter 1987): 36-45.

Spenser, Lyle. Calculating Human Resource Costs Benefits: Cutting Costs and Improving Productivity.
New York: Wiley, 1986.

Statistical Quality Control Handbook, 2d ed. Indianapo-
lis: AT&T Technologies, 1956.

Stebbing, Lionel. Quality Assurance: The Route to
Efficiency & Competitiveness. New York: Halsted
Press, 1986.

Stenkerd, Martin F. Productivity by Choice: The 20-1
Principle. New York: John Wiley, 1986.

Stevenson, H. H., and D. E. Gumpert. "The Heart of
Entrepreneurship." Harvard Business Review
(March/April 1985): 85-94.

Stewart, William T. "A Yardstick for Measuring Productivity." Industrial Engineering (February 1978):
34-37.

Strebel, P. "Organizing for Innovation Over an Industry
Cycle." Strategic Management Journal (March/April
1987): 117-24.

Strong, E. P. Increasing Office Productivity: A Seven-
Step Program. New York: McGraw-Hill, 1962.

Struthers, J. E. "Why Can't We Do What Japan Does?"
Canadian Business Review (Summer 1981): 24-26.

Swartz, G., and V. Constock. "One Firm's Experience
with Quality Circles." Quality Progress (September
1979): 14-16.

Sylwester, David L. "Statistical Techniques to Improve
Quality and Production in Non-Manufacturing Operations." Survey of Business (Spring 1984): 11-17.

Takeuchi, Hirotaka. "Productivity: Learning from the
Japanese." California Management Review (Summer
1981): 5-19.

Takeuchi, Hirotaka, and John A. Quelch. "Quality Is More Than Making a Product." Harvard Business Review (July/August 1983): 139-45.

Taylor, B. W., and K. R. Davis. "Corporate Productivity: Getting It All Together." Industrial Engineering (March 1977): 30-36.

Teece, David J. Strategy and Organization for Industrial Innovation and Renewal. Cambridge, Mass.: Ballinger, 1987.

Thompson, Harry, and Michael Paris. "The Changing Face of Manufacturing Technology." The Journal of Business Strategy (Summer 1982): 45-52.

Thompson, Phillip C. Quality Circles: How To Make Them Work in America. New York: AMACOM, American Management Association, 1982.

Tichy, Noel M. Managing Strategic Change: Technical, Political and Cultural Dynamics. New York: Wiley-Interscience, 1983.

Tichy, Noel M., Charles J. Fombrun, and Mary Anne Devanna. "Strategic Human Resource Management." Sloan Management Review (Winter 1982): 47-61.

Townsend, Patrick L. Commit to Quality. New York: John Wiley, 1986.

Twiss, Brian. Managing Technological Innovation. White Plains, N.Y.: Longman, 1986.

U. S. Congress, House Committee on Science and Technology. The Future of Science: Hearing Before the Task Committee on Science Policy of the Committee on Science and Technology. Washington, D.C.: U.S. GPO, 1986.

U. S. Department of Labor, Bureau of Labor Statistics. The Meaning and Measurement of Productivity. Washington, D.C.: U.S. GPO, September 1971.

Vesper, Karl H. Entrepreneurship and National Policy. Chicago: Heller Institute for Small Business Policy Papers, 1983.

---. "Entrepreneurs Affect People." Business Horizons (May/June 1985): 74-80.

Vicere, Albert A. "Managing Internal Entrepreneurs." Management Review (January 1985): 31-33.

Vough, Clair F., and Bernard Asbell. Tapping the Human Resource: A Strategy for Productivity. New York: AMACOM, 1975.

Wadsworth, Harrison M., et al. Modern Methods for Quality Control and Improvement. New York: Wiley, 1986.

Wait, D. J. "Productivity Measurement: A Management Accounting Challenge." Management Accounting (May 1979): 24-30.

Walsh, D. S. "Analyzing and Solving Productivity Problems." Training and Development Journal (July 1980): 70-74.

Walter, Craig. "Management Commitment to Quality: Hew-
    lett-Packard Company." Vital Speeches (August
    1983): 22-24.
Walters, Roy W., and Associates, Inc. Job Enrichment
    for Results: Strategies for Successful Implemen-
    tation. Reading, Mass.: Addison-Wesley, 1975.
Walton, R. E. "Quality of Working Life: What Is It?"
    Sloan Management Review (April 1976): 13-22.
Welch, J. L., and D. Gordon. "Assessing the Impact of
    Flextime on Productivity." Business Horizons (De-
    cember 1980): 61-62.
Werther, William B., Jr., et al. Productivity Through
    People. New York: West Publishing, 1986.
Wheelwright, S. C. "Manufacturing Strategy: Defining
    the Missing Link." Strategic Management Journal
    (January/March 1984): 77-91.
---. "Restoring the Competitive Edge in U.S. Manufac-
    turing." California Management Review (Spring
    1985): 26-42.
Wheelwright, S. C., and R. H. Hayes. "Competing Through
    Manufacturing." Harvard Business Review (January/
    February 1985): 99-109.
Williams, Kathy. "Enhancing Productivity Through Auto-
    mation." Management Accounting (July 1981): 54-55.
Wilson, A. H. "Engineering and Productivity." Engi-
    neering Journal (March 1977): 22-26.
Wise, J. "Setting Up a Company Productivity Program."
    Management Review (June 1980): 15-18.
Wood, Robert, Frank Hull, and Koya Azumi. "Evaluating
    Quality Circles: The American Application." Cali-
    fornia Management Review (Fall 1983): 37-53.
Wooten, Leland M., and Jin L. Tarter. "The Productivity
    Audit: A Key Tool for Executives." MSU Business
    Topics (Spring 1976): 31-41.
Wunnenberg, C. A., Jr. "Productivity in the Warehouse:
    Who Needs to Automate?" Management Review (Octo-
    ber 1977): 55-58.
Yager, Edward. "Japanese Managers Tell How Their System
    Works." Fortune (November 1977): 126-40.
Zeider, Joseph. Human Productivity Enhancement: Orga-
    nizational, Personnel and Decision Making Vol. 2.
    New York: Praeger, 1986.
Zeldman, M. "Moving Ideas from R & D to the Shop
    Floor." Management Review (December 1986): 24-27.

Index

Biographical Notes

Nancy Bancroft is the manager of Customer Management Consulting at Digital Equipment Corporation. She has extensive experience in computer systems development and management and in organization development. She has a B.A. from the College of William and Mary and completed an M.B.A. with a major in organizational development at Clark University, graduating with high honors. Her publications include articles in the <u>National Productivity Review</u> and in <u>Systems, Objectives, Solutions</u>, a North-Holland Press international journal.

Rodney D. Becker is the Vice President in Planning and Business Systems at Control Data. He received his B.A. at Wheaton College and his M.A. at the University of Illinois. He has been with Control Data since 1961 as personnel manager, director, general manager, and now vice president. His past experiences include employment supervisor at Magnavox, and service with the U.S. Army.

John R. Black is the director at the Quality Improvement Center at Boeing Commercial Airplane Company. He is responsible for assisting all levels of management to develop and implement processes leading to a cultural transition throughout the BCAC. He is a graduate of Gonzaga University with a B.A. in sociology, received an M.A. in human relations from the University of Oklahoma, and was granted an M.B.A. from City University in Seattle, Washington.

William Bradshaw is the manager of professional support services at the DMS-10 Division of Northern Telecom Inc. He holds a B.S. in social science from Campbell University and is a senior member of The Improvement Institute. He has conducted a variety of work methods improvement/simplification, work measurement, office automation, and productivity improvement studies throughout all functional areas of Northern Telecom's Research Triangle Park operations.

Michael A. Brewer is the Vice President/program general manager of Workstations/Office Systems at Unisys. He received a B.S. in economics from London School of Economics and Kings College. Since joining Unisys as an accounting trainee in London, he has held a number of increasingly responsible sales, promotion and product positions.

Vernon M. Buehler is a professor of business administration, assistant dean for business relations, and director of Partner's Program, College of Business, Utah State University, Logan, Utah. He holds the M.B.A. from the Harvard Graduate School of Business Administration and a Ph.D. in economics from George Washington University, Washington, D.C. He has been active in the field of government and business relationships, and his articles have been published in the Academy of Management Journal and other journals. He has taught public policy and business environment courses since 1972. He co-edited, with Y. K. Shetty, Productivity Improvement: Case Studies of Proven Practice (AMACOM, June 1981); Quality and Productivity Improvements: U.S. and Foreign Companies Experiences (Manufacturing Productivity Center, Illinois Institute of Technology, Chicago, March 1983); Productivity and Quality Through People: Practices of Well-Managed Companies (Quorum Books, 1985); and Quality, Productivity, and Innovation: Strategies for Gaining Competitive Advantage (Elsevier Science Publishing Co., 1987).

Paul F. Buller is an associate professor of management in the Department of Business Administration at Utah State University. He is director of the Masters of Business Administration (M.B.A.) degree program. Dr. Buller holds a B.S. degree in psychology and a master's degree in social work from the University of Utah, and an M.B.A. and Ph.D. in business administration from the University of Washington. He teaches courses in business policy, management, and business consulting and was selected as Professor of the Year in the College of Business in 1987. He has research, consulting, and management development experience in a variety of business and government organizations. His recent publications have appeared in Academy of Management Review, Group and Organization Studies, and Personnel Administration.

John Campbell is the personnel manager of the compensation, benefits, employment and personnel records functions in the Human Resources Group at National Semiconductor Corporation. He received his M.S. in human resource management at the University of Utah and his B.S. in psychology at Brigham Young University. He is a member of the board of trustees of Holy Cross Hospital's Preferred Providers Organization, the Holy Cross Care Executive Committee, and the American Society for Personnel Administration.

Ronald P. Carzoli is the senior Vice President in Human Resources at Mead Corporation. He received his B.A. from Wayne State University and his M.B.A. from the University of Akron. His other experiences include director of Personnel and Organization at Ford Motor Company, vice president of Industrial Relations at Ford of Europe, and vice president of Industrial Relations at Ford Motor Company.

Richard L. Chappell is a senior manager in the Management Information Consulting Division of Arthur Andersen & Co. His experience has been concentrated in the Financial Services Industry area. He received a B.S. and an M.S. from the University of Nebraska. He is a Certified Public Accountant and Registered Professional Engineer. He is also past president of the Institute of Industrial Engineers.

Michael Connors joined IBM in 1966. He has held executive positions in marketing, planning, and finance. He is now director of Information Systems & Applications. He received his B.S., M.S., and Ph.D. at Stanford University. He has been a member of the academic staff at UCLA and the University of California, Berkeley. He is the author of a book on mathematical theory and has authored over a dozen professional papers on business and systems control processes.

Roslyn S. Courtney is an executive at General Foods Corporation. She manages all aspects of personnel for corporate departments, an operating subsidiary, and an established business unit. She has written several articles on career management and strategic human resource planning. She holds a B.A. in political science from Penn State University and an M.B.A. from the Wharton School. She is a member of the Human Resource Planning Society of New York and serves on the Corporate Advisory Board of the NOW Legal Defense and Education Fund.

Des Cunningham is the chairman, CEO, and one of the two founding partners of Gandalf Technologies Inc. and is closely associated with the development and growth of the Canadian data communications industry. He is one of the founders and a past chairman of the Canadian Advanced Technology Association and is past president of the Ottawa Section of the Canadian Informations Processing Society. He is also director of IDEA Corporation in Ontario.

James R. Deters was named Vice President - finance of Borg-Warner Corporation in January 1985 after serving as Vice President - human resources, and controller. He has published articles on control and financial education, and he frequently lectures on management control. He received a bachelor's degree in finance from Ohio University, Athens, in 1959, and a master's degree in business administration from Ohio State University, Columbus, in 1963.

Rodney J. Falgout is the manager of Personnel Operations in Monsanto Company. He received his B.S. degree from the University of Southwestern Louisiana with a major in personnel management and a minor in psychology. He is involved in the American Society for Personnel Administration and the Human Resources Management Association of Greater St. Louis, Inc.

John J. Falzon is senior Vice President in charge of the Corporate Quality Department of Metropolitan Life Insurance Company. He is responsible for developing and supporting a corporate wide quality improvement process that engages all employees in a direct effort to exceed the expectation levels of their customers. He received his B.S. in management from Fordham University and is a member of the national business honor society, Beta Gamma Sigma. He served in the U.S. Army from 1950 to 1952.

Ed Finein is chief engineer and manager of Competitive Practices and the Product Delivery Process for Xerox Corporation. He is a member of the Strategic Business Office in the Business Products and Systems Group at Webster, New York. He is responsible for chairing the Chief Engineer's Council and for developing and maintaining the group's Product Delivery Process. He also manages all group competitive evaluation and analysis activities including benchmarking. He is a graduate of the University of Rochester with degrees in math and physics.

John Kenneth Galbraith is the Paul M. Warburg Professor of Economics Emeritus at Harvard University. He is a Ph.D. in economics from the University of California, was a Social Science Research Council Fellow at the University of Cambridge, and has taught at the

University of California, Princeton, and Harvard. He has written several books, the most recent of which is A View from the Stands.

Andrew S. Grove graduated from the City College of New York with a bachelor of chemical engineering degree and received his Ph.D. from the University of California, Berkeley. Upon graduation, he joined the Research and Development Laboratory of Fairchild Semiconductor and became assistant director of research and development in 1967. In July 1968, he participated in the founding of Intel Corporation, where, after serving as vice president and director of operations and executive vice president, he became president in 1979. He has written over forty technical papers and holds several patents on semiconductor devices and technology. His latest book is High Output Management.

Wayne L. Hanna graduated from Utah State University in 1965. He is director of telecommunications within McDonnell Aircraft Company. He served for two years as chairman on the MAP User Group. As chairman, he was responsible for the initial organization of the group co-sponsored by McDonnell Douglas Corporation and General Motors.

Thomas E. Helfrich is a vice president of the Travelers with companywide responsibility for Organization and Management Development. Prior to joining the Travelers in May of 1986, he worked with General Electric for fourteen years. He received an M.B.A. from DePaul University in 1977, a B.A. from the University of Notre Dame in 1972. He currently serves on the Corporate Linkage Model Development Program Advisory Committee, a Department of Labor funded effort to increase the upward mobility of women in corporations.

F. Kenneth Iverson has been chairman and CEO of Nucor Corporation since 1984. He received a bachelor's degree in aeronautical engineering from Cornell University and a master's degree from Purdue University. He is a director of Rexham Corporation, Wikoff Color Corporation., Cato Corporation., Southeastern Savings and Loan, S. H. Heist Corporation., the Council for a Competitive Economy, and the Greater Charlotte Foundation. In 1981 and 1982, the Wall Street Transcript named him the "Best Chief Executive in the Steel Industry."

Richard A. Jacobs is senior vice president of A. T. Kearney, Inc., and is a member of the firm's board of directors. He is a graduate of M.I.T., has done graduate work at M.I.T.'s Sloan School of Industrial

Management, and received an M.B.A. from Roosevelt University in Chicago. His areas of specialty include systems and data processing and management productivity.

Rosabeth Moss Kanter is an entrepreneur, business leader, best-selling author, and respected scholar. A magna cum laude graduate of Bryn Mawr, she earned both her M.A. and Ph.D. from the University of Michigan. She taught at Brandeis and Yale before joining the faculty at the Harvard Business School in 1986. She is the recipient of many national honors, including the Guggenheim Fellowship, four honorary doctoral degrees, and four "Woman of the Year" awards. She has written several books, the latest of which is <u>The Change Masters: Innovation and Entrepreneurship in the American Corporation</u>.

Dagnija D. Lacis is vice president and program general manager, Education for Corporate Program Management of Unisys. She received a B.S. in mathematics from Butler University in 1964 and a B.S. in chemistry the same year. She began her career with Unisys as a systems representative for the Business Machines Group. Since then she has had experience in senior management positions that have included such areas as programming, field marketing, and product management.

Theodore A. Lowe is a graduate of the University of Michigan, receiving both a B.S. in mechanical engineering and a master's degree in business administration. He is also an ASQC certified quality engineer. He is currently manager of quality improvement for General Motors Truck and Bus. In this position, he is responsible for leading the development and implementation of the GM Truck and Bus Quality Improvement Process.

Charles M. Lutz is an assistant professor of information systems management at Utah State University. He teaches both undergraduate and graduate courses in information systems analysis and design, business data communications, and information systems resource management. He is a former Army officer who worked in operations research and systems analysis. He is the author of numerous papers and major presentations.

Alexander MacLachlan is senior vice president of technology at the DuPont Company. He graduated from Tufts College in 1954 with a B.S. in chemistry. Three years later he received a doctorate in physical organic chemistry from the Massachusetts Institute of

Technology.   Numerous articles by him have been
published in technical journals. He also holds several
patents.

Glenn M. McEvoy is an associate professor of management
at the College of Business, Utah State University.   He
received his B.S. in Industrial Engineering from the
University of California at Berkeley, and his M.S. and
D.B.A. in Organization and Management from the
University of Colorado in Boulder. Dr. McEvoy has also
taught at the State University of New York and Loretto
Heights College.   His current research interests are
management training and skill evaluation, assessment
centers, and multiple rater performance appraisal systems.   His recent publications have appeared in the
Academy of Management Journal, Journal of Applied
Psychology, Personnel Psychology, Personnel Administrator, and the Journal of Business and Psychology.

John H. Moore became deputy director of the National
Science Foundation in June 1985.   Currently he is on
leave of absence from the Hoover Institution at
Stanford University where he was Associate Director and
Senior Fellow of the Institution.   He is also a member
of the Thomas Jefferson Center Foundation Board of
Directors.   He earned a B.S. in chemical engineering
and an M.B.A. from the University of Michigan.   He was
awarded a Ph.D. in economics by the University of
Virginia and was assistant chairman of that
university's Department of Economics from 1968 to 1972.

J. Tracy O'Rourke is the president and CEO of the
Allen-Bradley Company. He received a B.S. in mechanical engineering from Auburn University and served in
the U.S. Air Force. He is a member of the board of
directors and the compensation committee at the W. H.
Brady Company, and is a board member of Mrs. Fields
Company.

Wayne R. Pero is the manager of Quality Assurance and
Quality Performance at the Dow Chemical Company.   He
oversees development and execution of the quality performance effort at Dow divisions in the United States
and is responsible for maintaining a strong Quality
Assurance function.   He is a graduate of Bucknell
University, where he received a B.S. in chemical
engineering.

T. Boone Pickens, Jr., founded Mesa Petroleum Company
in 1964 and has served as its president and chairman of
the board since that time.   He received his B.S. in
geology from Oklahoma State University.   He has been
recognized for his management skills by the Wall Street

Transcript for the last four years. He currently
serves as a member of the board of directors of Texas
Commerce Bancshares, Inc., Houston. He is chairman of
the Board of Regents, West Texas State University,
Canyon; and Chairman, Texas Research League, Austin.

Hal Rosen is president of Lee Scientific, Inc. He
worked as a senior accountant with Deloitte Haskins &
Sells from 1977 to 1980 and as a small business consultant from 1980 to 1985. He earned his B.A. in
accounting at Brigham Young University in Provo, Utah,
in 1977. He is a member of the American Institute of
Certified Public Accountants, and the Utah Association
of Certified Public Accountants.

Robert Rubin is the director of information systems at
Pennwalt Corporation. He received his B.S. in physics
at Drexel University and his M.A. in physics at Temple
University. He is on the board of directors of the
Philadelphia Chapter of the Society for Information
Management. He is also the managing editor for the
National SIM Network, which is a bi-monthly newsletter.

Douglas A. Saarel is a senior vice president of the
Coca-Cola Company and director of Human Resources. He
received his B.A. in English from Rutgers University.
He completed post-graduate studies in business and
organizational behavior at San Jose State College and
received a J.D. degree from Loyola University in
Chicago. He spent some time in the U.S. military and
held a succession of command and staff positions.

Richard S. Sabo is the manager of Publicity and
Educational Services at the Lincoln Electric Company.
He is responsible for educational programs, publicity,
public relations, and technical publishing. He also
heads the Book Division and handles the editing, publishing, and sales of all books. He graduated from
California University and was awarded a master's degree
in education from Edinboro University. He has majors
in industrial arts, safety education, and counseling.

Lawrence Schein is a senior research associate in the
human resources program group of The Conference Board.
He holds a Ph.D. in sociology from the University of
Pennsylvania. His areas of concentration include
survey research, program evaluation, demographic analysis, medical sociology, and organization development.
He is the co-author of a major board research report on
innovations in human-resources management and is currently engaged in a series of studies of corporate
culture.